FOREST DYNAMICS

FOREST
DYNAMICS

An Ecological Model

Daniel B. Botkin

Oxford　　　　　　　New York
OXFORD UNIVERSITY PRESS
1993

Oxford University Press

Oxford   New York   Toronto
Delhi   Bombay   Calcutta   Madras   Karachi
Kuala Lumpur   Singapore   Hong Kong   Tokyo
Nairobi   Dar es Salaam   Cape Town
Melbourne   Auckland   Madrid

and associated companies in
Berlin   Ibadan

Copyright © 1993 by Oxford University Press, Inc.

Published by Oxford University Press, Inc.,
200 Madison Avenue, New York, New York 10016

Oxford is a registered trademark of Oxford University Press

Library of Congress Cataloging-in-Publication Data
Botkin, Daniel B.
Forest Dynamics: An Ecological Model / Daniel B. Botkin.
p. cm.   Includes bibliographical references and index.
ISBN 0-19-506555-7
1. Forest ecology—Computer simulation.
2. Trees—Growth—Computer simulation.
I. Title.
QK938.F6B66   1992
581.5′2642′0113—dc20
91-19555

2 4 6 8 9 7 5 3 1

Typeset by Thomson Press (India) Ltd., New Delhi, India
Printed in the United States of America
on acid-free paper

This book is dedicated to
James F. Janak and James R. Wallis,
with whom I first developed
the model described in this book.

Preface

This book is about a computer simulation that is a model of how forests grow and change over time—the JABOWA forest model, originally developed by my colleagues, James F. Janak and James R. Wallis, and me in the early 1970s (Botkin et al., 1970, 1972a, b). This model and others derived from it have been widely used around the world since then (Dale et al., 1985), including applications to forests of New Zealand (Develice, 1988), tropical forests (Doyle, 1981), and forests of Europe (Kienan and Kuhn, 1989; Leemans and Prentice, 1987, 1989; Oja, 1983), as well as boreal and temperate deciduous forests of North America (Pastor et al., 1987; Shugart, 1984; Solomon and West, 1985). In that sense the book is a technical discussion about a specific computer program, a program that is available as a companion to this book. However, because ecology is a relatively new science that deals with extremely complex phenomena that are yet little understood, any discussion of models, theory, and simulations in ecology leads us rapidly to a concern with some basic scientific concepts and some philosophical issues about the nature of understanding and tests of that understanding in the study of complex systems. For readers unfamiliar with ecology and the history of theories in this science, perhaps a few introductory comments might be useful to set the general context for the material that follows.

A model is a simplified representation of some aspects of the real world. It can take many forms—a picture, a map, a set of equations, a computer program. The kind of model used in science is defined as a simplified representation of a system, with hypotheses that describe and explain the system, usually expressed mathematically. Simulation comes from the Latin word *simulare*, meaning to mimic or imitate. In this sense a computer simulation is an imitation or a mimicking of reality. As most of us know today from television and movies, computer animation can be a striking and seemingly realistic mimicking of reality, and so computers provide us with a kind of technical power new in the last generation. Computer simulation in ecology is about 25 years old, and is itself a relatively new aspect of this science, one that remains controversial.

There are two fundamental uses of computer simulation in ecology. One is, as the term implies, to simulate some aspect of the real world and project changes, for example in the conditions of populations, ecological communities, ecosystems, even the entire global life-supporting and life-containing system, the biosphere. The other use is as an aid to understanding—a way of helping us learn the causes of what we see occurring in nature, the possible explanations for the complex systems that include and sustain life.

Among some users of computer programs in ecology, it has become customary to distinguish between these two uses—projecting future conditions and understanding causes and effects—by referring to a simulation as meaning simply the mimicking of reality, whether or not any attempt is made to understand the rules that underlie that mimicking, and by referring to a model as meaning a construct that is a set of hypotheses about causes and effects.

Although in science we seek understanding in preference to mere mimicking, in some practical problems the ability to make reliable projections, whether or not we understand the basis of the projections, can be important. Public opinion polls used to project how people might vote in an election is an example of how a projection is valued even if the causes of voter preferences are not understood. Forecasting weather from satellite images of cloud movements is another example of the utility of mimicking and projecting, even if we do not understand the physics that determines the dynamics of the atmosphere. Similarly, we can imagine cases where environmental projections are important but the overall goal of simulation is to increase our understanding.

The model described in this book has been applied to both uses—the projection of changes in the conditions of forests for practical purposes, and an aid to understanding. Because ecology is so new a science and its subject is so complex, our approaches to theories and models are incomplete. The model described here has been comparatively widely used and has been shown to be realistic, but could be improved on to great advantage. Part of the motivation for writing this book is to provide a complete documentation of the model so that people can better understand it. But an additional motivation is that some who read the book will see ways to develop theory in ecology and improve our ability to make projections and to understand the fascinating entities that contain and support life.

It is my perception that the complex systems that sustain and include life are inherently different from the systems usually analyzed by science, and that one of the limitations of ecological theory prior to 1970 is that most models, employing differential and integral calculus, mistakenly assumed that ecological systems were analogous to simple physical systems. At the time (1970) that my colleagues, James Janak and James Wallis, and I developed the original code for this model, few ecologists took computer models seriously, and fewer expected that successful computer models might be developed in the near future. Our original goal was to produce a model that would be as general as possible. In developing this model, I believe that we made three kinds of creative contributions: (1) an overall structure of the model (a "world view": a decision as to what fundamental, general qualities should be included and how these should be related one to another); (2) decisions as to how much to simplify and where simplification was necessary; and (3) specific algorithms for each process to be represented in the model.

This book provides the information necessary to understand and use the forest growth model. It describes some aspects of forested ecosystems and the ecology of trees. It sets the forest model within the broader context of the science of ecology and ecological theory and issues that confront society in the management of forests. It may serve as a prototype for approaches to other renewable natural resources issues. The current version of the model is keyed to this book; it is available on diskettes for use on microcomputers, so that the reader can use and test the model.

My purposes in writing this book are three: to place the model within the development of the science of ecology; to provide an explanation of the JABOWA model so that readers can understand the model, use it, test it, and develop their own extensions and modifications of it; and to suggest what might be and what is not likely to be possible in the development of theory for complex ecological systems. It is not my purpose to provide a complete literature review of all the derivatives of the original model. While this could be useful, it might confuse the reader who is trying to understand one version of the model. Therefore for the sake of clarity I have focused the discussion on the two major versions I have developed and used myself. This restriction is in no way meant to detract from the progress made by other investigators in their use of the model and their development of variants of it. In the last chapter, I attempt to relate my use of the model to the work of others and Appendix VII provides a list of references to many of these work as well as to related studies. I would like to acknowledge my appreciation to all the people who have used and extended the model and who recognize the connections between their versions and the original JABOWA model. The wide application of the model is a tribute to its strength and utility.

Because there has been widespread use and adaptation of the JABOWA model, some readers may be interested in a brief history and a family tree. The original model was developed in 1970 as a summer research project by James Janak, James Wallis, and me under a cooperative agreement between the IBM corporation and Yale University. By the end of the summer the first version of the model was operating, and our initial presentation was made that same year (Botkin, Janak, and Wallis, 1970). To my knowledge, most computer models developed since then that are known as "gap-phase" models of forests are derived from the original JABOWA model, mainly along one of two family trees. These have been given a variety of names, including CLIMACS, FIRESUM, FORET, FORENA, LINKAGES, SILVA, ZELIG.

John Aber, then a graduate student at Yale, worked with me to develop a version of the model that incorporated soil nitrogen effects; we applied this model to questions about forest harvesting practices and the effects of acid rain on forests near Brookhaven National Laboratory. John Aber then developed his own version, and subsequently a student of his, John Pastor, produced additional versions with a soil–forest feedback. A series of variants has developed from their work.

A second set of descendants began with software published in 1975 under the name FORET by Hank Shugart. This was a direct copy of the first version of JABOWA, recoded, but to my knowledge in its original form without changes in substance, operation, or algorithms. He, his colleagues, and students have since produced a number of applications and variants. Others, beginning either with the original JABOWA code or with Shugart's recoding, have produced additional versions; these versions include several for forests outside North America. Al Solomon was the first to suggest that the model might be applied to the study of pollen analysis and used to consider Holocene changes in forests, and he was the first to do this. Others, including Margaret Davis and myself, have also used the model for this application. Another group at the University of Washington, including Virginia Dale and Jerry Franklin, developed a version for western coniferous forests. At present there is a wide variety of applications and variations of the original model, some simplified, some more complex, some involving integration of this model with other software, such as geographic information systems.

The book begins with a discussion of the character and quality of ecological knowledge, then develops the context for a model, and finally develops the model itself. The second half of the book concerns tests of the model and applications to basic science and the management of natural resources. Over the years I have enjoyed using the model and learned much ecology from it; I hope the reader finds these same benefits.

Santa Barbara, Calif.   D. B. B.

Acknowledgments

I am first indebted to James F. Janak and James R. Wallis, with whom the original JABOWA model was developed. I would also like to acknowledge the contributions of Richard Levitan to the development of the water balance equations and the solution to the time-dependent water balance. I would especially like to thank Douglas A. Woodby and Jon C. Bergengren for reprogramming the model in C and the many hours they spent assisting me with development and modification of the program, especially in improvements in the water balance methods and the conduct of experiments using the program for this book. I thank Robert A. Nisbet for additional programming and for his cooperation in carrying out many of the simulations discussed in the book; Jon Bergengren, Susan Bicknell, James H. Fownes, MarkKamakea, F. Thomas Ledig, H. Jochen Schenk, and Eric D. Vance for helpful comments on early drafts of the manuscript. Tad Reynales helped in the first conversion of the FORTRAN version of the model for use on minicomputers and microcomputers. I would like to thank Dr. William F. Curtis, Donald Jackson, and Kirk Jensen of Oxford University Press for their interest in the manuscript and their insightful comments about how to make the best book out of the material.

The JABOWA model was first developed for use at the Hubbard Brook Ecosystem Study in the U.S. Department of Agriculture Forest Service's White Mountains Experimental Forest, New Hampshire, and many of the studies conducted there, especially those of F. H. Bormann and T. G. Siccama, have been helpful. I wish to thank them and all of their colleagues, whose empirical research has been valuable in tests of the model.

Some of the work reported here has been supported by grants from the U.S. Environmental Protection Agency, the National Science Foundation, and the Pew Charitable Trusts; related field work has been supported by the Andrew J. Mellon Foundation. Although results do not necessarily reflect views of any of these, and no official endorsement should be inferred from this publication, the support of these organizations is much appreciated.

Finally, I would like to express my appreciation to all those who have used the JABOWA model or variations of it over the years.

Contents

How to Use This Book

This book is designed for two related uses: as a technical manual for those who will use the JABOWA-II computer model of forest growth, which is available as a companion to this book for use on microcomputers; and as a description and discussion of the model, its background, its applications and its limitations and needs for additional work and improvement, for those who wish only to become familiar with the concepts and background of the model but do not plan to use it themselves. For the second use, the book is best read in a traditional manner from beginning to end, with the first chapter providing a general context and philosophy for the model, the second and third chapters describing the equations and algorithms that make up the model, and the rest of the book discussing the applications of the model in science and natural resource issues. Readers who wish to use this book as a technical reference for the model might simply refer to those parts of the book that contain material they want to know about as they use the software. However, readers who want to become serious users of computer models in ecology might do best to begin with Chapters 2 and 3 which explain the equations and algorithms that make up the model.

As an additional aid to those who use the software, the index contains terms noted in boldface that are of direct reference to the software, or seem to me to be of particular interest to those users. The manual that is provided with the software is also keyed to this book, to make cross-referencing more efficient.

FOREST
DYNAMICS

1

Nature and Knowing:
Theory and Ecology

Nature has ... some sort of arithmetical-geometrical coordinate system, because nature has all kinds of models. What we experience of nature is in models, and all of nature's models are so beautiful.

Buckminster Fuller (1966), quoted in "In the outlaw area," a profile of Buckminster Fuller by Calvin Tomkins, *The New Yorker*, January 8, 1966

The Importance of Forests and the Need
for Quantitative Theory of Forests

Forests have always been important to people. There is an ancient relationship between forests and people: forests provide many materials for civilization, and we continually clear forested landscapes as our civilization expands (Perlin, 1989). Deforestation was recognized as early as classical Greek times, when Plato wrote that the hills of Attica were a skeleton of their former selves (Thomas, 1956). For a long time, especially since the Industrial Revolution, the direct impact of human beings on forests has obscured slower, natural changes. For example, George Perkins Marsh, the first writer of the modern industrial era to point out the effects of civilization on environment, believed that forests remained constant except when suffering from human influence (Marsh, 1864). Today we have achieved a much longer perspective on the history of vegetation and of climate. We know that both have changed over the millennia, and that vegetation has responded to climatic change. Pollen records make this apparent (M. B. Davis, 1983; COHMAP, 1988). Faced now with potentially rapid and novel changes in forests, the effects of possible rapid climatic warming of our own doing, clearing of forests in the tropics, intensive logging of temperate and boreal forests, pollution of large areas of temperate and northern forests, fragmentation of forests and elimination of most older stands of trees, we need to evaluate what these changes will mean to the forests themselves and to the biological diversity they contain.

Forests are important sources of fuel, building materials, paper, and fiber. They are big business. In the three states of Michigan, Minnesota, and Wisconsin, forest products are estimated to add $20 billion annually to the economy. Forests have many other functions as well. They provide habitat for wildlife and endangered species of plants and animals; they control erosion; they provide water resources; and they offer opportunities for recreation of many kinds. People enjoy forests and value their beauty. In the 1970s the U.S. Supreme Court argued about whether trees should have legal standing (Stone, 1974). Environmental groups, such as Earth First!, have begun to take strong actions to protect forests.

Further, forests may affect the biosphere. Trees and forest soils are major storehouses of carbon, and the exchange by vegetation of carbon dioxide with the atmosphere can affect the rate of climate change. The influence on the biosphere of photosynthesis of terrestrial vegetation, mainly forests, can be seen in the famous Mauna Loa records of carbon dioxide concentration in the air. These records show a summer decrease and a winter increase, coinciding with growing-season photosynthetic uptake and winter respiratory output of carbon dioxide by vegetation (Ekdahl and Keeling, 1973; Keeling, Moss, and Whorf, 1989). Forests also influence climate locally by altering the surface energy balance through reflection of sunlight, friction caused by surface roughness, and evaporation of soil water. The rates at which these effects on climate occur depend on what species are present and the relative abundance of the species. These two biological factors in turn depend on the state of the environment and the history of the forest and its environment.

Because we have cleared so much of the forests and altered so much of what remains, most forests exist as patches and remnants. The few remaining extensive, continuous tracts lie primarily in the boreal forests of the north, in the most remote high mountains, and in some tropical regions. We would expect that a species with highly specific habitat requirements, requirements that are presently limited in geographic distribution, would be more greatly affected by rapid environmental change than species with broader environmental tolerances and whose seeds have the opportunity to migrate within a continuously forested area.

Forests and the species of trees that compose them have intrigued naturalists and scientists, and in the twentieth century have become one focus of ecological research. Because of their characteristics (including that trees do not move, are long-lived, and at higher latitudes have growth rings and can be readily aged), forests provide some of the best testing grounds for ecological theory. From a scientific point of view, the complexity of a forest is representative of the general level of complexity characteristic of ecological systems that contain and support life.

People have long believed that forests could be treated simply as products to be mined, as impediments to the progress of civilization, or, at best, as natural phenomena that could be understood simply from informal, casual observations and descriptions. In spite of notable instances of enlightenment, dominant twentieth-century attitudes have been like those ascribed in the eighteenth century to the people of Pennsylvania by the Swedish botanist Peter Kalm. "People are here (and in many other places) in regard to wood, bent only upon their own present advantage, utterly regardless of posterity," Kalm wrote in his journal about America. "(They take) little account of Natural History, that science being here (as in other parts of the world) looked upon as a mere trifle, and the pastime of fools" (Kalm, 1770). Now, as our forests disappear, management of the remaining areas can no longer be left to chance, whim, or informal beliefs. Forest management requires considerable knowledge and the ability to make projections into the future.* We need to develop reliable methods that help us understand the implications of the assumptions we make about forests and the effects of our actions on them. We need formal, quantitative models of forest growth.

The model discussed in this book is one of the oldest and one of the most widely used computer simulations of forest growth. This book is aimed at those who wish to use the model and are not very familiar with it, those who want to use it as part of a course, or those who wish to apply it to practical problems or use as an addition to their own research, either in its current form or with new modifications. The book is a companion to the JABOWA-II © 1989 software, prepared explicitly to be available with this book for use on microcomputers. The book can, in addition, provide a summary of the algorithms and assumptions and examples of applications for those already familiar with the model. As I explained in the Preface, but which is worth repeating here, it is not my purpose to provide a complete literature review of all the derivatives of the original model. While this could be useful, it might confuse the reader who is trying to understand one version of the model. Therefore for the sake of clarity I have focused the discussion on the two major versions I have developed and used myself. This restriction of the book is in no way meant to detract from the progress made by other investigators in their use of the model and their development of variants of it.

Natural History and Ecological Theory

Hiking up the Appalachian Trail in the Presidential Range of New Hampshire in the 1960s, it was easy to know when one reached an elevation somewhere between 760 m and 820 m above sea level. Within that range the forest abruptly changed. Sugar maple and beech, dominant below, gave way to spruce and fir. In the transition region itself, yellow birch, a species typical of lower elevations, seemed more important than it had been below. Then it too declined and disappeared as one walked to higher elevations.

Written accounts of forests of New Hampshire from the turn of the twentieth century told a slightly different story. Although the same transition was observed, spruce was reported to grow abundantly much lower down the

---

*The word *projection* is used here to mean a calculation by a model of a future state of a system, given the present state. In this sense, "present" could be time past or time future; it means the state that is known and defined. The word *prediction*, which has the connotation of an actual foretelling of future events, is avoided. The distinction here is the difference between the subjunctive (what might be if certain conditions were to hold), which is the use applied here to projection, and the reality of what "will" happen, which is a connotation of prediction.

mountain slopes than it does today (Chittenden, 1905). What accounts for changes in species composition with elevation? What information and concepts are necessary to predict these changes? And what accounts for the reported change in the distribution of spruce since the turn of the century? Was it simply the result of poor surveying long ago, with spruce growing no lower down the slopes then than it does not? Or has there been a real change in the distribution of spruce, perhaps as a result of intense logging at lower elevations?

These questions are specific examples of an underlying general question in ecology: How can we understand what determines the distribution and abundance of living things on the Earth? In the past, most attempts at explanation in ecology were strictly observational, describing what changes occurred and generating statistical correlations; or they were qualitative and informal and not open to direct scientific tests. Ultimately any explanation must derive from the development of theory, and in the broadest sense, a theory is a model. Any explanation of the distribution and abundances of trees in the Presidential Range of New Hampshire, or anywhere in the world, would have to account for past and present abundances and distributions.

There are several kinds of answers to the question: What determines the distribution and abundance of trees in a forest? Which kind we seek depends in part on what we want to explain. Do we want to explain the appearance or quality of the transition—simply what species are present and dominant in an area? Or do we want to predict the elevations at which takes place a quantitative change in some characteristic—biomass, or numbers of trees? This book explores an approach to explanation and theory in ecology. The pursuit of the simple question discovered on the trail in New Hampshire takes us to the heart of the idea of a science of ecology. It is simultaneously a technical and conceptual inquiry, one that penetrates to the essence of science yet connects us to many practical questions.

How We Can Assess the Response of Forests to Change

Three methods are open to us to assess the response of forests to change: present measurements, our knowledge of the past, and our ability to make projections using ecological theory. Present measurements and periodic monitoring are of great potential importance, but have been of limited utility in the past because, with rare exceptions, we have not measured forests for long enough to detect changes in forest quantities or to distinguish the causes of the changes we do observe.

Long-term monitoring of forests is difficult, is labor intensive, and requires great commitment over long periods of time from individuals and institutions. For example, Thomas Siccama conducted an excellent empirical study of the possible causes of the transition in species composition on Camel's Hump Mountain in Vermont, a study that occupied more than 5 years of intensive field measurements. Since the 1960s, with the beginning of the Hubbard Brook Ecosystems Study (Bormann and Likens, 1979), the Andrews Experimental

Forest Study in Oregon (Waring and Franklin, 1979), and similar projects at other U.S. Department of Agriculture Forest Service Experimental Forests, and with the initiation of the National Science Foundation project in long-term ecological research, there has been a great increase in governmental and institutional commitment to such studies. These are of great importance and need to be continued, as I have argued elsewhere (Botkin, 1989), but computer simulation can significantly shortcut certain aspects of these difficult investigations. As will be expanded on later, the model described in this book is a series of assumptions about forest dynamics, each of which can be changed. These assumptions, brought together in a computer program as a model, can help guide field research to make it more efficient. In addition, if used carefully the model can provide insights directly to its user, as readers of this book and users of the accompanying software can explore for themselves.

Some ecologists even argue that current forests are little different from those of presettlement times (Russell, 1983). That such a point of view can be put forward today suggests that we may have been unable to detect significant changes in forests in response to climate or other factors for three centuries. Even where measurements exist and show change, the causes of the changes can be confused. There are apparent correlations between climate change and forest growth, but changes in forest growth have also been attributed to the effects of acid rain and other forms of pollution. Observed current changes in forests might be responses to direct human actions, such as selective logging, that took place decades or centuries ago, the effects of which have been obscured until recently.

Even in instances where there might have been sufficient time to observe changes, such as at present geographic limits where a species might be most sensitive to change, we have rarely established proper programs to monitor changes over time, either to provide us with information leading up to the present or to provide measurements of change in the future. Such monitoring programs, made up of an array of permanent plots and measurements repeated at useful intervals, are needed, and it is hoped that more will be established in the future.

The past, especially information provided by fossil pollen and the correlation of this information with the history of climate, provides important insight into long-term changes in the distribution of trees. These help to establish the rates of seed migration across the landscape, and the kinds of changes in species composition that may occur within a forest, but not the rate at which existing forest composition will change.

The extensive work of M. B. Davis (see for example, Davis, 1983); Delcourt and Delcourt (see for example, Delcourt and Delcourt, 1977); the COHMAP group (COHMAP, 1988), and others has revealed the actual pathways and rates of migration of tree species in response to glacial retreat and climate change since the end of the last ice age. Using inferred potential rates of migration from these data to introduce or delete species in long-term runs of the JABOWA model, we can gain insight into aspects of long-term changes in forest composition not otherwise open to us (Davis and Botkin, 1985; Solomon et al., 1980;

Solomon and Shugart, 1984; Solomon and Webb, 1985), as I will discuss later in this book. The relationship between pollen deposits and climate or other environmental factors is a statistical correlation. A model, as an explicit set of assumptions, can help us gain insights into cause and effect of long-term processes. Thus a computer simulation of forest growth can augment the insights of the study of pollen records.

Soundly based theoretical models can provide insight into the rates of changes in forest composition, especially in response to novel aspects of environmental change. When ecology is at its best, theory is connected to observation, observations are connected to theory, and data are gathered systematically within a theoretical framework that tests major generalizations. This kind of ecology is very labor intensive, time consuming, and expensive to the point that discourages funding, and has rarely been conducted on the spatial and temporal scales necessary (Botkin, 1982, 1990). Computer models can significantly enhance our ability to address the issues of cause-and-effect relations on long temporal and broad spatial scales. For this to be accomplished in the best way, and for theory and observations to be best integrated, there is a need for a clear conceptual framework within which data are gathered systematically to test major generalizations.

To deal with the effects of human-induced environmental change on forests, it is essential that we develop a theory connected to observations, open to tests of validation and to calibration, and linking changes in populations and communities to environmental conditions. It is in this spirit that the JABOWA model has been developed, evaluated, and used.

Science, Theory, and Models

Every science involves theory, and every theory involves a model of how the object of study works. We think and explain science through models of reality. We use models in two ways: (1) In the *conceptual model*, the model is a metaphor, used to explain how something works. (2) In the *formal model*, the model is explicit in mathematical or numerical (e.g., a computer simulation) terms. In this chapter and throughout the book, a reference to models includes both kinds of models.

Ecological scientists applying mathematical models that they have not derived themselves should become aware of the assumptions and limitations of those models. Ecologists often use implicit models, without the benefits of using explicit assumptions. In ecology, it is common to separate "doing science" from "making models." Some ecologists are uncomfortable with theory and there has been a tendency to view "doing science" and "creating theory" as separate activities, a tendency that is decreasing today (Watt, 1962). In reality, there is a model behind every explanation. We gain when we make these models explicit, as I hope this book illustrates.

A model is an intentional simplification of reality, simplified so that phenomena of interest can be examined, analyzed, and understood. (For an interesting discussion of the role of models in science, see Schroedinger [1952].) While a model is necessary in science, it is also true that any model involves specific assumptions about how the world works (the world is assumed to be like the model). Thus the choice of a model is extremely important in the development of theory and of explanations. It is to our advantage to bring the underlying assumptions to the surface and make them part of active discussion.

Often scientists think of mathematics as free of assumptions that might restrict their perspective. But this is not the case. For example, differential and integral calculus, which was the foundation for most formal ecological theory during the first seven decades of the twentieth century, was derived to explain certain kinds of observations in physics. Calculus applied to real-world problems carries with it many strong assumptions, such as the assumption that events are continuous and deterministic. Ecologists using calculus must consider these assumptions explicitly. These assumptions are often inconsistent with ecological phenomena. The death of an elephant, for example, is a discrete event in the history of an elephant herd; the death of a large tree previously dominant in its local area is a discrete and important event for the remaining trees in that area.

Theory in Ecology: Some Earlier Analytical Approaches

Although the previous discussion might suggest that theory has played a minor role in ecology, on the contrary, theory has played a dominant role throughout twentieth-century ecology, shaping the very character of inquiry and conclusions. As Kenneth Watt (1962) has written, in the first half of this century ecologists tended to believe that their science lacked theory, whereas in reality it had "too much" theory—too much in the sense that the theory in use and influential was not thoroughly checked against or developed from observations. When it was so compared, the theory generally contradicted the facts. This has been true for a range of levels of phenomena, from single populations to the biosphere.

Many limitations and mistakes in ecological theory are illustrated by the development of theory in populations dynamics. During much of the first half of the twentieth century, there were three dominant formal models of population growth: (1) exponential, (2) logistic models of the growth of a single population, and (3) Lotka–Volterra equations for predator–prey interactions. Each was the simplest expression in the mathematics of calculus of an idea about population growth. The exponential model expresses the idea that a population grows at a constant percent increase per unit of time. The logistic model adds to this the idea that growth must be limited by competition among members of the population for limited resources (which are, however, available at a fixed rate). The Lotka–Volterra model begins with an exponentially growing prey, whose growth rate is reduced by interaction with its predator. The predator, in contrast, decreases exponentially in absence of prey, and increases according to the interaction of the two species.

The details of the mathematics of these models have been reviewed and analyzed in many books and articles (see for example, Goel, Maitra, and

Montroll, [1971]; May [1976]; Lotka's own book [Lotka, 1925] provides a clear exposition.) As theories of ecological phenomena, these models have serious limitations. They oversimplify, describing an entire population by a single number and describing the effect of one species on another by a single parameter. They ignore the influence of the environment and of time-varying stochastic events. They assume that events are continuous, not discrete; deterministic, not stochastic; that a sufficient description of a population is its size and rates of birth and death, so that age structure does not matter.[1]

When I began the work that led to the development of the forest model, it had become clear that simple, continuous, differential equation models based on the logistic and Lotka–Volterra models were no longer viable in ecology because their assumptions violated too much of what we knew from observations. It was necessary to develop new approaches. Some of these were made possible by computer simulation, which was initially looked down on by both theoreticians and empiricists. Most theoreticians in ecology preferred the elegance of formal analytical mathematics. This mathematics was simultaneously too limited to deal with the complexities of natural systems and too obscure to most field ecologists, so that theory and measurement tended to be isolated from one another rather than integrated. A theory was needed that was developed directly from an understanding of natural history and from the large wealth of observations about ecosystems and the growing understanding of the effects of the environment on physiological and ecological processes.

In recent years there has been considerable growth of interest in formal models for forests and other kinds of vegetation and in developing an approach to the analysis and evaluation of these models (Costanza and Sklar, 1985; DeAngelis, 1988; DeAngelis et al., 1985; French, 1990; Roberts, 1989; Smith and Huston, 1989).

Background to JABOWA: Other Models of Vegetation

Models of tree growth using differential equations were first developed in the nineteenth century (Greenhill, 1881). However, until the last several decades projections of forest conditions were based either on expert knowledge or on statistics derived from histories such as volume tables in forestry handbooks. Projections into the future assumed that future environments would be like the past. In the 1960s formal models of vegetation were initiated following several lines of investigation: (1) simple global models; (2) systems analysis models; (3) static stand structure models; (4) simple population dynamics models; (5) biophysical models; and (6) physiological models. There is no one model for all purposes. Ecology and its applications require a suite of models.

The Big Picture: Global Models of Chemical Cycles

Simple global models were early attempts to investigate the global carbon cycle and the role of vegetation in that cycle. Vegetation was characterized extremely simply as one or two "boxes" connected by a few lines of flux to the atmosphere and soils. For example, Bacastow and Keeling (1973) diagrammed the global carbon cycle as consisting of six storage compartments, two of which were land vegetation: long-lived and short-lived land biota. The state of the land biota was simply the content of carbon, and the only fluxes considered were those with the atmosphere. The change in carbon storage was calculated as simple, nonlinear functions of the carbon content in the biota and in the atmosphere. The purpose of this model was to consider the possible fate of carbon dioxide emitted to the atmosphere. The model did not deal with any of the details of vegetation or of ecological processes. The characterization of the biota was similar to the characterization of populations in earlier logistic and Lotka–Volterra population theory in that the state of the biota was defined by a single scalar quantity.

Systems Analysis and Analog Computers

Although not called such, the simple global model of Bacastow and Keeling (1973) was an example of a nonlinear systems model. Although most models of the biogeochemical cycles are simple systems analysis models, not all systems analysis models in ecology pertain to global processes. Some of the earliest ecosystem models used the methods of systems analysis (see for example Odum, 1957). Bernard Patten pioneered the conscious and formal application of engineering systems analysis to ecology (Patten, 1971). The application of systems analysis is a top-down approach. Phenomena are described by a series of boxes (representing the storage of a variable of interest, either matter or energy) and arrows (representing fluxes between the storage compartments). Part of the motivation behind the development of these models was the hope that formalisms developed in engineering sciences could provide a new basis to deal with the complexity of ecological processes. The perceived advantages were, on the one hand, the rapid development of analog computers that provided an electronic-physical model of the system, and, on the other, the development of formal, analytical mathematical methods for the solutions of linear systems. The success of engineering devices predicated on these formal methods in the real-world aspects of real engineering seemed to the proponents of systems analysis to offer a way to obtain a practical approach to ecological complexities. However, the mathematics of systems analysis had much in common with differential calculus of prior population models, and had the same limitations: the biota were characterized by simple (usually scalar) quantities, only a few flux pathways could be handled, and analytical mathematical solutions could not be extended in most cases beyond linear systems whereas real ecological systems were highly nonlinear. The complex processes and the true dynamics of ecological communities and ecosystems could not be characterized by these methods. Systems analysis played an important role, however, helping in the maturation of ideas in ecosystem ecology and forcing ecologists to consider the characteristics of stability of ecological systems. Systems analysis models are commonly used today in the analysis of chemical cycling in single ecosystems.

Static Structural Models. Models were also developed that combined a consideration of energy exchange with the exchange of water and carbon dioxide (in photosynthesis and respiration) for vegetation stands that were themselves static (the trees are fixed structures in time and space). These models developed first in agriculture and forestry, where they were helpful in projecting irrigation requirements and the limits of production and in considering the possible effects on primary production of increases in carbon dioxide concentration in the air near the vegetation (Waggoner and Reifsnyder, 1981). More recently, static models of vegetation stand structures have been refined to incorporate total water balance and rates of transpiration for trees (Running, 1984).

While each kind of model has had specific uses, none of the models described to this point allows projection of long-term changes over time so that dynamic processes such as ecological succession can be considered.

Simple Population Models. Although population dynamics dominated early theory in ecology, late nineteenth- and early twentieth-century work concentrated on animal populations, for example in the classic discussions by Verhulst (1838), Lokta (1925), Volterra (1926), and Gause (1934), and on analysis of population cycles (Elton, 1924), predator–prey interactions, competition, and, later, regulation of animal populations (Lack, 1967). This early population theory is characterized by definition of time-invariant birth, growth, and death rates, usually as scalar quantities but sometimes as age, sex, specific arrays, and the formulation of equations for population change using differential calculus or matrix algebra (the latter originated with the Leslie matrix model [Leslie, 1945], a major improvement of the mid-twentieth century in which each age class is explicit). Prior to the 1970s most population models were deterministic. Even in the original Leslie matrix model the number of births and deaths of each age class is calculated as a fixed percentage. However, a few models involved stochastic processes, such as Leslie's modification of his matrix model (Leslie, 1958). Most early population models did not consider the state of the environment in any explicit way. In general, the environment was considered to be constant or to have an important single variable that changed cyclically to drive population cycles. Since the 1980s there has been a growing number of age-class-specific matrix models (Wu, Rykiel, and Sharpe, 1987), some applied to trees (Buongiorno and Michie, 1980; Malanson, 1984). Most are deterministic and limited to comparatively static situations, with a constant environment and constant rates of birth and mortality. In one attempt to make the population age-structure approach more dynamic, Wu and I developed a stochastic model for an elephant population in which the entries in the matrix were transition probabilities rather than fixed rates, and these probabilities could be made functions of biological and environmental conditions (Wu and Botkin, 1980).

Biophysical Models. About the same time that the formulation of systems analysis and global models began, David M. Gates and a few others initiated another line of models at the opposite hierarchical extreme, that of individual leaves and individual plants and the exchange of energy between vegetation and the environment. The publication with the greatest initial impact in this field was Gates' original book, *Energy Exchange in the Biosphere*, published first in 1962. In contrast to the systems-analysis approach, these biophysical models began with "first principles"—the understanding of causes and effects— about the exchange of energy between surfaces and the atmosphere, and about the exchange of carbon dioxide and water during the processes of growth and respiration of vegetation. This work has developed into a rich and independent subdiscipline, with considerable insights about evolutionary adaptations of leaf structure, morphology, and pigments and considerable predictive power about energy and water conditions within which a given individual can persist. Advances in this field are summarized in Gates' more recent book, *Biophysical Ecology* (Gates, 1980).

Some static stand models apply the principles of biophysical models, and in this way the two modeling approaches are related. However, biophysical models are typically applied to a much smaller scale (a single individual or a component of an individual, such as a leaf) than are static stand models (which, as the name implies, are used for a stand of trees). In addition, biophysical models can be applied to situations where an organism is not fixed in time and space, as in successful applications to terrestrial animals. For example, Riechert and Tracy (1975) used such an approach to show that a spider making a web in a favorable energy environment could obtain more energy from its prey than a spider forming a web elsewhere. This approach also yielded insights that ranged from the role of posturing in thermoregulation of vertebrates (Tracy, Tracy, and Dobkin, 1975) to the behavior of toads (Bundy and Tracy, 1977). Biophysical models are useful in helping us understand the evolutionary advantages of structures. These models can project responses of an existing stand of vegetation, which is static in its structure over time, to energy exchange with its environment. Biophysical models have not been used to investigate temporal dynamics of vegetation structures themselves—production, growth, mortality of populations, or community dynamics—over time. Such processes were not the concern, nor were they within the analytical or computational capacities, of biophysical models in ecology.

The JABOWA model is a hybrid model. It is primarily a population dynamics model, but differs from most earlier models in that it deals with many species, does not assume that events are continuous, involves stochastic processes as a fundamental part of the dynamics, and relates growth and regeneration to several environmental variables. It bridges the old gap between animal and plant ecology because it has a foundation in population theory but incorporates vegetation—environment relations, time-varying environmental conditions, and competitive interactions among many individuals of many species.

Early Theoretical Concepts in Plant Ecology

Although plant ecology had little formal mathematical theory during the first half of the twentieth century, the study of plant communities was rich in concepts about the dynamic patterns of change in vegetation over time, and in change over space with changes in environmental gradients, such as changes in climate with elevation or changes in soil conditions that accompanied differences in bedrock. Plant ecology thus was characterized by a strong emphasis on qualitative or conceptual theory. This conceptualization, especially that which concerned the development of an ecological community or ecosystem over time, provided an important background for the JABOWA model discussed in this book.

As mentioned earlier, people have cleared forests since the earliest civilizations (Perlin, 1989), and the process of recovery of a forest from such disturbances has long been observed and reported. The early concept about forest dynamics was that a forest developed through a series of well-defined stages to a final steady-state condition, called the "climax," that was believed to persist indefinitely as long as there was no disturbance in the environment. This process was first called "succession" by Thomas Pownall in 1784 when he wrote that trees in a forest

> grow up, have their youth, their old age, and a period to their life, and die as we men do. You will see many a Sapling growing up, many an old Tree tottering to its Fall, and many fallen and rotting away, while they are succeeded by others of their kind, just as the Race of Man is. By this Succession of Vegetation the Wilderness is kept cloathed with Woods just as the human Species keeps the Earth peopled by its continuing Succession of Generations.   (Pownall, 1784)

Thoreau (1860) reintroduced this term in the nineteenth century to describe the development of pine woodlands following the logging of hardwood stands in New England, and soon after George Perkins Marsh used the term and discussed the idea and patterns of succession in great detail in his classic book, *Man and Nature* (Marsh, 1864).

As the twentieth century opened, there was a rich and rapid development of empirical studies of succession, including an excellent analysis of forest plots in the wilderness of Isle Royale in the second decade of the century by William Cooper (1913), studies of the succession of vegetation on sand dunes on the shores of Lake Michigan, and studies of the development of vegetation as bogs filled in, especially on glaciated terrain. For 100 years, the dominant theoretical concept was that succession led inevitably to a fixed climax condition, an idea that we find stated by George Perkins Marsh:

> Nature, left undisturbed, so fashions her territory as to give it almost unchanging permanence of form, outline, and proportion, except when shattered by geologic convulsions; and in these comparatively rare cases of derangement, she sets herself at once to repair the superficial damage, and to restore, as nearly as practicable, the former aspect of her dominion.   (Marsh, 1864)

One hundred years later, R. H. Whittaker, one of the most important plant ecologists of the twentieth century, reasserted the same idea, writing that where a climax community has been destroyed, vegetation communities "go through progressive development of parallel and interacting changes in environments and communities, a succession" and that "the end point of succession is a climax community of relatively stable species composition and steady-state function, adapted to its habitat and essentially permanent if its habitat is undisturbed" (Whittaker, 1970). At almost the same time, Eugene Odum, whom many consider the father of modern ecosystem ecology, wrote that succession is "an orderly process of community development that is reasonably directional and, therefore, predictable" and that succession "culminates in a stabilized ecosystem" (Odum, 1969).

Although generally accepted by plant ecologists until 1970, the hypothesis that a forest developed to a fixed climax was difficult to test, and in fact, as we will see later, was incorrect. Empirical studies, of which there were many, were at the time restricted to locating forest stands of known ages and comparing their present composition. Ages were known either because the time of the origin of the stand was known from historical records or tree rings of the oldest trees in the stand had been analyzed. The possibility that patterns of succession might be explored more fully through computer simulation, and new insight into these patterns thus be gained, was one of my primary motivations in developing a computer model of forest growth.

About the same time that the development of the JABOWA model began, there were other approaches to the development of theory of succession, such as that of Horn (1971). Equally important, a growing body of quantitative data began to shed new light on long-term patterns in vegetation. Among these were the development of new methods of measurements and analysis of data, and studies of many stands of known ages by a group at the University of Wisconsin, including Cottam and Curtis (1949), Loucks (1970), and McIntosh (much of which is reviewed in McIntosh, 1974, 1985). Also important were the beginnings of comparatively large studies of entire forested watersheds at USDA Forest Service Experimental Forests. Especially important for the JABOWA model were the studies at the Hubbard Brook Experimental Forest (see for example, Bormann and Likens, 1979), where the model was first applied, but also important were concomitant studies at the Andrews Experimental Forest in Oregon, and at Coweeta Experimental Forest in the Southeast. In these projects, larger resources than had been available in the past were applied to study many aspects of forest ecosystems. These empirical studies provided a new basis for comparison with theoretical projections and stimulated a reconsideration of the older idea of climax, an idea that has been largely abandoned today in its traditional guise—a fixed, constant condition and endpoint of succession.

Plant Physiology and Plant Anatomy as Background to Theory in Plant Ecology

There were other important antecedents that made possible the development of formal theory in plant ecology and the development of computer simulation. Especially crucial was a long history of laboratory research in plant physiology and the anatomy of woody plants, as well as field research in physiological ecology that took place mainly after the 1960s. Research in plant physiology and physiological ecology provided extensive knowledge about the relationship between photosynthesis, respiration, and growth of plants and key environmental conditions, especially light intensity, temperature, soil-water conditions, and the availability of chemical nutrients. An excellent recent review of much of this literature can be found in Kozlowski, Kramer and Pallardy (1991). The application of this knowledge in the development of the JABOWA forest model is discussed explicitly in Chapter 2 and Chapter 3, but it is important to emphasize here that without such a rich background in these fields the development of the forest model would have not been possible.

Studies of plant anatomy had led to an understanding of certain principles about the growth and form of plants, including a beginning understanding of what we might best refer to as the structural or engineering constraints on the shapes of plants. This had evolved into empirical studies of the relationships among the dimensions of trees, including the formulation of equations relating: diameter to height; diameter and height to total biomass and to leaf biomass. A wealth of equations were developed from which biomass of components of trees could be calculated from diameter and height, a method known as "dimension analysis" (Whittaker and Marks, 1975; Whittaker and Woodwell, 1968, 1969; Woods, Feiveson, and Botkin, 1991).

Another important empirical background was the development of large-scale ecosystems research projects, not only those at the USDA Forest Service Experimental Forests mentioned before, but also at Brookhaven National Laboratory and Oak Ridge National Laboratory, some of which was part of the International Biological Program. At Brookhaven, some of the earliest research was done to measure photosynthesis and respiration of trees under field conditions (Woodwell and Botkin, 1970; Botkin, 1969, 1970). My participation in that research was one of the beginnings of the formulation of the concepts that led to the JABOWA model.

Natural History and Ecological Theory

Another important part of the background of plant ecology was simply good natural history observation—the personal wisdom and experience of individuals who had spent years working in and studying forests. Their experiences, coupled with the wealth of research in plant physiology, plant anatomy, and physiological ecology and the increasing knowledge of community and the then-new ecosystem ecology, made it possible for experienced field scientists such as Murray Buell, with whom I studied at Rutgers University, to walk into a forest and give soundly based explanations of the distribution and abundances of the trees, and the relationships between the occurrences of trees and the conditions of the environment. What was lacking was a method by which these insights could be put to formal scientific tests in an efficient manner, and by which the rich background in knowledge could be applied to projections of future forest conditions, important both to basic science and to the applied problems of managing forest resources. The development of high-speed computers and high-level computer programming languages provided a means to make this transition, and allowed the development of the model described in this book.

It is interesting to contrast, in summary, the differences between the early development of ecological theory in animal and plant ecology during the first half of the twentieth century. Theory in animal ecology and theory in plant ecology had different and complementary strengths and weaknesses. Studies of vegetation emphasized the complex relationships between organisms and varying environmental conditions. Plant ecology emphasized environment–vegetation relations and focused on communities. Animal ecology emphasized the dynamics of a single population or two populations abstracted from the environment and formalized in classic nineteenth-century mathematics. The development of formal vegetation population theory, in the mode of the work on animal populations, did not begin to any degree until the late 1960s and early 1970s. In addition to the lines of theory I have described, Harper (1977) brought a new focus in plant ecology to the importance of the study of population dynamics of vegetation, and helped start a trend in that direction. As a result, by 1970 there was on the one hand a wealth of observations and individual understanding about the complexities of natural ecosystems, and on the other hand a simplified set of theories that were artificial and insufficiently connected to observation.

An Approach to Ecological Theory

How does a forest grow and change over time? How does it respond to changes in the environment, both intense and short-term, like a fire, and chronic and long-term, like climate change? A forested landscape forms a complex pattern that can be viewed hierarchically (O'Neill et al., 1986), which means as a set of phenomena, each connected to the others and each having its own characteristic scales of time and space. Because life on Earth is an interconnection of phenomena from the dynamics of the entire biosphere to the dynamics of populations of bacteria in a drop of mud in a lagoon, ecological phenomena are hierarchical. Most twentieth-century ecological theory focused on one level in this elaborate hierarchy, as I have discussed earlier, with animal ecologists restricting their interests to one or two populations abstracted from the environment, plant ecologists focusing on vegetation communities or associations. Only with the development of ecosystem ecology in the 1950s and 1960s was there a major attempt to break away from these restricted analyses, but for the most part attempts to deal with hierarchies have been limited to simple systems analysis models.

In terms of an ecological hierarchy, there are two approaches to the analysis of forests as ecological systems: top-down and bottom-up. The top-down approach begins with characteristics at the largest spatial scale and attempts to explain phenomena at each lower scale from the understanding achieved at the higher level. For example, characteristics of an ecosystem are used to explain the dynamics of an ecological community, whose characteristics in turn explain population dynamics, which in turn explain the responses of individual organisms. The bottom-up approach begins with a primary, lowest level. From this perspective, an ecological hierarchy is composed at the primary ecological level of individual trees of different species, sizes, ages, and vigor. Trees compete for resources: light, water, and nutrients. The success of a tree depends on inherited characteristics in response to the environment. With JABOWA, the primary level is the anatomy and physiology of the individual tree. The phenomena at each higher level are explained building on the understanding acquired about the processes at all lower levels. This book takes a bottom-up approach: the physiology and morphology of individual trees explain the response of the individual. The response of a group of interacting individuals explains the interactions among neighboring trees. Interactions among neighboring trees explain the dynamics of ecological communities, and communities, along with the status and dynamics of the local environment, explain forested ecosystems. Finally, the landscape's dynamics are the result of responses of individual ecosystems distributed spatially. A related discussion of a hierarchical perspective on derivatives of the JABOWA model can be found in Allen and Wyleto (1983).

This book is organized along the bottom-up path of explanation, beginning in a later section with an explanation of the dynamics of individual trees from a basis in physiology and anatomy.

Minimum Standards and Criteria for Models of Forest Stands*

One major and legitimate criticism of ecological models is the failure to test them adequately against observations, and prior to those tests, to establish minimum standards for model performance and criteria for evaluating that performance. Standards and criteria are analogous to setting up a confidence level before doing a statistical analysis. With criteria in hand, one could assess the quality and credibility of a model and its projections. The purpose of this section is to give guidance about standards and criteria.

In setting down criteria and standards, it is important to distinguish two uses of models: (1) to extend our scientific understanding; and (2) to help us deal with a specific, applied problem. The choice of criteria and standards can be more flexible (taking advantage of serendipitous results) for pure scientific research than for a specific applied problem.

A difference in emphasis between standards and criteria for models and those for statistical analysis lies with the importance of the conceptual basis— explanations in terms of causes and effects—in the criteria used to accept a model. Two analyses of the effects of climate change on forests illustrate the difference. Statistical correlations between the past distribution of fossil pollen

*This material was originally published, in a slightly different form, as Botkin, D. B., 1990 *Minimum standards and criteria for models of forest stands.* In B. Stout, ed., *Technical Bulletin,* National Countil of the Paper Industry for Air and Stream Improvement, Inc.

in pond sediments and past climate have been used to project future distributions of forest tree species under projected climatic conditions (Zabinski and Davis, 1989). This kind of approach is sometimes referred to as a "model" or a "statistical model," but I prefer to call it a "statistical correlation" to distinguish the method from models that have an explicitly conceptual basis. The acceptance of this method depends primarily on the strength of the correlations.

In the second approach, the JABOWA model and its descendants have been used to project forest composition and conditions based on a model in which cause-and-effect relationships are explicit (Solomon, 1986; Pastor and Post, 1988; Botkin, Woodby, and Nisbet, 1991). Such models are called "mechanistic," "conceptual," or "cause-and-effect" models. Each relationship between a biological variable and an environmental factor can be analyzed and modified so that the underlying explanation of the results can be made explicit. The acceptance of this method depends heavily on our belief that the cause-and-effect relations are correct.

Variables of Interest

There are many possible variables to consider in the formulation of a mathematical description of forest dynamics. Typically of interest are: amount of biomass, of production (both total and by species), and of merchantable timber; content of chemical elements; and shape and form of trees. When a model is to be used for an applied problem, one must choose at the outset, prior to the test, variables on which the performance will be determined. It is not legitimate to change the variables of interest when the intended test fails. For example, if a model selected to project below-ground production fails, one cannot then say, "Well, the model projects above-ground production successfully." On the other hand, when a model is to be used to extend our scientific understanding, it is legitimate to explore the response of different variables and to select these in an informal manner.

Structure and Choice of a Conceptual Basis

The choice of the structure and conceptual basis of a model is important. This choice is fundamental to the way that criteria and standards for models differ from those for statistical analyses. For example, one could establish as a criterion that the model be based on cause and effect, or that the model be founded at a certain hierarchical level of phenomena (Table 1.1). More specifically, one could establish as a criterion that the stand model considers growth of individual trees or begins with the physiology of individual leaves.

Hierarchical Criteria. Phenomena in a forest stand take place at various levels of detail. One of the most important decisions in the development of a model is the choice of the minimum level of detail to be considered. (An interesting discussion of hierarchical aspects of ecological phenomena can be found in O'Neill et al. [1986].) One scheme for classification of the level of detail is given

TABLE 1.1. Hierarchical levels for forest stands

Level	Phenomena
1	Cellular
2	Organ (leaf, root, etc.)
3	Individual plant
4	Population of plants of the same species
5	A plant association; all the species in the same local area
6	Ecological community
7	Ecosystem
8	Landscape element
9	Biome
10	Biosphere

in Table 1.1, for which the lowest level (highest detail) is physiological processes within cells, and the highest level (lowest detail) is the biosphere.

The kinds of models discussed earlier represent different hierarchical levels. Global models are at levels 9 and 10 in Table 1.1. Biophysical models are typically at levels 2 and 3. There is a tendency to require too low a level (too much detail) at the design stage in the development of a model. There is a trade-off in the selection. On the one hand, the lower the level (the greater the detail), the more satisfying the "explanation" of the phenomena seems. On the other hand, one rapidly reaches the limits of knowledge, the limits of the response of trees to details in their environment, and the limits of computer time and of money if one seeks always to build models from the lowest level. As an example, in the early 1970s some scientists attempted to develop forest-stand models of the growth of leaves and twigs in three dimensions. None of these models was successful; to my knowledge none was ever completed. One of the most important choices in the development of the model is the level of simplification, which involves the choice of hierarchical level. As a rule, one picks the level at which the model will operate to be one level lower than the level for which responses are required. For example, the response of populations is modeled from the level of individuals. In this way, concepts and assumptions agreed on for the individuals "explain" the response of the populations. This approach is consistent with the general character of explanation in science as discussed by Schroedinger (1952). In other words, I am claiming that the approach used in JABOWA is simply one that follows the mainstream of ideas about the nature of scientific explanation and the role of theory in that explanation. This provides a conceptual basis as an approach to a model but does not force details that are impossible to model, about which knowledge is lacking, and to which trees do not respond.

Generality, Realism, and Accuracy. Generality, realism, and accuracy concern the desired performance of the model (R. Levins, 1966). These must be distinguished one from another and a goal chosen.[2] *Generality* means that a model can be applied to many cases. The range of cases for which the model should provide projections is the *generality criterion* for model performance. *Realism* means that the projections of a model follow the same shapes of curves as the real trajectories. The acceptable average deviation from the shape of output for the variable of interest, in comparison with the real trajectory, is the realism criterion. I refer to this as the *qualitative criterion* for model performance. Most tests, however, seem to focus on accuracy, *accuracy* meaning the quantitative correspondence between the model's projection and the actual trajectory of the phenomenon. These distinctions may seem simple conceptually but are subtle in applications to ecological models. Consider the entire set of possible observations for variables of interest. Then choose a range for which the model should produce output. A completely general model would provide output for all observable cases (Mankin et al., 1975). The difficulty arises in choosing among models on the basis of their relative generality. Perhaps some observations are much more important than others, so that a model that included all important cases but was otherwise less general would be preferable to a model that included more cases but fewer of the important ones.

To illustrate this point, examine Figure 1.1, which shows hypothetical measurements of the growth rings of a single American beech tree in comparison with the projected growth of a tree from three models. Model one is an exponential curve, model two is a straight line, and model three is the fundamental growth equation of the JABOWA model, to be explained in Chapter two. The straight line, model two, approximates more points (lies near a greater number of observations) than either model one or three, and in this sense could be said to be the most general. However, if extended, the straight line leads to the absurdities that the beech tree will continue to grow at a constant rate forever and that it had no diameter until it was approximately 40 years old. Thus, the linear model is relatively general within the range of 60 to 160 years of age for the beech but is absurd on either side of that range.

The exponential curve, model one, goes exactly through the first five observations, and therefore could be said to be the most accurate within this range. However, this model also leads to the absurdity that the tree will grow indefinitely at an exponential rate, and the curve rapidly exceeds the hypothetical observed growth. Thus the exponential curve is accurate within a narrow range but is neither general nor realistic.

The sigmoid curve, model three, best approximates the general shape of the actual growth, showing a rapid increase in growth at early ages, a more or less linear range, and a slowing in growth as the tree ages. However, as illustrated, this curve intersects none of the hypothetical observations, so it is not accurate nor is it general in the sense expressed above.

The curves in Figure 1.1 have been drawn to suggest that realism is often to be preferred to either generality or accuracy, and this is a point I wish to make. However, there are occasions in which each of the three concepts will be the goal of choice. At this time there are no general rules for selecting among generality, realism, and accuracy; the selection must be made on an individual case basis.

In the case of realism, currently there are no general standards for a

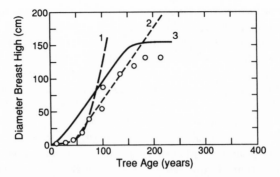

FIGURE 1.1. Generality, realism, and accuracy. Projections of three models are compared with hypothetical "observations" of the growth of an American beech tree, as measured from tree rings taken from a core. The hypothetical growth rings are shown as open circles. Curve 1 is an exponential, 2 is a straight line, and 3 is the fundamental growth equation of the JABOWA model (discussed in Chapter two). The exponential curve shows the most accuracy, because it passes exactly through the first four observations. However, it is unrealistic afterward, and leads to the absurdity that the tree will grow exponentially forever. The straight line curve approximates the largest number of points but passes through fewer than the exponential curve; hence the straight line is relatively general in comparison with the exponential but is less accurate. It leads to two absurdities: that the beech tree lived approximately 40 years with a zero diameter, and that its diameter would continue to increase at a constant amount each year indefinitely.

The sigmoid curve (three) is the most realistic in that it has the same general shape as the observations. However, it passes through none of the observed values. Hence the sigmoid curve is the most realistic, but it is least accurate. In the sense explained in the text, it is less general than curve two. It should be noted that some would argue, quite reasonably, that the realistic curve, as long as its accuracy is not too low, could also be viewed as the most general. The main conclusions to draw from this figure are that: (1) Realism is often to be preferred to accuracy and generality in the application of ecological models. (2) Realism can be taken as generality if the accuracy of the realistic curve is not too low. (3) However, whether generality, realism, or accuracy is preferred depends on the specific goals and application, and must be determined for each application of a model.

comparison between the shape of an output curve and the shape of the curve resulting from observations. Some more obvious choices that determine realism include: (1) It would not seem realistic for the model's output to lack an asymptote if the observations have one. (2) It would not seem realistic for the model's response to be concave when the observation is convex, and vice versa. (3) It would not seem realistic for the model's output to be linear when the observations are nonlinear, and vice versa, except when there is an explanation why a piece of a nonlinear response curve might appear linear. (4) It would not seem realistic for the model's response to be exponential if the observations were not, if the model's response was a positive exponential when a negative exponential was observed, and so forth.

As an example of realism, when we first applied the JABOWA forest model to forest stands in New England, the projections showed that biomass would peak in northern hardwoods forests in the middle of succession, between the first and second century following clearing, in contradiction to the previous dominant belief that biomass would peak in mature stands. An analysis of northern hardwoods forest stands in Wisconsin showed the same peak as projected by the model. The shape of the curve generated by the JABOWA model was realistic, while the older hypothesis was unrealistic (Botkin, Janak, and Wallis, 1972b).

This brings up the question: How do we deal with situations in which quantitative information is lacking? We can often devise a first approximation model, when the goal is a beginning of a realistic analysis, based on first principles. For example, suppose we lack specific information about the response of trees to temperature. To develop a first approximation of a model that is realistic, we can make use of the well-established relationship between cellular metabolic rates and temperature: within cells, the metabolic rate has a maximum at some temperature and decreases above and below that temperature. From this we can set down the general shape of a curve relating growth to average temperature, which is a curve with a maximum in the middle of the range of temperatures within which cells can survive. A parabola and a normal curve are examples of curves with the appropriate shape.

Following from this line of reasoning we can establish relative criteria for the evaluation of the realism of the model. The question is: How realistic is the model compared with the available information? If little quantitative information is available, a simple model based on first principles may represent a considerable advance, and may be valuable. On the other hand, in a situation where considerable quantitative information exists, the same simple model may be considered naive, oversimplified, and therefore unrealistic. For example, in the nineteenth century, when little was known about population growth and little information had been gathered from populations under natural conditions, the logistic growth curve for populations represented a considerable advance. Today, however, when we have abundant information about population growth, the logistic is naive, oversimplified, and unrealistic for most real situations.

A model's goal could be any one of the three criteria: generality, realism, and accuracy. Or it could be any pair or all three, although Levins suggested that for analytical models two but not three could be achieved (Levins, 1966). Even with modern computer simulation models, it is unlikely that all three could be achieved. At least one criterion must be given priority among the three, and ideally criteria ought to be selected prior to tests.

Tests of Accuracy. It would be desirable if ecologists could establish standards for acceptable accuracy. However, the approach will be different for deterministic and stochastic models. With a stochastic model, multiple runs of the model can be carried out and means, variances, and so on can be calculated. An acceptable level can be chosen a priori. With deterministic models, a standard can be chosen (e.g., the projections must be within 10% or 20% of the observed), but in this case the test is a comparison of a single value projected by the model against an observation.

Reasonable Standards of Accuracy. Because forests are intrinsically variable in space and time, there are practical limits to accuracy. As an example, wildlife managers know that it is generally impossible to distinguish the effects on plant growth of the removal by grazing or browsing of 10 percent of the leaves. (This is true for annual production and biomass, although the growth form of browsed woody plants is different from the growth form of unbrowsed plants.) As another example, let us consider a comparison of two stands thought to be identical. The stands might be plantations on the same soil type started in the same year. In my experience, those two stands might differ from one another by 10 or even 20 percent in any quantitative measure such as biomass or annual net production and still appear indistinguishable to the eye. In a study we have just completed in which we made the first statistically valid estimate of biomass for forests for any large area of the Earth, our goal was a 20 percent error of the mean (the observed for our number of samples was 23%) (Botkin and Simpson, 1990a).

In my experience with stochastic forest-stand models, two runs of the model, each with multiple trials and each beginning with the same initial conditions and subject to the same environmental trajectories and differing only in random number selection, can vary widely. The asymptotic level of variation for the JABOWA model of forest growth is about 10 percent. That is, as the number of iterations increases, the variance around the mean biomass, stem density, and basal area declines asymptotically, with the 95% confidence interval approaching about 10 percent of the mean as the iterations approach a large number. Thus for this model we do not require accuracy of better than 10 percent, and the number of iterations is chosen so that the 95-percent confidence level yields a 10-percent difference from the mean (i.e., if biomass is $50 \, kg/m^2$, the 95% C.I. is $5 \, kg/m^2$).

In basic scientific research it is unreasonable to demand that a model achieve an accuracy that exceeds the natural variability of forests as I have described it. Thus it is my opinion that a standard for accuracy need not be better than 10 percent and could be 20 percent. For example, if one seeks a 95-percent confidence level, then an acceptable value would be that the 95-percent confidence level differs from the mean by 10 or 20 percent. This applies to short-term measures of the current status of some factor or variable of interest. If concern is with economic issues, such as the changes in the value of timber as the level of air pollution changes, then a change much smaller than 10 or 20 percent may be of considerable interest.

Procedures to Establish Standards and Criteria

In developing a model, the following is a reasonable approach to a stepwise selection of standards and criteria.

1. Choose variables of interest. (What is it I want to project for the forest? Biomass? Merchantable timber? Production? Diversity? Transpiration?)
2. Determine the structure/conceptual level. Choose the hierarchical level

desired for the output. Then choose the next lower level (one level of detail greater for the minimum level at which the model operates) or write down explicit reasons that a finer level of detail is required and possible to achieve. For example, in many applications the population is the desired level for output (one wants to know the production and biomass by species). In this case, the minimum level at which the model operates is the individual.

3. Determine the generality criterion (the range of phenomena the model is required to reproduce).
4. Determine whether the goal is realism (the qualitative shape of an output curve) or accuracy (the quantitative difference between observed and projected values). If the goal is realism, then some (at this time in the development of the science) informal rules must be selected to determine acceptable correspondence between the shape of a curve and observations. There are no general rules at this time for such a selection.
5. Choose levels of accuracy. For forest-stand models, 10- or 20-percent accuracy in projections is reasonable.

Choice of Variables. Variables of interest include those that concern timber production: biomass, stem density, and merchantable timber; and those that concern multiple uses of forest ecosystems: biological variables including species diversity and dominance and amount of soil organic matter; biogeochemical variables: the content of chemical elements in the vegetation and soils and those lost to the atmosphere and to fluvial processes, as well as carbon-nitrogen ratios and other elemental ratios; and hydrologic variables: evapotranspiration and soil-water content. These are summarized in Table 1.2.

The State of the Forest

In attempting to understand a forest and to make projections about future conditions, we are concerned with identification of factors that define (or at least characterize) the state of the forest at a given time. The measure of the present state is taken to be the collection of information that is sufficient to project the next state of the forest. One of the obstacles to progress in ecology is the unclear definition of the state of an ecological system. One way that a computer model can aid scientific progress is by providing an explicit definition and characterization of the state of a forest expressed in a form (as a computer program) that can be readily modified as understanding changes or as assumptions are modified. The JABOWA model was conceived with this in mind as a technique to provide a dynamic repository of hypotheses about forest ecology that could be modified or replaced as knowledge increased.

Given the great complexity of natural ecological systems such as a forest, it is necessary to choose a state descriptor that is as simple as possible while capturing the essential dynamic qualities of the system. The approach to developing the model described here follows Occam's razor: always seek the simplest

TABLE 1.2. Properties and output of a forest-stand model

Minimum Dynamic Properties

1. Reproduce realistically growth to maximum age and size observed to occur in the region (and environment and habitat) for which the model will be operated
2. Simulate realistically competition among individuals for light
3. Simulate realistically the successional patterns in species composition, stem density, basal area, and biomass
4. Allow realistic response of growth, mortality, and regeneration of trees to major environmental factors including light, soil water content, and a measure of soil fertility

Desirable Dynamic Properties

1. Respond realistically to extreme climatic events
2. Respond realistically to seasonal climatic factors (e.g., spring or summer drought, unusual periods of spring warm weather)

Minimum Output Required

1. Stem density (total and by species)
2. Basal area (total and by species)
3. Biomass (total and by species)
4. Biomass for the stand divided by major components (total leaves, total bark, roots, bole)
5. Cubic feet (total and by species, for consideration of the amount of merchantable timber)

Output Desired but Not Necessary

1. Evapotranspiration
2. Carbon content (total and by species and by component)
3. Nitrogen content (total and by species and by component)
4. Content of other major chemical elements (total and by species and by component)
5. Components of dead organic matter: leaf litter, etc.
6. Age and size of trees at death

---

explanation consistent with observations. The JABOWA model was developed by beginning with the simplest case and simplest set of assumptions that explained available observations. Over time, the model has undergone changes to allow it to account for more observed qualities of a forest.

JABOWA has been intentionally developed "upward": knowledge about plant physiology and physiological ecology was used to explain the responses of individual trees. The responses of individuals in competition together explained community and ecosystem phenomena. First a model was developed that successfully mimicked the growth of individual trees. Next the model was extended to mimic a set of trees of different species competing under a constant environment. Then the model was extended to mimic the growth of a community of trees in a changing environment. Once the initial version of the model operated, it was tested and its implications examined. We considered the consequences of the assumptions to see if any were inconsistent with what was known to be true about forests.

A model of complex ecological phenomena like a forest can never be considered finished. The model is an integral part of the progress of science. Over time it serves as a progressively improving series of approximations. The

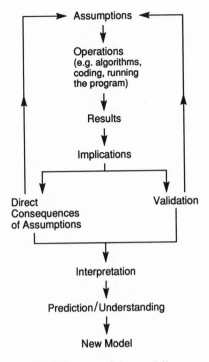

FIGURE 1.2. Diagram of the modeling process.

model must therefore be structured in an open way so that assumptions are altered easily and various assumptions can be tested and compared (Fig. 1.2).

Assumptions and Implications

The forest model demonstrates the implications of one's assumptions. Operation of the computer program forces one to confront these implications. There are several kinds or levels of assumptions, which I will refer to as parametric, primary, and secondary. The simplest kind is the quantitative value of a parameter in an equation, called parametric assumptions. Primary assumptions are those that define the biological response to environmental conditions. The equations or algorithms that represent these responses are called response functions or functional relationships. Secondary assumptions are those that determine the overall structure of the model (how functional relationships interact, for example, or how competition among individuals is expressed). The consequences of the model are its projections—its output for specific experimental conditions. Once available, these projections become hypotheses about nature. If they are verified they may become assumptions of a tertiary kind.

The process of model development is an aid to understanding. There are two kinds of understanding: one that arises from the development of the model, and another that results from using (sometimes called exercising) the model.

Computer modeling is a powerful tool when used properly. It is also dangerous when used blindly, especially when people who have not developed the model attempt to rely on it unquestioningly without understanding the assumptions. Even for the novice, models can serve to stimulate interest in the subject. One of my disappointments over the years is the way that JABOWA has been accepted and used by many investigators in its original form (sometimes recoded in a different computer language, but composed of exactly the same algorithms in the same structure), as if it were rigid and not open to improvement.

Notes

1. The original applications of these models to population dynamics occurred in the first decades of the twentieth century. They were carried out by Alfred Lotka in America, by Vernadsky in Russia, and by Volterra in Italy, with some elegant experiments conducted in the same period by the Russian G. F. Gause. The derivation of ideas from physics and chemistry was quite explicit. "If we cause to hesitate in defining life," wrote Lotka, let us "pass from legend to the world of scientific fact," and "borrow the method of the physicist." His fifth chapter was titled "The Program of Physical Biology," and he divided the discussion into statics and kinetics, exactly analogous to a text in mechanics. His chapter on kinetics investigated the solutions of simultaneous differential equations, which he applied to population growth.

The effect of predators on prey was studied in elegant and classic laboratory experiments by Gause (1934) using two single-celled *Protists, Paramecium caudatum* as the prey and *Didinium nasutum* as the predator. The experiments were conducted in small test tubes to which a nutrient solution containing bacteria (*Bacillus subtilis*), a food for the paramecia, was added at a constant rate. In one set of experiments, five paramecia were introduced into each tube; two days later three *Didinium* were added. The paramecia increased in abundance, reaching 120 individuals by the second day, and then declined rapidly when the predators were introduced. The predators increased to about 20 individuals. By the fifth day, the paramecia were completely eliminated by the predators, after which the predators died of starvation.

In other experiments, Gause repeatedly tried to obtain the oscillations predicted by the Lotka–Volterra equations, but could not. At best he was able to sustain several cycles before one or the other species became extinct, but the oscillations were not properly out of phase as predicted by the Lotka–Volterra equations. Gause concluded that the periodic oscillations were not a property of the interaction itself, as predicted by the equations, but resulted from other "interferences." Gause's analyses of theory are among the most scientifically complete in the first half of the twentieth century in that he considered concepts, formal theory, and experimental tests. In spite of the fact that his tests falsified the theory, the Lotka–Volterra equations continued to be used widely. This is what Kenneth Watt was referring to when he said that ecology had too much theory.

2. Many readers will recognize that these distinctions were first suggested by R. Levins (1966) in "The strategy of model building in population biology," *American Scientist* 54:421–431. However, he used the terms *generality, realism,* and *precision*. I have changed the last term to *accuracy*, to be consistent with the use of the terms *accuracy* and *precision* in biology, chemistry, physics, and statistical analysis. That is, I have followed the distinction given by Sokal and Rohlf (1969) that "*Accuracy* is the closeness of a measured or computed value to its true value; *precision* is the closeness of repeated measurements of the same quantity" (p. 13). In the discussion that follows, it is Sokal and Rohlf's idea of accuracy that is relevant.

2

The Forest Environment

This chapter begins the explanation of the forest model. The model has undergone a progression of changes over the years, but for simplicity I will refer to two versions of the model, the original version, referred to as version I or simply JABOWA, and the current version, version II or JABOWA-II. The major differences between versions I and II are the treatment of site conditions, reproduction, and growth of trees. Version II differs from version I in (1) species-specific differences in the shape of the fundamental growth equation; (2) improvements in the calculation of soil-water balance, including greater realism in the calculation of snow conditions and snow melting (the model now calculates a complete and comparatively realistic soil-water budget); (3) addition of two soil-water response functions, one for drought conditions and one for too-wet conditions; (4) addition of an index of soil fertility, represented by response to soil nitrogen; and (5) environmental conditions affecting both growth and regeneration (in version I, all response functions affected tree growth, but only some aspects of these functions affected regeneration in a restricted way).

From a technical point of view there are many other differences. Version I was written in FORTRAN for use on mainframe computers and lacked an integrated graphics display. (In the early 1970s, a version with a graphics display was produced, but it required the use of two computers, one to run the model, the second to take the model output and create a graphics display.) Version II, available for use with this book, is coded in C, has an integrated color graphics display, and runs on a personal computer.

Because two versions of the model are discussed, it is important to keep the following in mind when reading the book: *Unless otherwise stated, each algorithm described applies to both versions. Unless otherwise stated, values of parameters and results shown are for version II.*

In the model, a tree is characterized fundamentally by a diameter, height, and species. Its species is in turn defined by a set of parameters. Calculations are carried out in the following order: after initial conditions are selected by a user of the model and calculations required by these initial conditions are carried out by the computer, the population dynamics of trees on a plot are calculated in the order: regeneration of new stems, growth of existing stems, and determination of which live stems die. The model is coded so that it is easily modified. As explained in the first chapter, the code itself can be viewed as a set of hypotheses about forest dynamics, while the operation of the model on the computer demonstrates the implications of these hypotheses. In this sense the model is intentionally open to change; it is an open system in the sense explained by Soros (1990).

The Forest Environment

Like all living things, a tree must grow, reproduce, and die. The rate of growth of a tree varies with many factors, environmental and inherited. Primary environmental factors are the amount of sunlight, soil moisture, temperature, and availability of nutrients. Some risk of mortality always exists for trees; even when healthy they can be killed by fire or storms. The risk of death increases, however, when trees grow poorly.

Tree Growth

Consider a single tree growing under optimal conditions, when all the resources the tree requires are available. Growth is the net accumulation of organic matter and is therefore the difference between the amount of new organic matter produced by the leaves and the amount used by the rest of the living tissues. Of course the phenomenon is more complex. An individual leaf first carries out gross photosynthesis, which is the amount of organic matter produced before any utilization. Some energy fixed in the photosynthetic process is used in respiration by the leaves, and the net remaining—leaf net photosynthesis—is available for tree growth and is either stored within the leaves or exported via the phloem to the rest of the tree, to the stem and roots. Thus under optimum conditions, when no environmental resource is limiting, growth of a tree is directly proportional to the abundance of leaves, which can be measured as leaf weight or leaf area, and is inversely proportional to the amount of living but nonphotosynthetic tissue. There have been two alternative approaches to seeking an understanding of the shape and form of plants, one conceptual and the other strictly observational. Rashevesky, in his famous book *Mathematical Biophysics*, analyzed the shape and form of a tree as an engineering problem. He took the approach that a tree needs to support its leaves, branches, limbs, and twigs exactly as does any engineered structure, and he developed a generalized equation expressing tree shape. His equations are more detailed than required for our purposes, since they concern details about an individual tree that are not crucial to its competitive success, but his conceptual approach to tree shape and form is fundamentally correct.

The living nonphotosynthetic tissue is found in the xylem, phloem, and cambium, which lie in a narrow ring of essentially fixed width in the inner bark. The quantity of nonphotosynthetic tissue can therefore be approximated as proportional to the surface area of a simple geometric solid. The surface area of such a simple solid, such as a cylinder or cone, is proportional to the product of the diameter times height. From these simple statements we can set down a fundamental growth equation, which is stated first in terms of a change in tree volume, which is proportional to the square of the diameter times the height, so that:

$$\delta(D^2H) = R * \text{LA}\left(1 - \frac{DH}{D_{\max(i)}H_{\max(i)}}\right) * f(\text{environment}) \qquad (2.1)$$

where D = diameter at breast height
H = height of tree
R = a constant
LA = leaf area of tree
$D_{\max(i)}$ = maximum known diameter for trees of species (i)
$H_{\max(i)}$ = maximum known height for trees of species (i)

In this equation, D^2H represents the tree volume and DH represents the relative amount of living, nonphotosynthetic tissue. As the tree expands in size, the ring of nonphotosynthetic tissue, which in three dimensions forms a shell of fixed width, moves outward and increases in volume proportional to the surface area of a sphere or cylinder, that is, proportional to the production of diameter times height (Esau, 1977). Dividing DH by $D_{\max(i)}H_{\max(i)}$ sets as an upper limit to the size that a tree can reach the product of the maximum diameter times the maximum height. Hence this equation assumes that a tree cannot exceed the maximum diameter and height known to occur for individuals of its species, and that at this height and diameter the respiration of nonphotosynthetic tissue equals the photosynthetic rate of the leaves. In equation (2.1), f is a function of the effects of climate, shading, soil moisture, and fertility on tree growth. Algebraically R serves as a proportionality constant, and it disappears in the conversion of the equation for change in volume to the equation for change in diameter. However, as Fownes (personal communication, July 1991) points out, biologically, R is the maximum intrinsic net assimilation rate. There are two arguments to make this point. First, simply inspect equation (2.1) and consider when a seedling first begins to grow and $DH/D_{\max}H_{\max}$ (where D is very close to zero) and f (environment) is equal to 1. Then the change in volume is simply proportional to $R*\text{LA}$, and R can be interpreted to be at a maximum.

The second justification, provided by Fownes, is to consider the similarity between equation (2.1) and the classic growth analysis for plants, which assumes that change in biomass, dB/dt, is proportional to the total leaf area, LA, times the net photosynthetic or net assimilation rate per unit of leaf area, η, or

$$\frac{dB}{db} = \eta * \text{LA}$$

In a recent paper, Fownes and Harrington (1990) showed that growth trajectories for different spacings of trees in eucalyptus plantations on Hawaii could be summarized by a single empirical curve of η versus the leaf area index (LAI) for the stand, so that

$$\eta = \eta_{\max} * e^{(-\kappa*\text{LAI})}$$

where e is the base of natural logarithms. They also showed that the exponential factor, $e^{(-\kappa*LAI)}$, is the product of allometric factors (factors related to growth and form), which are analogous to $1 - DH/D_{max}H_{max}$ in equation (2.1), and competitive factors analogous to the shading component of f (environment). These two arguments lead one to conclude that R, from a biological point of view, is the maximum intrinsic net assimilation rate.

Change in Diameter. It is convenient and useful to convert equation (2.1) to calculate diameter, rather than volume, increment. Diameter is simple to measure (much simpler than tree volume) and to monitor directly, and is known to respond to changes in environmental conditions, as is reviewed in many basic books about silviculture, such as that by D. M. Davis (1985). The fundamental growth equation (2.1) can be solved for the change in diameter if we make the assumptions mentioned above concerning the relationship between diameter and height. It is well known that there is a strong correlation between tree height and diameter within a species (see for example, Crow [1978]), as this is a basis for many estimations, including merchantable timber volume and biomass of trees. To develop an equation for change in diameter from the equation for change in volume, we must make some additional assumptions. The first is that leaf area is proportional to leaf weight, W, and

$$W = C_i D^2 \tag{2.2}$$

where C_i is a constant.

A rationale can be used to justify the derivation of equation (2.2) relating leaf weight to the square of the diameter. Assuming that a unit of living but nonphotosynthetic tissue requires a unit of leaves to provide its metabolic needs, then, other things being equal, an idealized tree would add leaves in proportion to the amount needed to support the tissues of the rest of the tree. If this were true, a tree would add leaves with the square of the diameter. Equation (2.2) follows from this assumption. This was the rationale used in the original development of the model.

Another rationale to justify equation (2.2) is the pipe stem model, which gives the reverse causality—that the weight of leaves is limited by the supply of water transported through the xylem from the roots, and that the xylem cross-sectional area is proportional to the square of the diameter, as explained before.

At the time that the model was first derived, some authors suggested that leaf weight was proportional to several different powers of the diameter, and the second power lies in the middle of the observed range (Baskerville, 1965; Kittredge, 1948; Perry, Sellers, and Blanchard, 1969; Whittaker and Marks, 1975). More recently, Fownes (personal communication, July 1991) suggests that height is proportional to the square root of the diameter. If this were true, then DH would increase with $D^{1.5}$. Continuing to assume that leaf area is proportional to D^2 would imply that the photosynthesis/respiration ratio would continue to increase, rather than approach an asymptote, as assumed for equation (2.1).

Equation (2.2) assumes that the density of leaf tissue (grams per square centimeter of surface area of the leaf) is a constant. In fact, the density of leaves varies within a tree. "Sun leaves," those that have developed in bright sun, are smaller and thicker than "shade leaves." Sun leaves have a thicker waxy cuticle and more palisade cell layers. In some species the difference between sun and shade leaves is striking. Sun leaves at the top of a scarlet oak, for example, which are rarely seen by people, look more like leaves of scrub oak than the characteristic leaves of scarlet oak shown in field guides. The relationship expressed in equation (2.2) represents the average condition for a tree; for the purpose of the model this approximation is sufficient.

The function relating height to diameter is

$$H(D) = 137 + b_2 D - b_3 D^2 \tag{2.3}$$

where 137 is the height in centimeters at which diameter at breast height is measured. Hence that amount must be added to the height calculation.

Typically, relationships between height and diameter are determined by statistical regression analysis, fitting a set of data to a number of equations without restricting the choice of curves from allometric principles. At the time version I of the model was originally developed, Ker and Smith (1955) compared approximately 20 regression models, some of which gave quite similar fit to the data. More recently, White (1981) has observed that in these relationships the diameter is rarely raised above the third power. While these statistical relationships are useful as long as the correlation and significance are high, they leave one without a conceptual basis for a choice among many of them. In developing the model, we decided that it would be reasonable to restrict the relationship to a second-order equation, so that a product of height times diameter, with height a function of diameter, would convert to an equation of the third order consistent with the geometry that the volume of an object is a function of the cube of a linear dimension. This may seem a somewhat arcane justification to some, but it seemed better than no justification at all. The reader should note that this rationale does not assume that height is simply proportional to the square of the diameter, merely that the regression equation should not include a term greater than the second power. The exact form of equation (2.2) is a minor point in the model and could readily be changed if a change was thought useful.

Assuming that a tree of species i reaches its maximum height and maximum diameter at the same time, that is, when

$$D = D_{max(i)}$$

then $H = H_{max(i)}$ and $\dfrac{\delta D}{\delta t} = 0$, then it follows that

$$b_{2(i)} = \frac{2(H_{max(i)} - 137)}{D_{max(i)}} \tag{2.4}$$

$$b_{3(i)} = \frac{(H_{max(i)} - 137)}{D^2_{max(i)}} \tag{2.5}$$

(The derivations of b_2 and b_3 are given in Appendix I.)

The treatment of tree shape and form follows Occam's razor. It is the simplest explanation consistent with observation. Some of the first computer simulations of the 1960s and early 1970s treated shape and form in extreme detail, attempting to simulate the competition among individual leaves and branches. But the resulting three-dimensional detail was unnecessary to account for dynamics of a population of trees and was too complex to model successfully.

A strictly observational approach to tree shape and form, known as "dimension analysis," is the method by which the relationships between mass of the parts of a tree and its simple linear dimensions—diameter and height—are determined. In dimension analysis, individuals representing the range of sizes within a species are cut down. Components—leaves, twigs, bole, and so on—are either measured or subsampled for weight (see Bormann et al., 1971; Whittaker, 1966; Whittaker and Marks, 1975; Woods, Feiveson, and Botkin 1991). Statistical analyses of these measurements yield correlations between mass and height and diameter that can then provide a nondestructive method to calculate the mass of a tree. (There are numerous dimension analysis studies for individual species, for example, Aldred and Alemdag [1988], Evert [1985], Lieffers and Campbell [1984], Smith and Brand [1983]).

Dimension analysis is used in two ways in the model. In the dynamics of the model, the weight of leaves shading other trees is calculated from tree diameter. In the output from the model, dimension analysis relationships provide biomass of tree components (twigs, leaves, stems, roots).

Since the original development of the model many more measurements of the mass of the parts of a tree as a function of diameter and height have been made, and these could be used to modify the fundamental assumptions of equation (2.2) if this would improve our understanding or ability to make projections.

I have discussed in detail the rationales for our choice of relationships governing the shape of trees as an illustration of the overall approach we took in the development of the model. That approach is to search for generalizations that are the simplest consistent with understanding, that are derived from observations but where possible not blindly chosen from statistical analyses of observations. Ultimately, observations must be depended on, but where there are few observations of complex phenomena, one must seek to determine the simplest conceptual basis consistent with the observation. This seems to me to be one key to the practice of developing models for complex ecological systems. The term *conceptual basis* has a specific meaning in this context. It means that an understanding of phenomena at a lower level of organization is used to explain and project phenomena at a higher level of organization consistent with the discussion of hierarchy theory mentioned briefly in Chapter one (O'Neill et al., 1986). For example, an understanding of plant physiology (a "lower level of organization") is part of the information used to explain competition among trees (a "higher level of organization").

From equations (2.2), (2.3), (2.4), and (2.5), equation (2.1) can be rewritten as $\delta(D^2H) = 2DH\delta D$, so that

$$\delta D = \frac{\{G_i D[1 - (DH/D_{max}H_{max})]\} * f(\text{environment})}{274 + 3b_2 D - 4b_3 D^2} \tag{2.6}$$

$$\delta D = \frac{\{G_i D[1 - [D(137 + b_2 D - b_3 D^2)/D_{max}H_{max}]]\} * f(\text{environment})}{274 + 3b_2 D - 4b_3 D^2} \tag{2.6a}$$

where R (replaced here by G and C) is a proportionality constant and $G_i = R_i C_i$. Note that G, D_{max}, H_{max}, b_2, and b_3 are species specific. For simplicity of the formulation here, the subscript i to denote species specificity is omitted in some places, but should be implicit for the reader. Note also that the variables D and H are for each individual tree. The subscript n, for individual tree number, is also omitted for clarity, but should be implicit for the reader. (See Appendix IB for details.)

As an example, the parameters for sugar maple are

$G = 118.7$
$D_{max} = 170$
$H_{max} = 3350$
$b_2 = 37.8$
$b_3 = 0.111$

so that

$$\delta D = \frac{\{GD[1 - [D(137 + b_2 D - b_3 D^2)/D_{max}H_{max}]]\} * f(\text{environment})}{274 + 3b_2 D - 4b_3 D^2}$$

$$\delta D = \frac{\{118.7D[1 - [D(137 + 37.8D - 0.111D^2)/170*3350]]\} * f(\text{environment})}{274 + 3*37.8D - 4*0.111D^2}$$

The resulting graph of equation (2.6), when D is plotted against age (in years) or size is sigmoid, asymptotic to D_{max} (Fig. 2.1). The parameter G determines how early in its age (or size) a tree achieves most of its growth. The larger the value of G, the earlier a tree will reach one-half of its maximum size (the inflection point in the curve) (Figs. 2.1, 2.2). For example, sugar maple reaches one-half maximum size at age 117 with G set at the default value ($G_0 = 118.7$). (See Table 2.1, p. 41.) It reaches maximum size at year 30 when $G = 2G_0$, and at year 231 when $G = G_0/2$ (Fig. 2.2).

In JABOWA-I, all species were given the same shaped curve. G was chosen so that a tree reached two-thirds its maximum size at one-half its maximum age, a rather arbitrary decision based on natural history observations and experience, since detailed information was lacking (Botkin et al., 1972a, b). This sometimes resulted in unreasonable growth rates. In the calculation of values of G, the current version of the forest model differs from the original version. Currently the value of G is chosen according to the known maximum growth rates of each species, so that the ratio D/D_{max} varies among species from 2:9 to 8:9 at half-maximum age. Adjustments in the current version are meant to reflect what is known about the natural history of each species studied and whether its growth tends to be concentrated early or late in its life history.

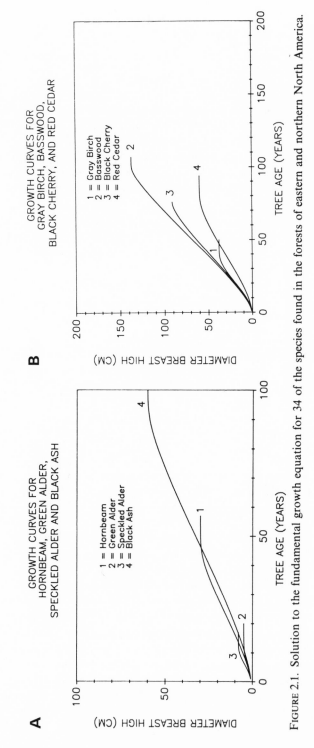

FIGURE 2.1. Solution to the fundamental growth equation for 34 of the species found in the forests of eastern and northern North America.

36

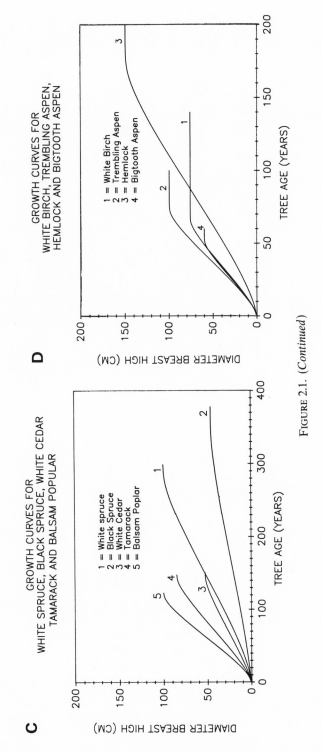

GROWTH CURVES FOR
WHITE BIRCH, TREMBLING ASPEN,
HEMLOCK AND BIGTOOTH ASPEN

1 = White Birch
2 = Trembling Aspen
3 = Hemlock
4 = Bigtooth Aspen

D

DIAMETER BREAST HIGH (CM)

TREE AGE (YEARS)

GROWTH CURVES FOR
WHITE SPRUCE, BLACK SPRUCE, WHITE CEDAR
TAMARACK AND BALSAM POPULAR

1 = White spruce
2 = Black Spruce
3 = White Cedar
4 = Tamarack
5 = Balsam Poplar

C

DIAMETER BREAST HIGH (CM)

TREE AGE (YEARS)

FIGURE 2.1. (*Continued*)

37

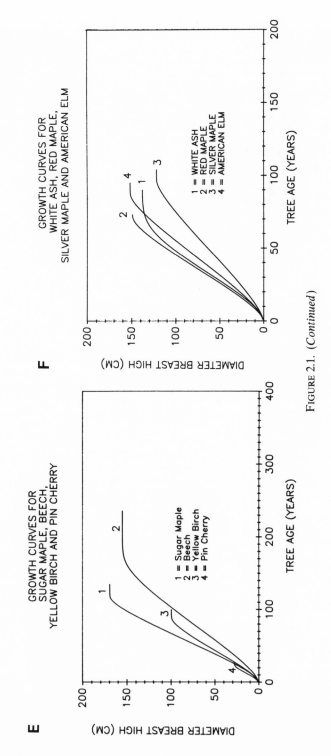

F

GROWTH CURVES FOR
WHITE ASH, RED MAPLE,
SILVER MAPLE AND AMERICAN ELM

1 = WHITE ASH
2 = RED MAPLE
3 = SILVER MAPLE
4 = AMERICAN ELM

DIAMETER BREAST HIGH (CM)

TREE AGE (YEARS)

E

GROWTH CURVES FOR
SUGAR MAPLE, BEECH,
YELLOW BIRCH AND PIN CHERRY

1 = Sugar Maple
2 = Beech
3 = Yellow Birch
4 = Pin Cherry

DIAMETER BREAST HIGH (CM)

TREE AGE (YEARS)

FIGURE 2.1. (*Continued*)

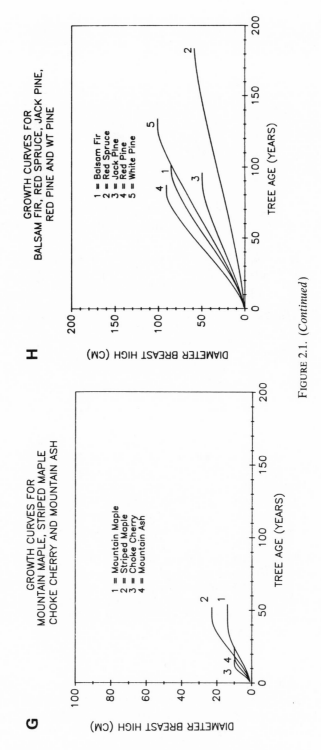

FIGURE 2.1. (*Continued*)

39

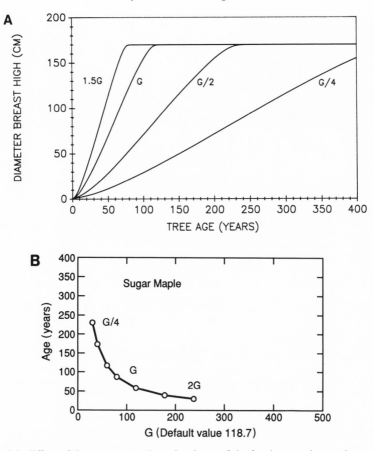

FIGURE 2.2. Effect of the parameter G on the shape of the fundamental growth equation.
(A) Solution to the fundamental growth equation for sugar maple with four different
values for the parameter G.

$G = G_d = 118.7$ (default value), $1/4G_d$, $1/2G_d$, and $1.5G_d$. The default value of G is
chosen so that a sugar maple tree growing under optimum conditions reaches two-thirds
its maximum size at one-half its maximum age. The growth curve is sensitive to the
value of G.

Under optimum conditions (when no environmental factor reduces growth), the age
at which a tree reaches one-half its maximum declines exponentially as G decreases. A
high G value means that a tree grows most rapidly when young. A low value means
that a tree gows most rapidly when comparatively old. Values for the parameters in the
fundamental growth equation are found in Table 2.1. (B) Age at which $D_{max}/2$ is reached
as a function of the parameter G.

In Botkin et al. (1972b), we proposed an alternative, more realistic method
for estimating G: Require that a tree obtain the maximum observed diameter
increment for its species, $\delta D_{max}/\delta t$, at some point during its lifetime, a measure
that can be obtained from field observations. Once $\delta D_{max}/\delta t$ has been so selected,

TABLE 2.1. Basic forest model parameters: intrinsic growth factors[a]

Species	Maximum Age (years)	Maximum Diameter (cm)	Maximum Height (cm)	B_2	B_3	G	C
Sugar maple	400	170	3350	37.8	0.111	118.7	1.570
Beech	366	160	3660	44.0	0.137	87.7	2.200
Yellow birch	300	100	3050	58.3	0.291	143.6	0.486
White ash	150	150	2440	30.7	0.102	147.5	1.750
Mountain maple	25	14	500	53.8	2.000	72.6	1.130
Striped maple	30	23	1000	76.7	1.700	109.8	1.750
Pin cherry	30	28	1126	70.6	1.260	227.2	2.450
Choke cherry	20	10	500	72.6	3.630	233.3	2.450
Balsam fir	200	86	2290	50.1	0.291	102.7	2.500
Red spruce	400	60	2290	71.8	0.598	50.7	2.500
White birch	140	76	3050	76.6	0.504	190.1	0.486
Mountain ash	30	10	500	72.6	3.630	155.6	1.750
Red maple	150	150	3660	47.0	0.156	213.8	1.570
Scarlet oak	200	30	3050	194.2	3.230	128.7	1.750
Hornbeams	150	30	1520	92.1	1.530	144.4	0.486
Green alder	30	5	300	65.2	6.520	143.3	2.000
Speckled alder	30	8	400	65.8	4.100	196.9	2.000
Chestnut	200	122	2740	42.7	0.175	195.2	1.750
Black ash	70	60	2130	66.4	0.554	96.2	1.750
Butternut	90	91	3050	64.0	0.352	192.2	1.750
White spruce	200	53	3350	121.2	1.140	91.8	2.500
Black spruce	250	46	2740	113.9	1.240	32.0	2.500
Jack pine	185	50	3050	116.5	1.160	142.0	2.000
Red pine	275	91	3050	64.0	0.352	156.4	2.000
White pine	450	101	4570	87.8	0.435	141.2	2.000
Trembling aspen	100	100	3050	58.3	0.291	173.7	0.486
White oak	600	122	3050	47.8	0.198	72.0	1.750
Red oak	400	100	3050	58.3	0.291	107.7	1.750
White cedar	400	100	2440	46.0	0.230	35.7	2.500
Hemlock	600	150	3660	47.0	0.156	86.0	2.000
Silver maple	125	122	3960	62.7	0.257	164.8	1.570
Tamarack	200	85	3050	68.5	0.403	86.3	2.000
Pitch pine	200	91	3050	64.0	0.352	86.5	2.000
Gray birch	50	38	910	40.7	0.535	119.5	0.486
American elm	300	152	3840	48.7	0.160	180.0	1.600
Basswood	140	137	4270	60.3	0.220	169.8	1.600
Bigtooth aspen	70	60	2130	66.4	0.554	176.7	0.486
Balsam poplar	150	100	2440	46.0	0.230	232.5	0.486
Black cherry	258	91	3050	64.0	0.352	166.7	2.450
Red cedar	250	60	1520	46.1	0.384	88.7	2.000

[a]See text for explanation and computation of B_2, B_3, and G.

the value of G is calculated as

$$G \cong 5H_{max}\left(\frac{\delta D_{max}}{D_{max}}\right) \qquad (2.7)$$

(The derivation of equation (2.7) is given in Appendix I.)

Although preferable because it is linked to observation, this method has not been employed by anyone to my knowledge. (A connection between *G* and classic botanical growth analysis is given in Note 1 to this chapter.)

Versions of the Model

Over the years JABOWA has undergone many modifications. The major conceptual differences between version I and version II have to do with

1. The number of environmental factors that affect tree growth. In version I, growth is affected by light and temperature. In version II, growth is affected by light, temperature, soil nitrogen, soil-water depletion, and soil-water saturation.
2. Regeneration. In version I, the program determined whether temperature conditions were within the range that would allow tree growth for a species, and for some species whether the soil moisture was *above* the minimum for the species. If these conditions were met, regeneration of a species could take place. The actual number of saplings added to a plot in any year was a stochastic function of available light. In version II, regeneration varies with the same functions of environmental conditions as does growth, except that the response to light is somewhat different, as described in Chapter 3.
3. Soil-water balance. Version II has a considerably improved and more realistic calculation of soil-water balance. Snow melt is treated in a physically more realistic manner. Depths of the water table and soil are separate parameters that influence tree growth and regeneration.
4. Species-specific differences. In version II, the fundamental growth equation can be modified for species-specific differences in morphology.

It is useful to repeat here that the algorithms given subsequently are found in both versions unless otherwise stated; values of parameters and results are for version II unless otherwise stated.

Competition among Neighboring Trees

Trees affect each other directly in local neighborhoods. A *forest neighborhood* is an area small enough so that a single large tree can shade all other trees in that neighborhood. In the model, the choice of the size of a plot should match the size of a forest neighborhood, and the plot size will change with the maximum size that mature trees of the forest can reach. The plot size should be small enough so that a large tree will shade every other tree on the plot at least for some time during an annual solar cycle. Thus where the characteristic tree size is large, a large plot should be used. Larger plots will be necessary in sequoia forests than in jack pine forests. In the input options of the original version of the model, the plot size could be altered by the user for each run. The default value was set at 10×10 m because (1) this is a common size in forest sampling,

(2) I have used this size in my field studies, (3) Bormann et al. (1971) used this plot size in the original sample of forest trees at the U.S. Forest Service Hubbard Brook Experimental Forest, for which the model was originally devised, and (4) this size is appropriate for the boreal and northern hardwoods forests of North America. Since few users of the first version opted to vary plot size, we discontinued this as a user-defined option in the interactive menus for the new program. Plot size is, however, simply and readily changed in the code.

In the model, environmental conditions at a site change over time at different rates. Soil fertility, expressed in terms of soil nitrogen content, changes most slowly. Soil moisture conditions change rapidly and are affected by monthly rainfall and evaporation. Light varies at many temporal scales, but the total quantity of light available on an annual basis is the primary measure of light affecting tree growth. In a real forest, the location of a site in relation to elevation, slope, and aspect can be important, as these affect temperature, precipitation, and soil characteristics—soil depth, average soil particle size, and percentage of rock in the soil—that influence the availability of moisture. The influence of these factors on growth and regeneration of trees in the model are discussed in sections that follow.

Calculations of Available Light

Anyone who has walked in a woodland on a breezy, sunny day knows that light beneath the trees changes all the time. Sun flecks, the splashes of direct sunlight, flash across the forest floor as the wind moves the branches. Where a tree has recently died, the forest is more open and the light brighter than elsewhere. In deep shade, light intensity may be one-tenth or one-hundredth or less of full sunlight. If we want to mimic competition among trees for light, how do we characterize this complex pattern that varies over time and location? There is a surprisingly simple solution, as long as our interest is in the overall growth of a tree on an annual or monthly basis. The solution lies in the observation that variations average out over time, so that a forest canopy acts over time as a uniform light-absorbing medium, as if it were a homogeneous solution of liquid pigments, chlorophyll, and accessory leaf pigments. That a forest canopy acts as a uniform light-absorbing medium has been shown in several studies (Kasanaga and Monsi, 1954; Loomis, Williams, and Duncan, 1967; Miller, 1969; Perry et al., 1969). In one of the most careful studies of the phenomenon, Baldocchi et al. (1984) found that photosynthetically active radiation was attenuated exponentially in an oak–hickory forest at Oak Ridge National Liboratory. These investigators used three towers with a moving tram system at eight levels in the forest. Regression analysis gave R^2 of 0.96 for the relationship between $ln(I_h/I_0)$ (the logarithm of light intensity at height h divided by light intensity above the forest canopy) and plant area index (the number of square centimeters of plant material above a square centimeter of ground surface).

This provides a key simplifying assumption of the model. Following from this assumption, AL(h), light intensity at any height h above the ground in a forest can be represented by

$$AL(h) = AL_0 e^{-k\int_h^{\infty} LA(h')dh'} \tag{2.8}$$

where AL_0 is the light available *above* the forest, h is the depth, measured down into the forest, $LA(h')$ is the distribution of leaf area or leaf weight with height per unit horizontal area and k is a constant called the light extinction coefficient (Botkin et al., 1972b). $LA(h')$ is the "shading leaf area" or "shading leaf weight." In the model, $LA(h')$ is leaf weight. This attenuation of light as a negative exponential is known as the Beer–Bouguer law. For each tree, the weight of leaves is represented as a scalar quantity with no vertical distribution—the leaves are assumed to be concentrated at the top of the tree. A walk in a forest will reveal to any careful observer that the distribution of leaves with height varies from species to species and roughly correlates with the successional stage characteristic of a species. Shade-tolerant species have leaves that extend from the top of the tree to near the ground, whereas the lowest leaves on a shade-intolerant tree are typically high up. Physiologically, this is the result of the semiautonomous role of leaves. For a leaf to be on a twig next year, this year's leaf must produce enough photosynthates to provide for its own needs, to export to the rest of the tree, and to produce next year's leaf bud. In the low level of light deep in the canopy, the photosynthetic response of a leaf of a shade-tolerant species can be sufficient for this to occur, whereas it is not sufficient for the leaf of a shade-intolerant species.

In contrast to this natural history observation, the model assumes that: (1) all of the leaf weight is concentrated at a point at the top of a tree; (2) therefore that all of the leaves of one tree shade any shorter tree, and (3) conversely, that none of the leaves of any shorter tree shade any leaf of a taller tree. These are among the key simplifying assumptions of the model.

The model was developed in stages, first a model of a single tree, then of a set of trees competing in a constant environment, then of a set of trees of different species competing in an environment that could be varied. At each stage, the assumption that leaves were concentrated at the top of a tree gave realistic results.

At various times I have considered revising this aspect of the model. At the time the model was originally developed, there were few data about the vertical distribution of leaves on a tree. Following Occam's razor and considering the lack of data, there was no reason to create a more detailed model. This may seem a curious result. But consider what is involved when one seeks a more detailed model. One needs quantitative information about the depth of the crown (i.e., height to the lowest branch that contains a live leaf must be measured in addition to total tree height), as well as the shape of the function describing the vertical distribution of leaves, determined as a function of tree diameter, height, and species.

The species-specific information would seem important because, as mentioned earlier, shade-tolerant trees have leaves extending farther down their trunks than shade-intolerant species.

Taking such an approach, one might assume a simple vertical distribution, such as a cylinder of leaves around the trunk, or leaves distributed following a normal curve with the leaves reaching a maximum density at a point midway between the top of the tree and the lowest leaf. Without detailed empirical studies, one would be trapped into picking parameters with little real basis, either conceptual or observational.

Once one begins to follow this train of thought, other details come to mind and seem relevant, such as the position of each tree relative to the others in the horizontal plane of the plot, so that shading as a function of sun angle could be considered; and perhaps even the three-dimensional positioning of leaves relative to one another. At this point one becomes involved in a level of detail far exceeding knowledge and several levels down in the hierarchy from population dynamics of whole trees.

In the early development of the model, the simple approach taken to leaf location seemed key to making development of the model practical. That it has also turned out to be sufficient to the present is intriguing.

Some might conclude that competition among trees is not a precise phenomenon, but instead that stochastic events that remove leaves and alter the detailed shape of the crown, including herbivory, diseases, and physical events such as wind and fire, are sufficiently important and common to prevent the evolution of very precise competitive interactions, and that overall rank ordering of competition, weighted quantitatively by species, is all that is possible.

Others may wish to explore this aspect of the model in more detail. Present high-speed microcomputers and minicomputer work stations have the speed and storage to allow the construction of much more detailed models. Twenty years of dimension analysis provides more data about the depth of the crown, if not the distribution of leaves with height. Selecting such an approach should not violate Occam's razor, or exceed what we know observationally about trees in a forest.

The light extinction coefficient, k, was chosen originally to approximate known absorption of forest canopies, typically resulting in deep shade in a mature forest where light intensity is one-hundredth or less that of full sunlight. (See, for example, Botkin, 1969.) The value of $k = 1/6,000$ was found to give good agreement with these observations and to give good results using units of leaf weight. There is no reason that the value of k could not be changed, and it was our expectation when the model was first developed that later field measurements would lead to a more accurate determination of k, which could then be incorporated into the model. Baldocchi et al. (1984) found that k was 0.732 (SD = 0.141) for photosynthetically active radiation, measured against leaf area index. Interestingly, those who have used variants of the forest model, however, have not found it necessary to add precision to this parameter.

The light available to a tree, AL, is calculated according to equation (2.9):

$$AL = PHIe^{-k*SLA} \qquad (2.9)$$

where PHI is the light incident above the canopy, SLA is the shading leaf weight, and k is defined as above.

PHI is normalized to 1, but it would be a minor change to use real values or to correct this value for changes among years or within seasons if one wanted to model growth on a shorter time basis. We expected such modifications of the model to be made by users of the model, but such details have not been found necessary. Corrections are made for differences in annual day-length and sunlight intensity as a function of latitude according to values given in Thornthwaite (1948, Table V, p. 93); the model interpolates between values given in that table (reproduced in Appendix III). This correction affects the calculation of E_0 described later. Appendix III reproduces Thornthwaite's table and gives the method for calculating the latitudinal correction factor.

Temperature Conditions

Temperature affects all metabolic processes. Internally, at the cellular and biochemical level, temperature affects tree growth by changing rates of enzyme reactions and thus changing rates of photosynthesis and respiration. Heat energy from sunlight and air masses provide the energy that drives evaporation of water and makes possible the transport of water from the roots to stems and leaves. An excess of soil and leaf temperature dries the soil and leads to drought effects on trees.

Net growth of trees occurs only above some temperature minimum at which photosynthetic processes proceed at rates greater than those for respiratory processes and both are greater than zero. While the minimum temperature for growth is a species-specific factor, the average thermal properties of the environment can be represented by the choice of a baseline below which growth is assumed to be negligible. For northern temperate forests this minimum is generally taken to be about 40°F (approximately 4.4°C) and this is the minimum used in the version of the forest model discussed in this book. Other baselines might be appropriate for forests of quite different environments, such as tropical rain forests. In many applications, a baseline of 0°C is used, but the 4.4°C baseline is also common.

The important thermal characteristic of the environment for trees is not temperature but rather the heat sum: the product of temperatures above the zero-growth level and the duration of those temperatures. This is the integral under the curve (and above the minimum growth temperature) of temperature plotted against time, and is known as "growing degree-days."

$$\text{DEGD} = \int T_t dt \tag{2.10}$$

where T_t is the instantaneous temperature.

Where more specific information is lacking, DEGD can be approximated as the product of the mean monthly temperature times the number of days in the month (Fig. 2.3). Where available, growing degree-days for a site can be calculated from actual daily site weather, summed over monthly intervals. Lacking these data, one can obtain weather records from the nearest weather station. When a site lies at a known elevation above or below a weather station, standard temperature lapse rates can be used to extrapolate climate records

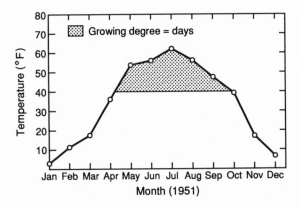

FIGURE 2.3. Growing degree-days for 1951 for Mount Pleasant, Michigan, weather records. This graph illustrates the definition of growing degree-days (see text for explanation).

to represent those at the site. The standard temperature lapse rates are 3.6°F/1,000 ft in the summer (6.5°C/1,000 m) and 2.2°F/1,000 ft (4.01°C/1,000 m) in the winter, as given in Sellers (1965). In the version of JABOWA-II available as a companion to this book, the summer lapse rate is used if the average temperature for a month is above 0°C, otherwise the winter lapse rate is used.

Where actual local weather records are not available, one can assume that the temperature profile is sinusoidal and can estimate degree-days for each month from January and July average temperatures, as discussed in Botkin, Janak, and Wallis (1972a, b) by the equation

$$\text{DEGD} = \frac{365}{2\pi}(T_{\text{July}} - T_{\text{Jan}}) - \frac{365}{2}\left(40 - \frac{T_{\text{July}} + T_{\text{Jan}}}{2}\right) + \frac{365}{\pi}\left[\frac{\left(40 - \frac{T_{\text{July}} + T_{\text{Jan}}}{2}\right)^2}{T_{\text{July}} - T_{\text{Jan}}}\right]$$

(2.11)

where temperature is in Fahrenheit. (Examples of the degree-day calculation are given in Tables 2.2 and 2.3.)

TABLE 2.2. Effect of two different elevations—100 m and 500 m—on site conditions: average of 22 years of climate data from watershed 6 of Hubbard Brook Experimental Forest[a]

| | (A) Temperature and basic water balance at 100-m elevation[b] | | | | | | | |
Month	Mean Temp. (°C)	Precip. (mm)	Water-S (mm)	Snow-S (mm)	Snowmelt (mm)	Snow-D-D (degree-days)	I	A
Jan.	−7.00	84.08	70.48	185.65	0.00	0.00	37.97	1.10
Feb.	−4.76	80.63	70.48	266.28	0.00	0.00	37.97	1.10
March	0.72	85.72	70.48	0.00	266.28	127.67	37.97	1.10
April	7.04	92.41	70.48	0.00	0.00	313.29	37.97	1.10
May	13.76	121.64	56.99	0.00	0.00	532.09	37.97	1.10
June	18.25	120.86	40.77	0.00	0.00	649.49	37.97	1.10
July	20.93	106.08	32.63	0.00	0.00	754.38	37.97	1.10
Aug.	19.79	116.87	38.00	0.00	0.00	718.74	37.97	1.10
Sept.	14.77	114.11	48.71	0.00	0.00	545.13	37.97	1.10
Oct.	8.36	104.25	65.63	0.00	0.00	364.63	37.97	1.10
Nov.	2.73	106.34	70.48	0.00	0.00	183.92	37.97	1.10
Dec.	−4.06	101.57	70.48	101.57	0.00	0.00	37.97	1.10
Totals		1234.56				266.28		

| | (B) Derived water balance calculations at 100-m elevation[b] | | | | | | | |
Month	dw/dt	P (mm)	U (mm)	E_0 (mm)	E (mm)	S (mm)	ρ (mm)	DW±
Jan.	0.00	0.00	0.00	0.00	0.00	0.00	0.00	0.00
Feb.	0.00	0.00	0.00	0.00	0.00	0.00	0.00	0.00
March	17.28	351.99	0.00	2.63	2.63	322.09	17.28	−0.00
April	2.45	92.41	0.00	35.52	35.52	54.44	2.45	0.00
May	13.50	121.64	0.00	83.77	83.77	51.37	0.00	−13.50
June	16.22	120.86	0.00	115.10	104.86	32.22	0.00	−16.22
July	8.14	106.08	0.00	135.92	95.86	18.37	0.00	−8.14
Aug.	5.37	116.87	0.00	118.12	87.41	24.09	0.00	5.37
Sept.	10.71	114.11	0.00	74.01	67.86	35.54	0.00	10.71
Oct.	16.92	104.25	0.00	35.96	35.95	51.38	0.00	16.92
Nov.	14.68	106.34	0.00	8.89	8.89	82.77	9.83	4.85
Dec.	0.00	0.00	0.00	0.00	0.00	0.00	0.00	0.00
Totals	29.56	1234.56	0.00	609.92	522.75	682.25	29.56	0.00

| | (C) Temperature and basic water balance at 500-m elevation[c] | | | | | | | |
Month	Mean Temp. (°C)	Precip. (mm)	Water-S (mm)	Snow-S (mm)	Snowmelt (mm)	Snow-D-D (degree-days)	I	A
Jan.	−8.98	101.95	23.49	232.36	0.00	0.00	30.25	0.98
Feb.	−7.01	97.21	23.49	329.67	0.00	0.00	30.25	0.98
March	2.11	110.20	23.49	331.53	108.23	40.09	30.25	0.98
April	4.14	102.60	23.49	0.00	331.53	226.22	30.25	0.98
May	11.26	123.61	19.16	0.00	0.00	454.56	30.25	0.98
June	16.01	125.28	14.35	0.00	0.00	582.33	30.25	0.98
July	18.71	110.63	11.82	0.00	0.00	685.46	30.25	0.98
Aug.	17.69	119.58	13.64	0.00	0.00	653.65	30.25	0.98
Sept.	13.12	116.60	18.35	0.00	0.00	495.58	30.25	0.98
Oct.	7.17	110.48	23.49	0.00	0.00	327.78	30.25	0.98
Nov.	0.89	125.21	23.49	0.00	0.00	128.81	30.25	0.98
Dec.	−5.83	130.41	23.49	130.41	0.00	0.00	30.25	0.98
Totals		1373.75				439.77		

TABLE 2.2. (*Continued*)

		(D) Derived water balance calculations at 500-m elevation[c]						
Month	dw/dt	P(mm)	U(mm)	E_0(mm)	E(mm)	S(mm)	ρ(mm)	DW±
Jan.	0.00	0.00	0.00	0.00	0.00	0.00	0.00	0.00
Feb.	0.00	0.00	0.00	0.00	0.00	0.00	0.00	0.00
March	12.46	108.23	0.00	0.00	0.00	95.77	12.46	0.00
April	17.18	434.13	0.00	24.53	24.53	392.43	17.18	0.00
May	−8.18	123.61	0.00	74.06	74.06	57.73	0.00	−8.18
June	−17.47	125.28	0.00	105.47	102.94	39.80	0.00	−17.47
July	−8.96	110.63	0.00	124.84	96.98	22.61	0.00	−8.96
Aug.	4.77	119.58	0.00	109.22	86.95	27.86	0.00	4.77
Sept.	10.29	116.60	0.00	70.35	67.19	39.12	0.00	10.29
Oct.	17.29	110.48	0.00	35.28	35.28	57.90	0.00	17.29
Nov.	14.81	125.21	0.00	3.85	3.85	106.54	12.55	2.26
Dec.	0.00	0.00	0.00	0.00	0.00	0.00	0.00	0.00
Totals	42.19	1373.75	0.00	547.60	491.79	839.77	42.19	0.00

[a]Other site information: Adjusted root depth = 0.50 m; K (snow evaporated) = 2.70 mm/day/°C; water table depth = 1.75 m; snow-temperature = − 3.40°C; soil texture = 150 mm/m; soil nitrogen = 79 kg/ha. See text and glossary for further information about column headings. The meanings of the symbols (the column headings) not given below appear in the glossary and in the text. I and A are from the Thornthwaite equations. K is the millimeters of snow evaporated per day per degree of temperature.

[b]W_{max} = 70.48 mm; degree-days = 2,199.98; W_k = 49.34 mm. Runoff (S) is the amount explicitly calculated by the equations described in the text. ρ is the additional amount that must run off to keep the actual water storage less than or equal to the maximum.

[c]W_{max} = 23.49 mm; degree-days = 1,756.79; W_k = 16.45 mm.

Data courtesy of T. C. Siccama; calculations courtesy of J. Bergengren.

TABLE 2.3. Effect of soil texture on site conditions at 500-m elevation in White Mountains of New Hampshire: average of 22 years of climate data from watershed 6 of Hubbard Brook Experimental Forest[a]

		(A) Sandy soil (soil moisture-holding capacity 50 mm/m depth of soil)						
Month	dw/dt	P(mm)	U(mm)	E_0(mm)	E(mm)	S(mm)	ρ(mm)	DW±
Jan.	0.00	0.00	0.00	0.00	0.00	0.00	0.00	0.00
Feb.	0.00	0.00	0.00	0.00	0.00	0.00	0.00	0.00
March	5.73	108.23	0.00	0.00	0.00	102.51	5.73	−0.00
April	5.78	434,13	0.00	24.53	24.53	403.83	5.78	−0.00
May	−4.33	123.61	0.00	74.06	74.06	53.88	0.00	−4.33
June	−4.81	125.28	0.00	105.47	95.29	34.80	0.00	−4.81
July	−2.53	110.63	0.00	124842	91.77	21.40	0.00	−2.53
Aug.	1.82	119.58	0.00	109.22	89.18	28.58	0.00	1.82
Sept.	4.72	116.60	0.00	70.35	69.34	42.54	0.00	4.72
Oct.	7.56	110.48	0.00	35.28	35.28	67.64	2.41	5.14
Nov.	5.75	125.21	0.00	3.85	3.85	115.60	5.75	0.00
Dec.	0.00	0.00	0.00	0.00	0.00	0.00	0.00	0.00
Totals	19.67	1373.75	0.00	547.60	483.31	870.77	19.67	−0.00

[a]The meanings of the symbols (the column headings) are given in the glossary and in the text. I and A are from the Thornthwaite equations. The water balance calculations are given in Table 2.2(B).

Data courtesy of T. C. Siccama; calculations courtesy of J. Bergengren.

TABLE 2.3. (*Continued*)

Month	dw/dt	P (mm)	U (mm)	E_0 (mm)	E (mm)	S (mm)	ρ (mm)	DW ±
(B) Clay-loam soil (soil moisture-holding capacity 250 mm/m depth of soil)								
Jan.	0.00	0.00	0.00	0.00	0.00	0.00	0.00	0.00
Feb.	0.00	0.00	0.00	0.00	0.00	0.00	0.00	0.00
March	15.32	108.23	0.00	0.00	0.00	92.92	15.32	0.00
April	27.14	434.13	0.00	24.53	24.53	382.47	27.14	−0.00
May	−9.56	123.61	0.00	74.06	74.06	59.10	0.00	−9.56
June	−24.15	125.28	0.00	105.47	105.47	43.96	0.00	−24.15
July	−20.87	110.63	0.00	124.84	106.63	24.87	0.00	−20.87
Aug.	4.37	119.58	0.00	109.22	87.25	27.96	0.00	4.37
Sept.	14.12	116.60	0.00	70.35	64.91	37.57	0.00	14.12
Oct.	21.94	110.48	0.00	35.28	35.28	53.26	0.00	21.94
Nov.	24.41	125.21	0.00	3.85	3.85	96.95	10.26	14.15
Dec.	0.00	0.00	0.00	0.00	0.00	0.00	0.00	0.00
Totals	52.72	1373.75	0.00	547.60	501.97	819.06	52.72	−0.00

Data courtesy of T. C. Siccama; calculations courtesy of J. Bergengren.

The Soil-Water Balance and Soil Moisture Conditions*

> With the disappearance of the forest, all is changed ... precipitation becomes as regular as the temperature; the melting snows and vernal rains, no longer absorbed by a loose and bibulous vegetable mould, rush over the frozen surface, and pour down the valleys seaward, instead of filling a retentive bed of absorbent earth, and sorting up a supply of moisture to feed perennial springs.
>
> George Perkins Marsh, 1864

Water becomes available for tree growth primarily through the soil (in some rare exceptions, such as the coastal redwoods of California, trees intercept fog and water condenses into larger droplets that then enter the soil; this provides an additional input not accounted for by rainfall records). Because water is so important to tree growth, it is necessary that soil-water relationships be considered in a realistic manner. JABOWA-II calculates a complete soil-water balance, which begins with rainfall input. This is a major modification in JABOWA-II compared with the original model. The following description is for version II, except where noted. Additional technical details of the soil-water balance calculations appear in Appendix III.

Water Balance Equations

For the purpose of calculating the soil-water budget, the soil can be viewed vertically (Fig. 2.4). The depth of the soil is defined in the model as the depth to which tree roots can penetrate or that material can be transported directly

---

*The description of the soil-water balance follows closely that found in Botkin, D. B. and R. E. Levitan, 1977, *Wolves, moose and trees: An age specific trophic-level model of Isle Royale National Park*. IBM Research Report in Life-Sciences RC 6834. Richard Levitan did much of the work to develop the details of this improved treatment.

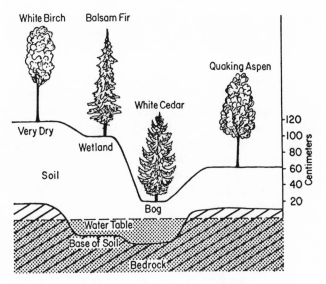

FIGURE 2.4. Plant soil-water relations.

to tree roots. Below is a subsoil that may be bedrock or unconsolidated particles where chemical elements are not available to the roots. At some depth, a water-saturated zone begins in which all capillary pore spaces between soil particles are filled with water and there is essentially no gaseous atmosphere. The top of the saturated zone is the *water table*, and the depth to the water table (the distance from the surface of the soil to the water table) is denoted as DT. Above the water table is the unsaturated zone, where the pore spaces between soil particles are in part occupied by air and in part by water. The relationship between the depth to the base of the soil (denoted as DS in meters) and the depth to the water table is important to tree growth in the model. (As a technical note and for the sake of completion, I will point out that JABOWA-II also contains another parameter, root depth, which is not invoked in any of the results reported here and is disabled in the version of the model available as a companion to this book, but is available with minor changes in the computer code for cases where roots might penetrate into the water table and obtain nutrients below that surface, or where a user might wish to add more complexity to the soil moisture responses.)

The relationship between depth to the soil base and depth to the water table is used in the model to distinguish trees adapted to wet ground and trees adapted to dry areas (see Chapter 3). If the water table is higher than the base of the soil (if DT < DS), some tree roots will be in a saturated zone, and I will refer to the soil as *water saturated*. If the saturated zone extends to the soil surface, I will refer to the soil as *completely saturated*. In a completely saturated soil, tree roots do not have oxygen and cannot respire without specialized adaptations that allow air to pass from the surface of the soil down to the roots, such as aerenchyma tissue found in trees characteristic of swamps, such as cypress. Tree species vary in their ability to survive water saturation, either periodic flooding (complete soil-water saturation) or a continuous shallow depth to the water table. Tropical trees in the flooded forests along the Amazon River are flooded for months at a time and survive; some temperate-zone trees can withstand little flooding.

The amount of water available to a tree for growth is the amount stored in the soil, not the amount simply received from rainfall. Water may become available when transferred from other areas by subsurface runoff or through upward capillary action. Some rain that falls evaporates immediately or runs off the soil surface and is never available to the trees. Once in the soil, water can be lost through downward percolation, horizontal runoff, or evaporation from the soil, being raised to the top by capillary action. The water taken up by trees is eventually transferred to the atmosphere by transpiration from the tree leaves.

The generalized fundamental water balance equation is

$$\frac{dw}{dt} = P + U - E - S \tag{2.12}$$

where w = amount of water held in capillary storage in root zone

P = precipitation per unit time (rainfall plus snowmelt) at site

U = upward capillary transport per unit time from water table to root zone

E = evapotranspiration (evaporation + transpiration) per unit time

S = water loss via runoff and downward percolation per unit time.

(The water balance equations are those originally given by Thornthwaite [1948], as stated in Sellers [1965]). In past and present versions of the JABOWA model, the water balance has been calculated monthly. This time interval was chosen for convenience, because monthly weather records are common, and a daily or hourly calculation would increase the computer time required, a consideration more important when the model was first developed than it is today. There is nothing intrinsic in the model requiring a monthly interval, and with the high speed of modern microcomputers and computer work stations, another time interval could be used if it were advantageous or necessary.

Precipitation

Ideally, temperature and rainfall should be measured at a site of study, but this rarely happens, and long-term weather records from remote forest sites are even rarer. Thus a way must be found to use weather records from the nearest weather station. For temperature there are standard lapse rates (changes in temperature with elevation) mentioned earlier of $3.3°F/1,000\,ft$ $(0.65°C/100\,m)$ elevation in the summer and $2.2°F/1,000\,ft$ $(0.40°C/100\,m)$ in the winter. In other cases, other topographic corrections may be required. For example, in an application of the model to Isle Royale National Park, a correction was introduced for the distance of the site from the shore of Lake Superior, because the cold lake waters produced a colder local climate near the shore than was found in the interior (Botkin and Levitan, 1977). The detail required for these corrections is an empirical matter.

Just as with temperature, precipitation ideally would be measured on site, but this is rarely the case. Weather station records of actual monthly precipitation averages are used and are called base precipitation data, $BASEP_j$. Version I adjusted precipitation with elevation according to the following linear lapse rate equation, consistent with information provided by the U.S. Forest Service's Hubbard Brook Experimental Forest:

$$PP_j = BASEP_j + RLAPSE*(ELEV_{site} - ELEV_{base}) \tag{2.13}$$

where RLAPSE equals the average lapse rate for the difference in elevation between the base weather station and the plot, $BASEP_j$ is the rainfall recorded at that station in month j, and PP_j is the calculated rainfall in month j at the site.

In version I, the value of RLAPSE was 0.03589 cm per meter rise in elevation. The code for this linear relationship exists in version II, but is disabled in the software available as a companion to this book. In recent years, we decided that changes in rainfall with elevation in mountainous terrain was too site specific—too dependent on slope, aspect, proximity to an ocean or large lake—to be represented by a simple general curve. For specific applications, where the change in rainfall with elevation is known, this correction factor can be added readily.

The relationship between elevation and rainfall is more complex than the relationship between elevation and temperature. As a consequence, the relationship between elevation and actual evapotranspiration also can be complex (Reiners, Hollinger, and Lang, 1984). In general, rainfall increases with elevation, as an air mass cools and the amount of water that can be stored in the gaseous state decreases.

For example, in Santa Barbara, California, the mountains rise above 1,200 m (4,000 ft) within 15 km of the ocean. Rainfall at the summit and just on the leeward side of the summit averages about twice that at the shore, about 90 cm/yr versus 45 cm/yr. As air descends on the leeward side, it warms and expands, and friction between the air and ground warms the air more. The capacity for moisture storage in the air increases, and the inland valley air thus is dry. Near Santa Barbara the inland valley rainfall declines to 30 cm/yr.

Snow Accumulation and Snow Melt[2]

In JABOWA, when the average site temperature is below freezing, all precipitation accumulates as snow, and when the average monthly temperature (T_j) is above $-3.4°C$, which means that some parts of some days are above $0°C$, snow melts uniformly over the month. In version I, all of the winter's accumulated snow melted in the first month with an average temperature above zero. This method often led to an artificially high loss to soil-water storage of much of the winter's accumulation. In version II, snowmelt (μ) (in millimeters melted per day) is a function of the energy available to melt the snow, so that

$$\mu_j = K_s(T_j - \tau) \tag{2.14}$$

where τ is the threshold temperature ($-3.4°C$ in version II of the model) at which snowmelt begins to occur and $K_s = 2.7$ (Dunne and Leopold [1978]). Note that in the first year of the simulation, which begins on January 1, in the northern hemisphere, the computer will not know how to do the soil-water and snow-depth calculations. Therefore, year 1 is started with an accumulation of the snow from year 1, months 7 through 12. Thus the program assumes that year 0 (before the simulation begins) is identical in snow conditions to year 1. K_s, a snowmelt degree-day factor, converts temperature to the millimeters of snow melted per day per degree of temperature.[3] Its units are

$$K_s = 2.7 \, \text{mm/day/}^\circ\text{C}$$

which is a value chosen to give realistic snowmelt dynamics during the spring for Hubbard Brook Experimental Forest. There is also an iterative solution to establish year 1's initial soil-water storage. The iteration proceeds until the initial value converges.

Evapotranspiration

Evapotranspiration is the sum of the water evaporated from the soil surface to the atmosphere and that transferred by transpiration from leaves. These are calculated as one process. Both processes ultimately depend on sunlight as the source of energy. Evapotranspiration increases with air temperature, with the amount of direct sunlight, and with the amount of water available. Evapotranspiration can be limited either by the amount of energy or by the amount of water. *Potential evapotranspiration* is defined as the amount of water that could be transferred from the soil to the atmosphere given the amount of thermal energy present. *Actual evapotranspiration* is defined as the amount that could actually evaporate given the water available and the energy available. Note that in this terminology, actual evapotranspiration is a theoretical estimate and not a measured value. Where water is not limited, actual and potential evapotranspiration are the same.

The rate of evapotranspiration is a complex function of many factors. Classically these have been estimated by empirical relationships. The most widely used method is that developed by Thornthwaite, and this has been used with the forest model (Thornthwaite, 1948; Thornthwaite and Mather, 1957). As with other aspects of the environment as defined in the model, there is nothing intrinsically necessary about this method; other methods could be readily substituted if concepts and data to determine parameters were available. Thornthwaite's method was chosen originally because it was in wide use, gave reasonable results, and the parameters required could be readily obtained. As with other aspects of the model, it was not expected that users would stay entirely with this original choice, and the model was written so as to make modification or substitution of this method by another straightforward. There are other methods such as Penman's that are preferred by some climatologists, but these require data that are difficult to obtain and generally unavailable except in specific experiments (Monteith, 1965). The calculation of evapotranspiration is one topic in which further consideration would be most helpful, at least from a theoretical point of view. The basic determinants for the choice of a method are time scales and questions of interest. For study of forests over decades and centuries under climatic regimes for which the Thornthwaite method was developed, this empirical approach seems adequate.

Actual and potential evapotranspiration are calculated following a modified Thornthwaite water-balance calculation (Thornthwaite, 1948; Sellers, 1965). Potential evapotranspiration, E_0, is calculated in millimeters per standard month as

$$E_{0j} = 16(10T_j/I)^a \tag{2.15}$$

where j = the month (30 days with 12 hours of daylight)

T_j = average temperature for month j, so that T_j = max (0, mean temperature (°C) for month j)

I = heat index for month, where

$$I = \sum_{j=1}^{12} \left(\frac{T_j}{5}\right)^{1.514} \tag{2.16}$$

$$a = (0.675I^3 - 77.1I^2 + 17{,}920I + 492{,}390) \times 10^{-6} \tag{2.17}$$

This equation is considered valid between 0°C and 26.5°C. In the original Thornthwaite equations, evapotranspiration increases exponentially with temperature. Clearly, this is unrealistic beyond a certain range. A modified Thornthwaite method (Sellers [1965]) is used.

Above 26.5°C and up to 38°C, the potential evapotranspiration follows a parabolic function expressed by

$$E_{0j} = -41.947 + 3.246(T_j) - 0.0436(T_j)^2 \tag{2.18}$$

Above 38°C, potential evapotranspiration becomes asymptotic at 185 mm of water (Thornthwaite, 1948; R. Nisbet, personal communication 1990). Factors to correct for length of month and length of daylight at selected northern latitudes are tabulated in Thornthwaite (1948). Examples of a and I for specific conditions are given in Tables 2.3 and 2.4. The modification above 26.5°C has the effect of reducing the evapotranspiration rate relative to the original basic formula.

The rate of actual evapotranspiration, E, is determined by the potential rate and by the availability of water stored in the soil. Above some value of soil moisture, w_k, actual evapotranspiration proceeds at the potential rate. At this level energy, not water, limits the rate of evaporation. Below w_k, water also limits evaporation. This rate of evapotranspiration depends on the soil moisture:

$$E = E_0 \quad \text{if } w \geqslant w_k \tag{2.19a}$$

$$E = E_0\left(\frac{w}{w_k}\right) \quad \text{if } w < w_k \tag{2.19b}$$

Following Sellers, the crossover occurs by definition when $w/w_k = 1$. w/w_k is 0.7 w_{max}, where w_{max} is the maximum capillary storage or field capacity. For those not familiar with soil characteristics, it may be useful to explain that particles in the soil can hold on to a certain amount of water against the pull of gravity.

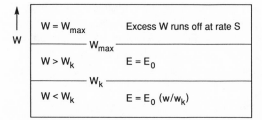

FIGURE 2.5. The relationship between water held in the soil and evapotranspiration (91).

Field capacity is the maximum amount of water that can be held against gravity by the soil particles. Any water added beyond this amount will flow downward, however slowly.

The time-dependent solution for E is given in Appendix III. The relationship between W, the amount of water stored in the soil, and E, the evapotranspiration rate, can be visualized as in Figure 2.5. Thornthwaite's method was determined for mid-latitudes and is simply extrapolated to high latitudes without empirical tests.

It is instructive to compare the projections of evapotranspiration from the modified Thornthwaite (as used with the model) and the Penman method, especially because there is much current interest in global climate change and therefore in what would be the appropriate method to calculate evapotranspiration for a rapidly changing, possibly novel climate. Figure 2.6 shows a comparison of the two methods for a single site; the results are reasonably similar for our purpose, especially at higher temperatures.

Percolation and Surplus Rainfall

When $w = w_{max}$, runoff, S, is at a maximum rate that increases with $[P^2/(E_0 + P)]$, where P is precipitation in millimeters per month and E_0 is the potential evapotranspiration (Sellers, 1965).

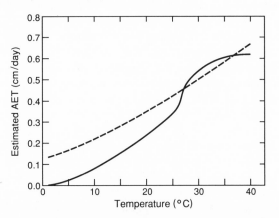

FIGURE 2.6. Comparison of the Penman (dashed line) and modified Thornthwaite (solid line) methods for the calculation of evapotranspiration as a function of temperature for a single site.

Continuing to follow the empirical relationships given in Sellers (1965), it is assumed that the surplus amount of rainfall that runs off S is

$$S = \frac{[0.8P^2][w]}{[E_0 + P][w_{max}]} \qquad (2.20)$$

or $\varsigma = \beta \times \Omega$, where $\beta = 0.8P^2/(E_0 + P)$ and

$$\Omega = \frac{w}{w_{max}}$$

This is another empirically based relationship. It provides more runoff and/or percolation as rainfall intensity increases or the ground becomes saturated (the harder it rains and the wetter the ground, the greater percentage of the water runs off). Upward capillary transport was not calculated in version I, except for a version developed for Isle Royale National Park (Botkin and Levitan, 1977); it is calculated in version II provided as software that is a companion to this book.[3]

If the water table (DT) is above the soil depth (DS), then obviously there is no net upward capillary flux of water from DT to DS, and the value of U, upward capillary water transport, is zero. If the water table is below the soil depth, then the upward capillary flux of water is calculated as

$$U_j = \frac{K*(w_{max} - w)}{w_{max}*DT} \qquad (2.21)$$

where K is a coefficient, set to 15 mm/month of water transport.

Tables 2.2 and 2.3 show sample water balance calculations for the Hubbard Brook Experimental Forest control watershed (watershed 6), using the average of weather conditions for a 22-year period as measured on site at different elevations within the watershed (Siccama, personal communication, 1990). Table 2.3 compares site conditions under the same monthly precipitation and temperature and soil nitrogen for a sandy and a clay-loam soil.

Evapotranspiration is greater in the clay-loam soil and the runoff is smaller, showing that plants use more water in the clay-loam. The differences are not great under these conditions, with evapotranspiration differing by less than 20 mm/yr. This is because the actual evapotranspiration is nearly as large as the potential evapotranspiration, a result of a comparatively cold climate with relatively high summer precipitation characteristic of northern New Hampshire. Both soils decline in water storage during May, June, and July.

Table 2.2 compares basic water balance data and calculations for two elevations, 100 (Table 2.2(A) and (B)) and 500 m (Table 2.2(C) and (D)). The second and third columns, temperature and precipitation, are read-in variables—mean monthly temprature and total precipitation. The columns labled Water-S and Snow-S are the millimeters of water storage and snow storage, respectively, in the soil for each month. If the temperature (column 2) is below 0°C, then precipitation is stored as snow. For example, in (A) temperature drops below zero in December and the 101.57 mm of precipitation becomes snow storage.

To this is added the storage in the next January (84.04 mm precipitation) and in February (80.63 mm), to total 266.28 mm of snow. (Remember that Table 2.3 is for a steady-state climate calculated as the 22-year average Hubbard Brook Experimental Forest weather conditions, so that monthly temperature and precipitation in one year represent the conditions in all years. Calculations can cycle from December to January.)

When the average monthly temperature climbs above $-3.4°C$ in the spring, snowmelt begins. As mentioned earlier, this average monthly temperature implies that some parts of some day light hours will be above 0°C. With the rapid warming that occurs in New Hampshire, all the accumulated snowmelts in the first month of warming (providing 266.28 mm into column 6, Snow-M, in March), but this is not necessarily the case in JABOWA-II. The column labled Snow-D-D is the number of snow degree-days that is summed from a base of 0°C for the snowmelt calculation (see Appendix III). The factors I and A in the last columns are variables of the Thornthwaite evapotranspiration equations (2.15, 2.16, and 2.17).

Table 2.2(B) and (D) presents derived water balance calculations for Hubbard Brook sites at 100- and 500-m elevations, respectively. Remember that the change in water storage, dw/dt, is the sum of precipitation input and upward capillary input minus loss to evapotranspiration and runoff, or

$$\frac{dw}{dt} = P + U - E - S \qquad (2.12)$$

The amount stored is only that which can be held by the soil particles up to field capacity, so that dw/dt can overestimate the storage. The change in storage is later corrected for the amount of water remaining above field capacity at the end of the month. This water runs off (given in column 8 under variable ρ) and decreases the amount stored. The corrected amount is variable DW± where

$$\text{DW}\pm = \frac{dw}{dt} - \rho \qquad (2.22)$$

where ρ is water lost via runoff and downward percolation.

With this explanation, we can now compare sites for the effects of elevation (Table 2.2) and soil texture (Table 2.3). In Table 2.2, temperature is lower but rainfall is higher at 500 m than at 100 m. Water storage is much greater at 100 m; snow storage greater at 500 m. At both elevations there is a water deficit during May, June, and July (the net storage declines). Note that the rainfall recorded on the site varies in a complex way, with rainfall greater in April at 500 m than at 100 m, but greater in March at the lower elevation than at the higher. This complexity makes one cautious about using simple precipitation lapse rates, as I discussed earlier. As expected, potential evapotranspiration is greater for all months at the lower elevation. Actual evapotranspiration is slightly greater for most months at 100 m, but is higher in July at the higher elevation due to the larger water deficit at 100 m during that month. Evaporation proceeds at the potential (maximum) rate through May and again from October through December at both sites. These tables show that the calculations of site conditions in the present version of the model respond to changes in temperature and rainfall in the soil-water balance.

There has been some misunderstanding of the soil-water calculations in the model. Some have mistakenly believed that since the model uses Sellers' modified Thornthwaite method to calculate potential and actual evapotranspiration, the model stops with that calculation and does not contain a complete water budget. As the equations discussed in this section make clear, JABOWA-II does calculate a complete water budget that is a function of soil storage capacity as well as availability of water throughout the months and energy available to evaporate water. Stephenson (1990) has suggested that the distribution of vegetation is better correlated with water balance than with the calculation of Thornthwaite's evapotranspiration index. JABOWA-II provides a complete water balance and its methods are consistent with that conclusion.

Table 2.3 shows the effect of soil texture on the soil-water balance for two sites at 500-m elevation otherwise identical with the site in Table 2.2(C) and (D). Table 2.3(A) gives results for a site with a coarse sandy soil (50 mm/m water-holding capacity) and Table 2.3(B) gives results for a clay soil (250 mm/m). The sandy site has a much smaller water-holding capacity, and the increase in storage during the spring and fall is one-third or less the increase in storage in the clay soil. The loss of water during the summer is also much less from the sandy soil than from the clay soil for the simple reason that the amount stored was much less. Potential evapotranspiration is of course the same for both sites, because that value does not depend on stored water. Actual evapotranspiration is the same from January through May for the two sites. During these months there is ample water, and actual evapotranspiration equals potential evapotranspiration. In June and July, actual evapotranspiration is greater on the clay soil than on the sand, but in August and September actual evapotranspiration is slightly greater on the sandy site. Total runoff is greater from the sandy site and total evapotranspiration is less. These tables show the complexity of the soil-water budget in the model.

Soil Nutrient Conditions and Chemical Cycling

Version I of the model did not consider any aspect of soil chemical fertility. Of course, the fertility of soil in terms of the availability of chemical elements required for vegetation growth and regeneration is of great importance to the growth, survival, and regeneration of trees. There is a wealth of laboratory research on mineral nutritional needs of plants, and there is no need to emphasize here the use made of fertilizers in agriculture and the extensive agricultural and fundamental research that has taken place to determine what elements are required, in what concentrations, and in what ratios. More will be said about the response of trees to the fertility of the soil in the third chapter; here I only discuss how one aspect of that fertility is represented.

Soil nutrient conditions are expressed as the concentration by weight of a chemical element in the soil, in kilograms per hectare. This is simply empirical information. In existing versions of the model, only nitrogen is considered explicitly, in part because of the importance of nitrogen in tree growth, and in part because more field measurements exist for this chemical element than for others. At the time the original model was developed, there were few field studies of the effects of experimental changes in soil nutrient conditions on trees.

The first approach to developing a nitrogen response function was in Aber, Botkin, and Melillo (1978, 1979). The key that allowed a beginning of the development of the nitrogen response function discussed in Chapter 3 was an old but classic study by Mitchell and Chandler (1939), who added several levels of measured amounts of nitrogen fertilizer to a New England forest and measured the responses of trees of various species. The gap in their study was a direct measure of the amount of nitrogen already available in the soil. This raises two issues, one empirical and the other technical. The technical issue is that, using the Mitchell and Chandler study as a basis, one can only derive a quantity for relative nitrogen content of the soil. In the version of the model available as a companion to this book, the relative available nitrogen is scaled from -100 to $+100$, and can be thought of as an approximation of the kilograms per hectare of nitrogen actually available for tree growth in a year, in comparison with a baseline, unfertile, New England soil (whose available nitrogen on this scale would be zero). The "available" nitrogen value that is provided as input to the model is scaled to create a relative available nitrogen:

$$v = -170 + 4.0 * \text{AVAILN} \qquad (2.23)$$

where AVAILN is the amount of available nitrogen, represented as kilograms per hectare per year.

The empirical issue is that most measurements of soil nitrogen give the *total* nitrogen concentration in the soil, much of which is tied up in organic compounds that decompose slowly and are not readily available to trees. The amount that is important to tree growth is not the total, but the amount that is available directly for uptake by vegetation. Trees can take up nitrogen either as nitrate or ammonia, both volatile and reactive compounds with a short residence time in a soil, and for which there have been relatively few studies (Waring and Schlesinger, 1985). Nitrogen becomes available through bacterial fixation, through recycling from leaf fall and decomposition, and from rainwater input. Existing measurements suggest that available nitrogen is typically less than 100 kg/ha/yr. The amount returned from litter and leaf fall has been measured to range between approximately 14 and 30 kg/ha/yr (Kozlowski, Kramer, and Pallardy, 1991). Nitrogen fixation by bacterial symbionts in root nodules of a successional shrub, *Ceanothus velutinus*, in forests of Oregon contributed as much as 100 kg/ha/yr (Binkley, Cromack, and Fredriksen, 1982), but this probably represents an extreme value.

I hasten to point out that JABOWA-II does not include a feedback between the trees and the soil. Soil nitrogen content affects tree growth, but tree growth does not change soil nitrogen content. The return of chemical elements to the soil by trees, or the addition of chemical elements over time by soil symbionts, is not explicit in the model. However, Pastor and Post have developed a version that includes chemical recycling and a feedback between the forest and the soil (Pastor and Post, 1985, 1986).

Summary

This chapter began the explanation of the model, first describing the basis for the fundamental growth equation and the way that this equation is modified by environmental conditions. The balance of the chapter discussed the rationales, concepts, and equations for the model's calculations of environmental conditions. The fundamental growth equation was developed, as much as was possible at the time, from basic understanding of the physiology, anatomy, and morphology of trees, subject always to Occam's razor that an explanation should be the simplest consistent with observation. Assumptions about shape and form and the relationship between photosynthetic production of organic matter and utilization of this organic matter by respiring, nonphotosynthetic tissues allowed the development of an equation describing the change in the volume of a tree over time, assuming no environmental conditions were limiting. It is interesting to note that the resulting equation, converted to calculate diameter rather than volume, yields a sigmoid curve characteristic of the growth of many individual organisms, although we did not plan or anticipate that outcome when we began to formulate the equation.

The projected "actual" growth of a tree is calculated by reducing the maximum growth calculated from the fundamental growth equation by functions that represent suboptimal environmental conditions. Thus it is necessary to develop methods to calculate the state of the key aspects of the environment. In the original version of the model, only light and temperature were primary environmental variables; soil water was calculated and used to affect some regeneration of trees. The present version adds to light and temperature a consideration of the effects of soil drought, too much moisture in the soil, and soil nitrogen concentration on tree growth and regeneration. With the greater emphasis on soil-water effects, it was important to increase the realism and completeness in the calculation of the soil-water balance. The present version of the model calculates a complete water balance, including input from precipitation and upward percolation, and losses due to evapotranspiration and runoff.

The calculations of environmental conditions include some of the most intricate and complex parts of the model. Much of the discussion about environmental conditions in this chapter concerned physical aspects of the environment rather than biological responses, effects, or interactions between the physical environment and trees. The assumptions behind the calculations of environmental conditions depend largely on knowledge of physical and chemical processes, the understanding of which has been developed in large part by scientists outside biology, and the ecological appropriateness of these calculations is limited substantially by developments in those fields. The environmental calculations of the model can be readily modified as knowledge develops in those fields. The computer code for the forest model is designed to facilitate those modifications.

With an understanding of the fundamental growth equation and the calculation of environmental conditions, we can now turn to a consideration of how the model calculates the response of trees to suboptimal environmental conditions; this is the subject of Chapter 3.

Notes

1. Fownes (personal communication, July 1991) points out the similarity between equation (2.7) and some other classic growth analyses. In this equation, G is proportional to maximum diameter increment divided by maximum diameter, which is diameter relative growth rate. Biomass relative growth rate, dB/B, is classically

$$E*LAR$$

where E is unit leaf photosynthetic or assimilation rate and LAR is leaf area per unit biomass. Since $G = RC$ as well, this observation reinforces the conceptual connection between R and maximum net assimilation rate.

2. I thank J. Bergengren for contributions to improvements in the water balance part of the model.

3. Some addtional comments may be useful to readers of the book and users of the program, particularly those interested in the influence of snow as a supply of water and snowmelt on tree growth. In JABOWA-II, these calculations have been extensively modified and greater realism added. However, some readers might want to explore the effects of even greater realism. For example, in reality the rate of snowmelt is also a

function of sunlight intensity and air temperature, and in the model snowmelt additionally could be made a function of insolation. I caution the reader, however, not to violate Occam's razor or develop details that extend beyond data to test the results.

4. When upward capillary movement was considered by Botkin and Levitan (1977), it was considered to be a minor factor.

3

The Tree in the Forest:
The Response of Populations of Trees
to the Environment

The special conditions required for the spontaneous propagation of trees
are ... exemption from defect or excess of moisture, from perpetual frost, and
from the depredations of man and browsing quadrupeds. Where these
requisites are secured, the hardest rock is as certain to be over grown with
wood as the most fertile plain.

George Perkins Marsh, 1864

The Influence of the Environment
on an Individual Tree

In the previous chapter I described the growth of an individual tree as a function
of inherited characteristics under nonlimiting environmental conditions. Under
those conditions growth rate changes only with age or size, as illustrated in
Chapter 2, Figure 2.1. Actual growth can range from zero to the value calculated
from the fundamental growth equation, depending on competition with other
trees, which determines the light each tree receives, and on other environmental
conditions, that is, on the availability of resources (soil water and soil nitrogen)
and the thermal conditions of the environment. A tree responds to all environmental conditions, and a fundamental question is how this total response
integrates the individual effects of separate environmental factors. Does each
factor act independently on the tree or do factors act together?

In the history of the study of plants, the classic answer is *Liebig's law of the
minimum*, which states that the single factor in least supply will limit plant
growth. In this sense, there are no interactions among limiting factors. For
example, if a soil is low in phosphorus and this limits plant growth, the addition
of nitrogen to the soil will have no effect on tree growth. An alternative is that
resources interact. For example, since enzymes require nitrogen, addition of
nitrogen might allow a tree to produce more enzymes involved in the uptake
of phosphorus, improving its efficiency in the use of phosphorus. Additional
nitrogen fertilizer would lead to an increase in tree growth even if phosphorus
were the limiting factor. Although the latter is possible, most plant ecologists
and agricultural scientists have accepted Liebig's law.

Mathematically, the response of a tree to each environmental factor can be expressed as a function that takes on values between 0 and 1 (or in some cases slightly more than 1), which affects the fundamental growth equation (2.1). Each of these relationships will be called a *response function*. To follow Liebig's law of the minimum, one would calculate the value of each response function and then choose the one with the smallest value and multiply the fundamental growth equation by that value, ignoring all others. Formally,

$$\frac{\delta D}{\delta t} = \left(\frac{\delta D}{\delta t}\right)_{opt} * MIN\{f_1, f_2, \ldots, f_j\} \qquad (3.1a)$$

where $\delta D/\delta t$ is the change in diameter (the actual diameter growth); $(\delta D/\delta t)_{opt}$ is the fundamental growth equation (2.1); $f_1, \ldots, f_j$ represents the response functions; and MIN{ } is the function that becomes the minimum of the values contained in the brackets. The approach expressed in equation (3.1a), which is an expression of Liebig's law of the minimum, is called an *additive approach*; the functions are said to be chosen *additively*.

A second approach assumes that the functions continue to influence the growth simultaneously and independently. The simplest expression of this is to assume that the fundamental growth equation is multiplied by each response function, so that

$$\frac{\delta D}{\delta t} = \left(\frac{\delta D}{\delta t}\right)_{opt} * f_1 * f_2 * \ldots f_j \qquad (3.1b)$$

In this case the tree responds to each factor independently, but responds to all independent factors simultaneously. The tree's calculated response to a specific level of nitrogen is the same whatever may be the level of phosphorus, the temperature of the air, or soil moisture conditions. Its total growth is the product of these independent responses.

The reader should understand that the resulting total response may resemble synergistic curves. I define a synergistic relationship as one where one environmental factor influences another and then the combination of two factors acts on tree growth as a single function to produce a new, single response curve.

The important point here, worth repeating, is that the total growth response of the tree is the product of all environmental factors, though each factor operates independently through its own response function on the tree. I emphasize this point because some colleagues have urged me to apply multiple regression analysis to data about the response of trees to two environmental factors, and then to use this integrated response in the model. As an example, tree ring analysis (the study of the correlation between past climates and diameter increments in trees) often employs multiple regression analysis. A single function is generated relating diameter increment to air temperature and rainfall. With the structure of JABOWA, such analysis cannot be used directly. The relationship must be decomposed into two separate ones: diameter increment as a function of precipitation and diameter increment as a function of temperature. Then these studies can be used to improve JABOWA.

The assumption that the response functions operate independently means that they are graphically *orthogonal*; each plot of growth versus a response function will lie 90 degrees from each other plot (Fig. 3.1). This approach is called *multiplicative*. JABOWA operates under the multiplicative assumption (equation 3.1b), but it is simple to convert it to a model that operates under the additive assumption (equation 3.1a).

In the original formulation of the model, we chose to use multiplicative interaction to capture the possible interactions among factors and because it seemed intuitively preferable. It may interest the reader to know that we compared the additive and multiplicative interactions in a model of an algal community within a freshwater ecosystem. We found that both methods ordered

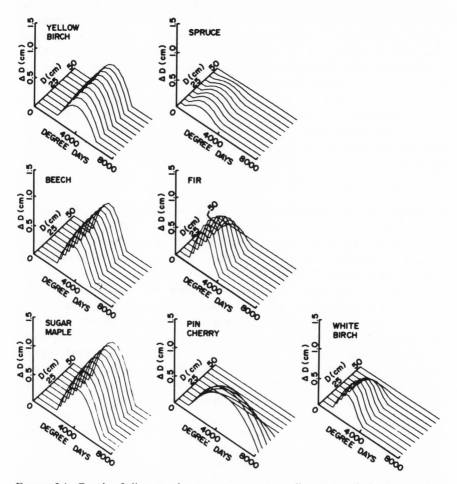

FIGURE 3.1. Graph of diameter increment versus tree diameter and the temperature response functions in three-dimensional space. These graphs show aspects of the fundamental niche of tree species. The graphs are part of the assumptions of the model. (Reproduced from Fig. 5.2b in D. B. Botkin, 1981, Causality and succession, Chap. 5, pp. 36–55, In D. C. West, H. H. Shugart, and D. B. Botkin, eds., *Forest Succession: Concepts and Applications*. Springer-Verlag, New York.)

species dominance correctly, but the multiplicative approach was quantitatively closer (the multiplicative model was more accurate) than the additive model (Lehman, Botkin, and Likens, 1975a, b).

An obvious question raised about multiplicative interactions is, what happens when there are many factors and all of them are at suboptimal levels? Can any tree grow? For example, if four factors—temperature, light, soil drought, and soil nitrogen—are at 50 percent of maximum, then the actual diameter growth will be $(0.5)^4 \Delta D_{opt}$ or $0.0625 \Delta D_{opt}$; the tree would grow at a little more than 6 percent of its maximum. This suggests that, if multiplicative interactions exist in nature, then only a few factors actually determine tree growth, or only a few are much less than optimal at any one time, or trees can withstand very suboptimal conditions. Operation of the forest model (that is, using the forest model), as discussed in Chapter 4, provides some insight into a possible answer.

From the point of view of mineral nutrition of plants, one might believe that many factors could limit tree growth. On the order of 20 chemical elements are required by green plants. If ten were characteristically below 50 percent of their maxima and interactions were multiplicative, most of the time plants would grow little. As I will discuss later, there is a connection between the concept of multiplicative interactions and niche theory (see Chapter 4). This provides additional insight into the question I have raised here.

Individual Response Functions

The totality of the effects of environmental conditions on tree growth, also referred to throughout as totality of "environmental response functions" and denoted in the general form by f(environment) in equation (2.6) is a product of several factors:

$$f(\text{environment}) = f_i(\text{AL}) * Q_i * s(\text{BAR}) \qquad (3.2)$$

for a tree of the ith species, where $f_i(\text{AL})$ is the light response of the species and $s(\text{BAR})$ is a function of the maximum basal area that the plot can support (similar to the forester's idea of a site index). For each site, the value of each response function is computed for each species. In JABOWA-II, Q_i is the product of response functions for thermal properties of the environment, drought (wilt factor), soil-water saturation (soil wetness factor), and soil fertility. (This approach follows Botkin and Levitan, 1977.) Q_i will be referred to as the "site quality" for species i. For each species i,

$$Q_i = \text{TF}_i * \text{WiF}_i * \text{WeF}_i * \text{NF}_i \qquad (3.3)$$

where TF_i is the temperature function. WiF_i is the "wilt" factor (an index of the drought conditions that a tree can withstand). WeF_i is a "soil wetness" factor, an index of the amount of water saturation of the soil a tree can withstand. NF_i is an index of tree response to nitrogen content of the soil. The calculation of each environmental factor was described in Chapter 2. The next sections describe how these factors change growth and regeneration. (In JABOWA-I, Q_i consisted only of the temperature response function.) Note that parameters for the response functions are given in Tables 3.5 and 3.6.)

Light and Tree Growth

Earlier I described some aspects of photosynthesis and growth. During the twentieth century, hundreds of experiments have been conducted to measure photosynthesis as a function of light (see, for example, Botkin, 1969; Botkin, Woodwell, and Tempel, 1970; Kozlowski, Kramer, and Pallardy, 1991; Vowinckel, Oechel, and Boll, 1975; Woodwell and Botkin, 1970). The general shape of the curve relating net photosynthesis to available light is an upward-sloping asymptotic curve (Fig. 3.2). At low light intensities there is a rapid increase in net photosynthesis as light increases. Eventually other factors, including intrinsic (genetic) capacities, become limiting. The photosynthetic rate becomes "light saturated"—additional light has a smaller and smaller effect on growth. Eventually increases in light fail to increase net photosynthesis.

The general form of the light response function can be written as

$$f(\text{AL})_L = A_1\{1 - e^{(A_2\text{AL} - A_3)}\}\tag{3.4a}$$

where A_1, A_2, and A_3 are empirically derived constants, and AL is the light available to a tree, and L is the light tolerance class (class 1 is intolerant of low light, 2 is intermediate in tolerance, and 3 is tolerant of low light).

Although there have been hundreds of studies of the response of growth and photosynthesis of vascular plants to light intensity, few exist for tree species and fewer still for individual mature trees under natural conditions. When the

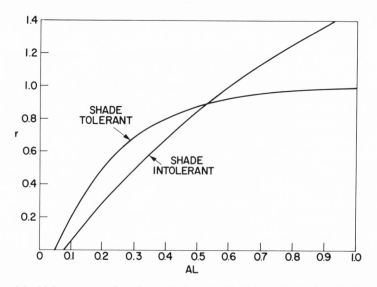

FIGURE 3.2. Light response functions. AL is available light and r is the relative growth rate. The curves are solutions to equations (3.4a) and (3.4b).

model was first developed, very few measures were available, and in the original version of the model, species were put into one of two light response categories, tolerant or intolerant, whose response functions were

$$f(AL)_1 = 2.24\{1 - e^{-1.136(AL - 0.08)}\} \qquad \text{Shade intolerant} \qquad (3.4b)$$

$$f(AL)_3 = 1 - e^{-4.64(AL - 0.05)} \qquad \text{Shade tolerant} \qquad (3.4c)$$

(In versions I and II, light tolerance class 2 has the same equation as class 1). The parameters in these equations were obtained from data in Kramer and Decker (1944) and others (Kramer and Kozlowski, 1960). These continue to be the light response functions in the model, but the computer code allows three categories of response functions: tolerant, intolerant, and intermediate. The code also allows each species to have a unique light response. Some evidence suggests that each does have a unique light response curve (Botkin, 1969). It is even conceivable that genotypes within a species might have distinctive curves. However, experience with the model shows that, at least for trees with C_3 photosynthetic pathways, we can limit the responses to two categories, shade tolerant and shade intolerant.[1] A shade tolerant species is comparatively efficient in using light when little is available, but it becomes light saturated at comparatively low light intensities. A shade-intolerant species has comparatively low growth rates under low light conditions but responds more rapidly to increases in light and reaches saturation at a much higher light intensity than a shade-tolerant species (Fig. 3.2). Foresters sometimes talk about a third category, intermediate, with light response curves between the tolerant. As I mentioned earlier, JABOWA allows for three shade tolerance classes, but in the existing version intolerant and intermediate species share the same light response function. Regeneration algorithms separate shade intolerant and intermediate species from each other, as I will discuss later. The accuracy of such shade-tolerance classification has been discussed (Lorimer, 1983).

The relationships shown in Figure 3.2 are for near-instantaneous measurements of net photosynthesis of tree leaves. Of course the net photosynthesis of individual tree leaves will vary with changes in light, temperature, and soil-water availability during the day, and there are differences in the instantaneous photosynthetic rates of different leaves in different parts of the same tree, some of which will be shaded and some sunlit. Thus one cannot expect a one-to-one correspondence between instantaneous net photosynthesis of a single leaf and total tree growth. However, measurements of net photosynthesis over an entire growing season from a surprisingly small sample of leaves give a reasonably good estimate of net forest production (Botkin, Woodwell, and Tempel, 1970). The assumption of the model is that the shape of the leaf photosynthetic curve (either shade tolerant or shade intolerant) will correspond reasonably well to the response of a whole tree to annual insolation.

To summarize, parameters in light response equations derive from laboratory measurements of shade-tolerant and shade-intolerant trees (Kramer and Kozlowski, 1960), and the simplifying assumption was made that two categories could represent the growth of all species. It was expected that, as more information became available for specific tree species, species-specific parameters would be substituted. I have been surprised through my own use of the model and use by others to find that the general categorization has been adequate.

Temperature, Degree-Days, and Tree Growth

There is a strong empirical relationship between tree growth and climate, especially between tree growth and temperature or degree-days. This relationship holds for long-term records, over thousands of years. It is used as one basis for reconstruction of past climates, and the relationships between tree growth and climate are used to determine ages of archeological sites and past climate–vegetation relationships. The study of these long-term relationships is so soundly based empirically as to develop into its own field of study, dendrochronology (Fritts, 1976; Jacoby and Hornbeck, 1987).

We can put these relationships on a conceptual and theoretical foundation. The rate of an enzyme reaction within a cell increases with temperature to a maximum and then decreases. Such unimodal curves are biologically universal and typically fall within a range between 0°C and 50°C. The sum of metabolic activities within an organism also leads to a similar unimodal response of total growth rates as a function of temperature. The temperature of a tree depends on external environmental conditions mediated by the ability of the tree to increase evaporation of water to cool leaves under warm conditions and to reduce this evaporation under cooler conditions, and by other factors as discussed in Kramer and Kozlowski (1979) and in Gates (1980).

The general relationship between net photosynthesis of woody plants and temperature is shown in Figure 3.3(A). Trees from widely different habitats and environments have temperature response curves of the same shape, roughly a parabola with the peak response at higher temperatures for tropical trees than for temperate ones. For example, *Pinus cembra* reaches a peak net photosynthetic rate at approximately 18°C, whereas tropical *Acacia carspedocarpa* reaches a peak at 40°C (Larcher, 1969). It has been suggested that this parabolic curve is the result of three interacting physiological responses: an exponential increase of gross photosynthesis that occurs at low temperatures, an increase in dark respiration that at first lags gross photosynthesis but increases at a more rapid rate as temperature rises; and an inactivation or destruction of the photosynthetic apparatus at high temperatures (Kozlowski, Kramer, and Pallardy, 1991; Kramer and Kozlowski, 1979). Regardless of the specific form of the function, there is a strong empirical relationship observed between tree growth and temperature or tree growth and degree-days. For example, Fuller, Reed, and Homes (1987) reported that the most important variable related to tree growth was cumulative air temperature degree-days.

From this we can propose that the general form of the temperature response of a tree is parabolic, as shown in Figure 3.3(B). However, that response is to the instantaneous temperature. The actual response over a longer period—a month or a year—is a response to the integrated thermal properties of the environment. These can be represented by growing degree-days, discussed in Chapter 2, which

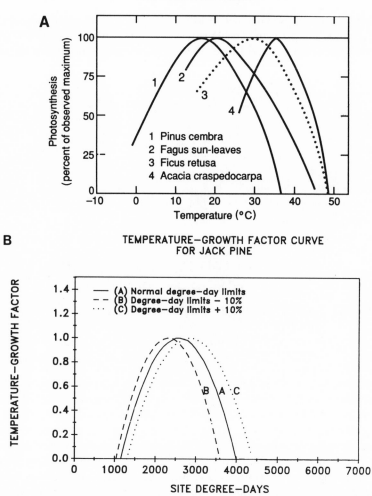

FIGURE 3.3. Growth as a function of temperature. (A) Observed photosynthesis as a function of air temperature. (Fig. 5.17, p. 196, from P. J. Kramer and T. T. Kozlowsi, 1979, *Physiology of Woody Plants*, Academic Press, New York; adapted from Larcher [1969]). (B) JABOWA temperature-growth factor as a function of degree-days. The temperature derating factor, $TF_{(i)}$, is a function of the growing degree-days, as explained in the text. (contd. Fig. 3.3)

I remind the reader is the sum above some minimum of the average temperature of a day over a month or a year. The minimum growing degree-days is the value below which tree growth ceases. This may be a species-specific factor that has evolved over long periods of time as an evolutionary adaptation of trees to regional climate. For trees in boreal and northern hardwoods forests, the minimum growing degree-days value is assumed to be 4.4°C (40°F). To my knowledge, this base has been used in all versions of the model. As discussed in Chapter 2, where actual local weather records are not available one can assume that the temperature profile is sinusoidal and estimate degree-days for each month from January and July average temperatures, as discussed by Botkin et al. (1972a, b) from equation (2.11)

$$DEGD = \frac{365}{2\pi}(T_{July} - T_{Jan}) - \frac{365}{\pi}\left[40 - \frac{(T_{July} + T_{Jan})}{2}\right]$$
$$+ \frac{365}{\pi}\left\{\frac{[40 - ((T_{July} + T_{Jan})/2)]^2}{(T_{July} - T_{Jan})}\right\} \tag{2.11}$$

where temperature is in Fahrenheit.

The temperature derating factor, TF_i, is a parabolic function of the growing degree-days. For species i,

$$TF_i = max(0, TDEGD_i) \tag{3.5}$$

$$TDEGD_i = \frac{4(DEGD - DEGD_{min(i)})(DEGD_{max(i)} - DEGD)}{(DEGD_{max(i)} - DEGD_{min(i)})^2} \tag{3.6}$$

where DEGD = growing degree-days during current year at site as discussed in Chapter 2

$DEGD_{min(i)}$ = value at thermal minimum (northern end of range of species i)

$DEGD_{max(i)}$ = value at thermal maximum (southern end of range of species i)

In theory, the values of $DEGD_{min(i)}$ and $DEGD_{max(i)}$ could be determined from laboratory experiments in which the rate of photosynthesis was measured over a range of temperatures, as has been done in some cases, primarily for seedlings (Ledig and Korbobo, 1983). This is a preferred method. However, there may be considerable differences in this response for seedling, sapling, and mature stages in an individual tree and among individuals and ecotypes within a species, so that the determination of an average value for these parameters for a species might require measurements of many individuals known to be representative of the variability within the species. Unfortunately, such extensive measurements are generally not available. Lacking such measurements, $DEGD_{min(i)}$ and $DEGD_{max(i)}$ can be estimated from species range maps and lines of temperature isotherms for January and July, as shown for yellow birch in Figure 3.4. The development of a program for direct estimation of these parameters following methods of Ledig and Korbobo (1983) and others, and using standardized environmental conditions, would be useful.

The parabolic curve representing the response of tree growth to temperature imposes certain stringent assumptions, the most important of which are: (1) photosynthesis and growth become zero at specific degree-day values, and (2) the temperature response curve is symmetric, falling off at equal rates above and below the peak. In reality the curve could be asymmetric, especially with a steeper decline toward higher temperatures above the peak than the rise at lower temperatures before the peak, which might result from rapid destruction or deactivation of the photosynthetic apparatus as proteins and therefore

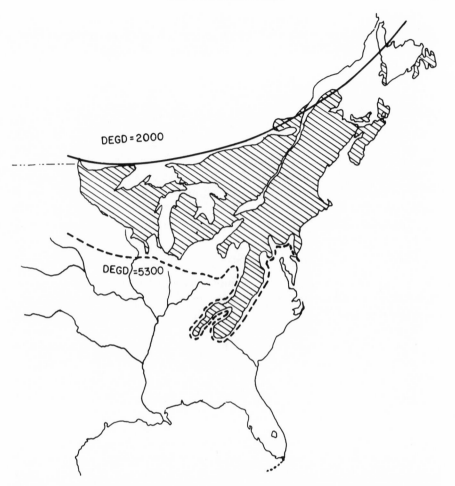

FIGURE 3.4. Temperature isotherms correspond to geographic limits of yellow birch. The January minimum temperature isotherm that gives 2,000 growing degree-days closely approximates the northern limit of yellow birch, while the July maximum that gives 5,300 growing degree-days closely approximates the southern limit. This gives an approximation of $DEGD(i)_{min}$ of 2,000 and $DEGD(i)_{max}$ of 5,300 for this species. Because of the lack of detailed experimental data, this method has been used as the general estimation procedure for these parameters. (From Botkin et al., 1972b.)

enzymes became denatured. Such asymmetries have been observed for algae, and we applied that asymetric relationship in a model of a phytoplankton community (Lehman et al., 1975a). However, available data do not support this choice of a curve for woody plants.

It is possible that growth declines gradually toward the geographic limits of the range of a species, so that the growth would decline asymptotically with temperature toward zero instead of ceasing abruptly at one temperature. This gradual approach to the thermal limits can be represented by a bell-shaped or

Gaussian curve. One could justify the use of this curve for a species with high genetic diversity. Local genotypes near the limit of the range could have different temperature response functions than genotypes in the center of the range. Although individual trees would reach zero growth at specific temperatures, net population growth would decline gradually because individuals would reach zero growth at different temperatures.

Such a Gaussian response curve is represented by

$$\mathrm{TF}_i = e^{\{-(\mathrm{DEGD} - \gamma)^2/2\sigma^2\}} \tag{3.7}$$

where TF_i is the temperature response function for species i, DEGD is the celsius degree-days as defined in Chapter 2, γ is the average of the maximum and minimum degree-day limits, and σ is an estimate of the standard deviation of degree-days (Woodby, 1991). It is straightforward to substitute the Gaussian curve for the parabolic and find which gives the more realistic results.

Tree Responses to Soil Moisture Conditions

The growth of a tree changes with too much and too little water. Unfortunately, studies of the response of total biomass increment of an entire tree to soil water have been qualitative and do not provide a solid basis from which water response functions can be developed. Existing studies suggest that tree growth is insensitive to the exact value of soil water except near the extremes of drought and saturated soil (Kozlowski, 1968, 1970, 1972, 1982a, b). Although much is known about anatomical, morphological, and physiological details of water transport in a tree, including the forces, pressures, and sources of energy and the cellular processes in leaves, xylem, phloem, and roots (Kramer and Kozlowski, 1979), and although models exist that project water use and transport by individual trees, such as the models of Jarvis (1981), Running (1984), and Waring, Schroeder, and Ore (1982), less is known about the response of total net biomass increment to soil-water conditions. The observed, general shape of the response to drought is shown in Figure 3.5.

Drought conditions for a tree occur when energy available for evapotranspiration greatly exceeds the amount of water available in the soil for that evaporation. Remember that the amount of water that could be evaporated and transpired given the available energy is known as potential evapotranspiration. When potential evapotranspiration is much greater than the calculated actual evapotranspiration, the net amount of water stored in the tree declines. If this continues long enough, leaves wilt and eventually the tree dies. There are two ways in which a drought effect has been calculated in the model. In the first method, too little water is represented by the difference between potential evapotranspiration, E_0, and actual evapotranspiration, E, normalized by the potential evapotranspiration, so that

$$\mathrm{WILT} = \frac{(E_0 - E)}{E_0} \tag{3.8a}$$

WILT is a dimensionless quantity that ranges from zero in swampy sites to

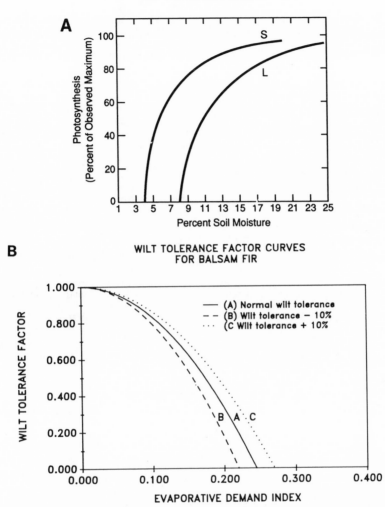

FIGURE 3.5. Growth as a function of decreasing soil moisture. (A) Net photosynthesis of sweetgum (S) and loblolly pine (L) with decreasing soil moisture. (Fig. 5.21, p. 203, from P. J. Kramer and T. T. Kozlowski, 1979, *Physiology of Woody Plants*, Academic Press, New York. Data on sweetgum from Bormann [1953]; on loblolly pine from Kozlowski [1949].) (B) Model water response function. Effect of soil drought on balsam fir growth. This is a plot of the function WiF for balsam fir, showing the effect of an error of plus and minus 10 percent in the estimation of this parameter on the response function.

about 30 percent in thin interior soils with a low till depth (Botkin and Levitan, 1977). The wilt factor measures the lack of a positive property of soil moisture for plant growth. Note that since E can never exceed E_0, WILT can never be less than zero. WILT can also never be greater than one. (For an example of the size of the wilt factor for two different soil conditions, see Table 3.1).

In the second method, the wilt factor is calculated as the difference between

TABLE 3.1. Effect of soil texture on site quality for two locations near Mount Pleasant, Michigan, weather station—a coarse sandy site and a clay soil[a]

(A) Environmental conditions for a clay soil (soil moisture-holding capacity 250 mm/m)

Month	Temp. (°C)	Precip. (mm)	Water-S (mm)	Snow-S (mm)	Snowmelt (mm)	Snow-D-D (degree-days)	I	A
Jan.	−7.59	54.86	100.00	161.80	0.00	0.00	31.16	1.00
Feb.	−6.88	42.16	100.00	203.96	0.00	0.00	31.16	1.00
March	−1.65	60.20	100.00	117.96	146.20	54.15	31.16	1.00
April	3.66	102.62	100.00	0.00	117.96	211.80	31.16	1.00
May	12.69	65.02	63.19	0.00	0.00	498.89	31.16	1.00
June	15.73	66.04	42.79	0.00	0.00	573.80	31.16	1.00
July	19.29	97.54	43.89	0.00	0.00	703.49	31.16	1.00
Aug.	17.23	90.93	47.74	0.00	0.00	639.60	31.16	1.00
Sept.	12.78	118.11	69.28	0.00	0.00	485.47	31.16	1.00
Oct.	8.90	143.26	95.58	0.00	0.00	381.44	31.16	1.00
Nov.	−2.69	47.75	100.00	0.00	47.75	21.40	31.16	1.00
Dec.	−4.95	59.18	100.00	59.18	0.00	0.00	31.16	1.00
Totals		947.67			311.91			

Month	dw/dt	P (mm)	U (mm)	E_0 (mm)	E (mm)	S (mm)	ρ (mm)	DW ±
Jan.	0.00	0.00	0.00	0.00	0.00	0.00	0.00	0.00
Feb.	0.00	0.00	0.00	0.00	0.00	0.00	0.00	0.00
March	17.24	146.20	0.00	0.00	0.00	128.96	17.24	−0.00
April	19.05	220.57	0.00	21.41	21.41	180.11	19.05	0.00
May	−36.81	65.02	0.00	84.62	83.64	18.19	0.00	−36.81
June	−20.40	66.04	0.00	106.36	76.28	10.16	0.00	−20.40
July	1.10	97.54	0.00	131.87	82.00	14.44	0.00	1.10
Aug.	3.85	90.93	0.00	108.15	71.67	15.41	0.00	3.85
Sept.	21.54	118.11	0.00	68.56	59.97	36.61	0.00	21.54
Oct.	26.30	143.26	0.00	42.36	42.35	74.60	0.0	26.30
Nov.	9.34	47.75	0.00	0.00	0.00	38.41	4.92	4.42
Dec.	0.00	0.00	0.00	0.00	0.00	0.00	0.00	0.00
Totals	41.21	995.43	0.00	563.33	437.32	516.90	41.21	0.00

Species site qualities—clay soil

Species	Q_i (Site Quality)	W_iF	W_eF_i	TDEGD$_i$	NF$_i$
Sugar maple	0.171	0.592	1.000	0.848	0.342
Beech	0.173	0.592	1.000	0.857	0.342
Yellow birch	0.054	0.166	1.000	0.958	0.342
White ash	0.006	0.166	1.000	0.377	0.093
Mountain maple	0.102	0.334	1.000	0.892	0.342
Striped maple	0.097	0.334	1.000	0.848	0.342
Pin cherry	0.060	0.650	1.000	0.991	0.093
Choke cherry	0.139	0.650	1.000	0.626	0.342
Balsam fir	0.041	0.166	1.000	0.450	0.545
Red spruce	0.057	0.166	1.000	0.628	0.545
White birch	0.234	0.650	1.000	0.660	0.545
Mountain ash	0.119	0.405	1.000	0.860	0.342

TABLE 3.1. (*Continued*)

Species	Q_i (Site Quality)	W_iF	W_eF_i	TDEGD$_i$	NF$_i$
Red maple	0.181	0.753	1.000	0.441	0.545
Scarlet oak	0.000	0.753	1.000	0.000	0.545
Hornbeams	0.071	0.753	1.000	0.276	0.342
Green alder	0.000	0.000	1.000	0.000	0.545
Speckled alder	0.000	0.000	1.000	0.926	0.545
Chestnut	0.000	0.753	1.000	0.000	0.545
Black ash	0.000	0.000	1.000	0.989	0.093
Butternut	0.005	0.444	1.000	0.131	0.093
White spruce	0.007	0.166	1.000	0.479	0.093
Black spruce	0.000	0.000	1.000	0.552	0.545
Jack pine	0.406	0.822	1.000	0.906	0.545
Red pine	0.409	0.800	1.000	0.938	0.545
White pine	0.352	0.753	1.000	0.857	0.545
Trembling aspen	0.255	0.753	1.000	0.993	0.342
White oak	0.047	0.753	1.000	0.182	0.342
Red oak	0.114	0.753	1.000	0.443	0.342
White cedar	0.000	0.000	1.000	0.581	0.093
Hemlock	0.061	0.166	1.000	0.678	0.545
Silver maple	0.000	0.000	1.000	0.547	0.093
Tamarack	0.000	0.000	1.000	0.517	0.342
Pitch pine	0.000	0.822	1.000	0.000	0.545
Gray birch	0.313	0.753	1.000	0.762	0.545
American elm	0.007	0.166	1.000	0.481	0.093
Basswood	0.030	0.405	1.000	0.795	0.093
Bigtooth aspen	0.000	0.000	1.000	0.857	0.342
Balsam poplar	0.050	0.650	1.000	0.839	0.093
Black cherry	0.000	0.650	1.000	0.000	0.342
Red cedar	0.075	0.753	1.000	0.182	0.545

(B) Environmental conditions for a sandy soil (soil moisture-holding capacity 50 mm/m)[b]

Month	dw/dt	P(mm)	U(mm)	E_0(mm)	E(mm)	S(mm)	ρ(mm)	DW±
Jan.	0.00	0.00	0.00	0.00	0.00	0.00	0.00	0.00
Feb.	0.00	0.00	0.00	0.00	0.00	0.00	0.00	0.00
March	4.99	146.20	0.00	0.00	0.00	141.22	4.99	0.00
April	4.76	220.57	0.00	21.41	21.41	194.40	4.76	0.00
May	−10.93	65.02	0.00	84.62	63.55	12.39	0.00	−10.93
June	−1.40	66.04	0.00	106.36	59.52	7.93	0.00	−1.40
July	1.13	97.54	0.00	131.87	81.97	14.44	0.00	1.13
Aug.	0.88	90.93	0.00	108.15	74.11	15.94	0.00	0.88
Sept.	6.64	118.11	0.00	68.56	66.9	44.56	0.00	6.64
Oct.	6.41	143.26	0.00	42.36	42.36	94.49	2.74	3.67
Nov.	4.26	47.75	0.00	0.00	0.00	43.49	4.26	−0.00
Dec.	0.00	0.00	0.00	0.00	0.00	0.00	0.00	0.00
Totals	16.74	995.43	0.00	563.33	409.82	568.86	16.74	0.00

Footnotes on page 78.

TABLE 3.1. (*Continued*)

Species site qualities—sandy soil

Species	Q_i (Site Quality)	W_iF	WeF_i	$TDEGD_i$	NF_i
Sugar maple	0.114	0.394	1.000	0.848	0.342
Beech	0.115	0.394	1.000	0.857	0.342
Yellow birch	0.000	0.000	1.000	0.958	0.342
White ash	0.000	0.000	1.000	0.377	0.093
Mountain maple	0.003	0.011	1.000	0.892	0.342
Striped maple	0.003	0.011	1.000	0.848	0.342
Pin cherry	0.044	0.480	1.000	0.991	0.093
Choke cherry	0.103	0.480	1.000	0.626	0.342
Balsam fir	0.000	0.000	1.000	0.450	0.545
Red spruce	0.000	0.000	1.000	0.628	0.545
White birch	0.173	0.480	1.000	0.660	0.545
Mountain ash	0.034	0.117	1.000	0.860	0.342
Red maple	0.152	0.633	1.000	0.441	0.545
Scarlet oak	0.000	0.633	1.000	0.000	0.545
Hornbeams	0.060	0.633	1.000	0.276	0.342
Green alder	0.000	0.000	1.000	0.000	0.545
Speckled alder	0.000	0.000	1.000	0.926	0.545
Chestnut	0.000	0.633	1.000	0.000	0.545
Black ash	0.000	0.000	1.000	0.989	0.093
Butternut	0.002	0.175	1.000	0.131	0.093
White spruce	0.000	0.000	1.000	0.479	0.093
Black spruce	0.000	0.000	1.000	0.552	0.545
Jack pine	0.363	0.736	1.000	0.906	0.545
Red pine	0.359	0.703	1.000	0.938	0.545
White pine	0.296	0.633	1.000	0.857	0.545
Trembling aspen	0.215	0.633	1.000	0.993	0.342
White oak	0.039	0.633	1.000	0.182	0.342
Red oak	0.096	0.633	1.000	0.443	0.342
White cedar	0.000	0.000	1.000	0.581	0.093
Hemlock	0.000	0.000	1.000	0.678	0.545
Silver maple	0.000	0.000	1.000	0.547	0.093
Tamarack	0.000	0.000	1.000	0.517	0.342
Pitch pine	0.000	0.736	1.000	0.000	0.545
Gray birch	0.263	0.633	1.000	0.762	0.545
American elm	0.000	0.000	1.000	0.481	0.093
Basswood	0.009	0.117	1.000	0.795	0.093
Bigtooth aspen	0.000	0.000	1.000	0.857	0.342
Balsam poplar	0.037	0.480	1.000	0.839	0.093
Black cherry	0.000	0.480	1.000	0.000	0.342
Red cedar	0.063	0.633	1.000	0.182	0.545

[a]Elevation = 468 m, soil depth = 0.4 m; water table depth = 0.6 m; soil nitrogen content = 60 kg/ha; degree-days = 1840.14. Symbols in column headings not given above are explained in the glossary and text. The wilt factor is calculated according to the first method:

$$WILT = \frac{E_0 - E}{E_0} \qquad (3.8a)$$

and evapotranspiration is calculated as discussed in Chapter 2:

$$E_0 = 16\left(\frac{10T_m}{I}\right)^a \qquad (2.15)$$

[b]Other conditions as in (A).

the field capacity and the normalized actual water storage, so that

$$\text{WILT} = \frac{w_k - w}{w_k} \tag{3.8b}$$

In either case, the effect of drought on tree growth, WiF_i, is

$$\text{WiF}_i = \max\left\{0_j, 1 - \left(\frac{\text{WILT}}{\text{WLMAX}_i}\right)^2\right\} \tag{3.9}$$

where WLMAX_i is the maximum wilt tolerable by species i. WILT and WLMAX_i are dimensionless quantities. In practice, the value of WLMAX_i has been estimated as a relative factor, with species known to be characteristic of very dry or very wet sites given extreme values and other species given intermediate values consistent with the known habitats. This approach to the derivation of WLMAX_i is less satisfactory conceptually than the derivation of other parameters in the model, and remains a subject for which improvement is desirable, both from experimental studies to specify the shape of the relationship between tree growth and drought, and from theoretical analysis of tree–soil-water flux. As discussed earlier, available information suggests that trees grow reasonably well in relation to soil moisture except at the extremes—that the curve relating tree growth to soil moisture is level except at the extremes, where a rapid decrease occurs. If WiF_i were made a function of $\text{WILT}/\text{WLMAX}_i$ to the first power, the slope would be too gradual. Raising this ratio to higher powers steepens the slope. The ratio $\text{WILT}/\text{WLMAX}_i$ is raised to the second power to provide a realistic steepness to the slope of the decrease in tree growth with soil dryness.

Unless otherwise stated, results presented in this book use the first method, but both methods are available in the model which is available for use with this book. The shape of WiF_i is shown in Figure 3.5(B) for several species. This function has the general shape observed for trees (Fig. 3.5). Note that when $\text{WLMAX}_i = \text{WILT}$, that is, when the maximum drought tolerable by a species equals the current site conditions, then WiF_i is zero and individuals of the species cannot grow.

Soil texture affects the amount of water stored, thus affecting site quality (the function Q_i in equations [3.2] and [3.3]). The effect of soil texture on site quality can be seen in Table 3.1 for a clay-loam soil (soil moisture-holding capacity 250 mm depth of water/m depth of soil) (Table 3.1(A)) and for a sandy soil (soil moisture-holding capacity 50 mm/m) (Table 3.1(B)). Texture has a large effect, which varies with species. The wilt factor for sugar maple drops from 0.592 to 0.394 and the site quality, Q_i from 0.171 to 0.114. Here the site is dry—the soil is comparatively shallow and the water table deep, so that the site quality favors upland species adapted to dry conditions, including jack pine. The difference between sites is sufficient to allow yellow birch to persist marginally in the clay soil (site quality = 0.054), but to fail (site quality = 0) in the sandy soil. Inspection of Table 3.1 shows which species will be favored during competition for specific site conditions. Users of the software available as a companion to this book can inspect the table, list species that they believe should dominate, and then run the model using site conditions in Table 3.1 or for other cases and observe the quantitative projections of the model for these two habitats. Symbols in column headings not given below are explained in the glossary and text. The wilt factor is calculated according to the first method:

$$\text{WILT} = \frac{E_0 - E}{E_0} \tag{3.8a}$$

and evapotranspiration is calculated as discussed in Chapter 2:

$$E_0 = 16(10\text{T}_m/\text{I})^a \tag{2.15}$$

Effect of Elevation on Site Quality. Change in elevation affects both evapotranspiration and temperature, but the effect is much larger on the temperature factor (Table 3.2). For example, at 500 m elevation, boreal species including balsam fir, red spruce, and white birch have a sufficient site quality to allow growth, but at 100 m the temperature quality is 0.0 for fir and spruce and 0.048 for white birch; fir and spruce cannot grow at 100 m elevation and white birch will do poorly. In contrast, the northern hardwoods, such as sugar maple and beech, have a considerably improved temperature quality at 100 m in comparison to 500 m. Although these species can grow at both elevations, their growth will be much better at 100 m.

Soil-Water Saturation and Tree Growth. Too much water in the soil has negative effects on tree growth, principally because water-saturated soil lacks oxygen that is necessary for root tissue respiration. Cessation of root growth has several effects. Uptake of chemical elements takes place at the growing, actively metabolizing ends of roots, and a lack of oxygen suppresses nutrient uptake. Suppression of growth renders roots more vulnerable to fungal or bacterial diseases. Trees with a poor root structure are more vulnerable to physical damage from wind. Again, although much is known about the details of these mechanisms at the cellular level, less is known about how biomass increment of an entire tree responds quantitatively to soil-water saturation. As a first approximation of these effects, we have chosen as our measure of site wetness simply the reciprocal of the water table depth (Botkin and Levitan, 1977).

The factor for site wetness is calculated as

$$\text{WeF}_i = \max\left[0, 1 - \left(\frac{\text{DTMIN}_i}{\text{DT}}\right) \right] \tag{3.10}$$

where DTMIN_i is the minimum distance to the water table tolerable for species *i* and DT is the depth to the water table. The shape of this response curve is shown in Figure 3.6. In the absence of numerical data in the literature, we provided values for parameters reflecting the hydrological environments that each species is known to tolerate (Table 3.5, pages 90–91).

TABLE 3.2. Effect of elevation on site quality $(Q_i)^a$

(A) At 100-m elevation (from conditions given in Table 2.3A and B, Chapter 2)

Species	Q_i	W_iF	WeF_i	$TDEGD_i$	NF_i
Sugar maple	0.499	0.776	1.000	0.992	0.648
Beech	0.501	0.776	1.000	0.998	0.648
Yellow birch	0.339	0.542	1.000	0.965	0.648
White ash	0.139	0.542	1.000	0.593	0.431
Mountain maple	0.410	0.634	1.000	0.998	0.648
Striped maple	0.408	0.634	1.000	0.992	0.648
Pin cherry	0.339	0.808	1.000	0.972	0.431
Choke cherry	0.415	0.808	1.000	0.793	0.648
Balsam fir	0.000	0.542	1.000	0.000	0.718
Red spruce	0.000	0.542	1.000	0.000	0.718
White birch	0.028	0.808	1.000	0.048	0.718
Mountain ash	0.031	0.673	1.000	0.072	0.648
Red maple	0.380	0.864	1.000	0.612	0.718

(B) At 500-m elevation (from conditions given in Table 2.3C and D, Chapter 2)

Species	Q_i	W_iF	WeF_i	$TDEGD_i$	NF_i
Sugar maple	0.427	0.835	1.000	0.789	0.648
Beech	0.429	0.835	1.000	0.793	0.648
Yellow birch	0.392	0.664	1.000	0.913	0.648
White ash	0.092	0.664	1.000	0.320	0.431
Mountain maple	0.400	0.731	1.000	0.844	0.648
Striped maple	0.374	0.731	1.000	0.789	0.648
Pin cherry	0.361	0.859	1.000	0.975	0.431
Choke cherry	0.323	0.859	1.000	0.581	0.648
Balsam fir	0.281	0.664	1.000	0.589	0.718
Red spruce	0.362	0.664	1.000	0.760	0.718
White birch	0.467	0.859	1.000	0.758	0.718
Mountain ash	0.464	0.760	1.000	0.943	0.648
Red maple	0.257	0.900	1.000	0.397	0.718

[a] Soil texture 150 mm/m; soil nitrogen content 79 kg/ha.

Exchange of Chemical Elements between an Individual Tree and Its Environment

For readers not familiar with chemical cycling and vegetation, the following background material may be helpful. Every organism exchanges chemical elements with its environment. Each organism is made up of many chemical compounds; as it grows it must accumulate more chemical elements to make more compounds. In addition, some chemical elements inevitably are lost, and these must be replaced. Some loss occurs in the elimination of wastes—for example, green plants give off oxygen to the atmosphere as a waste product of photosynthesis. Some occurs through loss of tissue as, for example, when roots or the inner, living part of the bark dies and the material is sloughed off.

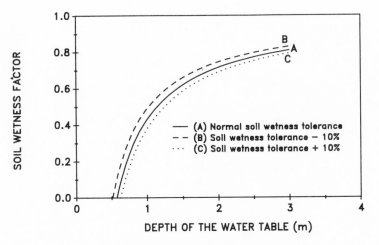

FIGURE 3.6. Water response function. Effect of soil-water saturation on sugar maple growth. This is a plot of WeF_i for several species (see text).

There are 24 chemical elements required by at least some kinds of organisms, and each must be available at the right time, in the right concentration, and in the right radio to other elements. Trees can store a certain amount of some chemical elements for later use, but storage has limits, which vary seasonally. In deciduous angiosperm trees, nitrogen can be transported out of leaves in the autumn before leaf fall and transported to roots, where it is stored until spring. A tree must obtain the right amounts of all the elements needed though these elements vary in abundance and availability over time—with seasons and among years—and they are distributed in a patchy way throughout their environment.

Chemical elements required by living things can be separated into three categories: (1) *macronutrients*, nutrients required in large amounts by all life forms; (2) *micronutrients* that are used in small amounts by all life forms; and (3) *micronutrients* that are used in small amounts by some, but not all, life forms. Macronutrients include the "big six" elements: carbon, hydrogen, nitrogen, oxygen, phosphorus, and sulfur. These are the basic building blocks of organic compounds. Carbon, hydrogen, and oxygen make up carbohydrates, fats, and oils. These three plus nitrogen form proteins. Phosphorus and sulfur are required for compounds that control many cell functions. Phosphorus is the energy element, acting in the compounds adenosinetriphosphate (ATP) and adenosinediphosphate (ADP) to store and transfer energy within cells. Sulfur occurs in enzymes that control important cell activities.

Chemical cycling in ecosystems has been the subject of considerable study during the last two decades, as exemplified by analyses of eastern deciduous forests of North America (Bormann and Likens, 1979; Borman et al., 1974; Likens et al., 1977). There has been an emphasis on changes in nutrient storage at the ecosystem level (Schlesinger, 1991; Waring and Schlesinger, 1985), as discussed for example by Vitousek and Reiners (1975) and Gorham, Vitousek, and Reiners (1979). Volumes of research exist about the effects of limitations of each chemical element on plant growth, (Kozlowski, Kramer, and Pallardy, 1991) especially for crops (Kramer and Kozlowski, 1979).

These studies provide a picture of the general response of a plant to the concentration of a required chemical element in the soil. The shape of this curve is like the growth response of a tree to available light (equation 3.3) (Figs. 3.1, 3.2). This is known as a *dose-response curve*. Dose-response curves characteristically show sensitivity to low concentrations and insensitivity to high concentrations. At low concentrations of a chemical element in the soil, a unit increase causes a large increase in growth; at high concentrations, the same unit increase causes a small increase in growth, and when concentrations are sufficiently high, additional increases in the soil lead to no growth increase and eventually to a suppression of growth (although this final negative effect occurs at levels not usually found for macronutrients and most micronutrients in nature). From this background we know that the general shape of the growth response of a tree to the concentration of any required chemical element will resemble that shown in Figure 3.2.

We need to consider another factor before we can arrive at a model of growth response to chemical nutrients. Plants do not grow in direct response to the abundance of a chemical element in the soil, but in response to the internal stores of that element. There are two steps in the utilization of molecules of a nutrient: uptake and growth. We might try to simplify the representation of these processes by devising a dose-response curve that relates growth directly to external concentration. Experience with models of other photosynthetic organisms, including freshwater algae (Lehman, Botkin, and Likens, 1975a, b) and salt marsh grasses (Morris, 1982; Morris, Houghton, and Botkin, 1984) suggests that greater realism is obtained in model output if the two stages are retained. With this background, we can set down the nutrient response function of the forest model.

In theory, one could introduce into the model a nutrient response function for each required chemical element. This would require approximately 40 additional equations (one each for uptake and one each for growth from internal stores for approximately 20 elements). This is unnecessary in practice because (1) some elements vary together so that the variation in one would represent, for purposes of a model, the variation in the other; and (2) some elements are rarely limiting, and where they are a specific response function could be added. Data are insufficient to develop a complete set of nutrient response functions. It is not an overstatement to say that nitrogen may be the only element for which the response of trees under natural conditions is sufficiently well studied to develop a function. Even for this element data are sketchy. A model with 20 or more chemical elements would be computationally clumsy, requiring parameters for which data are poor or lacking. Such a model would provide little additional realism at the cost of great effort. In the long term there are situations for which additional nutrient response functions will be important. However, given the paucity of data for any chemical element, we have developed algorithms only for nitrogen. In Chapter 4 the success of this simplification of the model can be examined.

Soil Nitrogen Tolerance

Green plants cannot use molecular nitrogen directly, but they can use either nitrate or ammonia, and some evidence suggests that these are used equally efficiently (Morris, 1982; Morris, Houghton, and Botkin, 1984). As is well known, nitrate and ammonia are made available by various bacterial reactions. Readers unfamiliar with the nitrogen cycle and interested in these reactions can refer to a number of standard references, such as Waring and Schlesinger (1985), Schlesinger (1991), and other recent papers discussing nitrogen cycling in forests, such as Aber, Melillo, and McClaugherty (1990); Stevens et al. (1990); Hendrickson (1990); Van Miegroet, Johnson, and Cole (1990); Nohrstedt (1989). When nitrogen is low in concentration in the soil, the addition of a small amount produces a relatively great increase in plant growth, if other factors are not limiting. As the concentration of nitrogen in the soil increases, the addition of each new unit of nitrogen results in a smaller increase in plant growth. As a result, the generalized response of a green plant to soil nitrogen concentration is a curve that increases to an asymptote with an increase in soil nitrogen, a curve of the same form as the light response function (equation 3.3a).

Although hundreds of studies have been done on the response of plants in greenhouses to changes in concentration of a specific chemical element, there are few studies of the response of trees in a forest to changes in the soil concentration of a single element. Such studies require long periods—for forest trees results might not be evident in fewer than 5 or 10 years. There are many confounding variations in natural forest soils. And until recently there were few funds for such studies. One of the few classical studies of the response of a forest to fertilization was that done in New England by Mitchell and Chandler (1939). This study provides the basic shapes of the curves and the definition of the gradient. Mitchell and Chandler measured the growth response of trees against the amount of nitrogen added in fertilizer. The study has one serious limitation: the authors did not measure the nitrogen concentration in the soil prior to the addition of fertilizer, so the absolute concentration of nitrogen is not known.

Mitchell and Chandler were not alone in having difficulty in dealing with the concentration of nitrogen in soils. Although there have been a number of recent studies of the nitrogen cycle and of concentrations in the soil, the connection between a measured amount of nitrogen in a soil and the amount actually available for uptake by trees remains unclear.

Most measurements of soil nitrogen give the total amount (Federer, 1984; Huntington et al., 1988), and suggest that total nitrogen can be as mush as 1,000 to 10,000 kg/ha, but most of that is bound in complex organic compounds not available to plants. In a review of nitrogen cycling, Waring and Schlesinger (1985) reported that nitrogen concentrations in the forest floor ranged from approximately 600 to 700 kg/ha in boreal and temperate forests, were less than 400 kg/ha in temperate deciduous forests, and were approximately 200 in tropical forests. Total nitrogen concentration in organic and inorganic fractions of the soil, typically measured to a depth of 20 to 100 cm, ranges up to more than 8,000 kg/ha. Plants can take up nitrogen as nitrate or ammonia, but these are produced by bacterial action, through the fixation of atmospheric molecular nitrogen, or the decomposition of organic compounds, processes that can vary comparatively rapidly with changes in soil temperature and moisture. Nitrate and ammonia are reactive and relatively volatile so that they have a short residence time in the soil, which means that even a perfect measure of the status of these compounds would have to be done frequently (or monitored continuously) to provide an exact measure of available nitrogen. In spite of these limitations, considerable progress has been made in recent years in obtaining measurements that correspond to nitrogen levels that can be taken up by plants, and these suggest that nitrogen becomes available for plant uptake at a rate generally less than 100 kg/ha/yr (Aber, Melillo, and McClaugherty, 1990; Clark and Rosswall, 1981; Hendrickson, 1990; Pastor et al., 1984, Van Miegroet et al., 1990). Available nitrogen is highest in forests with a high abundance of trees that have symbiotic nitrogen-fixing bacteria in root nodules; in these, nitrogen fixation can approach 100 kg/ha/yr (Bormann and Gordon, 1984). Nitrogen fixation by free-living bacteria seems to be much less, on the order of 0 to 3 kg/ha/yr (Roskoski, 1980; Waring and Schlesinger, 1985). Further field research is needed to resolve the question of the relationship between nitrogen measured in the soil and nitrogen available to a tree. Thus in part the problem is one of measurement; it is still not clear what measurement method corresponds to the level of nitrogen available to trees. In spite of these limitations, Mitchell and Chandler's study provides a basis from which a dose-response curve can be derived, and their study is a basis for the following discussion. The factor for nitrogen tolerance, NF, is based on work by Aber, Botkin, and Mellilo (1978, 1979). NF is a function of the available nitrogen in the soil, AVAILN, and the nitrogen concentration in the leaves λ. Different curves for responses to the nitrogen gradient are used for each of three classes—tolerant (1), intermediate (2), and intolerant (3). Here, "tolerant" means tolerant of low levels of available nitrogen.

This analysis leaves one question unanswered: What is the correspondence between the quantities derived from the Mitchell and Chandler study and measured amounts of nitrogen reported in the literature? Woodby (1991) followed Aber et al. (1979) and normalized the results from the Mitchell and Chandler study to represent kilograms per hectare to give amounts of the same order of magnitude as nitrogen available by biological fixation on an annual basis. The nitrogen content of the leaves is computed for each tolerance class, N, using the following equation:

$$\lambda_N = \alpha_1 [1 - 10^{-\alpha_2(\text{AVAILN} + \alpha_3)}] \tag{3.11}$$

and the response of the trees calculated as:

$$\text{NF}_i = \frac{(\alpha_4 + \alpha_5 * \lambda_N)}{\alpha_6} \tag{3.12}$$

where λ is the concentration of nitrogen in leaves of tolerance class N; AVAILN

TABLE 3.3. Coefficients for the nitrogen response function

Nitrogen Class	α_1	α_2	α_3	α_4	α_5	α_6
Intolerant	2.99	0.00175	207.43	−5.0	2.9	3.671
Intermediate	2.94	0.00234	117.52	−1.2	1.3	2.622
Tolerant	2.79	0.00179	219.77	−0.6	1.0	2.190

is the concentration of available nitrogen in the soil, NF_i is the nitrogen response function of species i, and the coefficients (α_1 through α_6) are given in Table 3.3 and some resulting nitrogen values are given in Figure 3.7.

Conceptually, the two equations (3.11) and (3.12) create a two-step nitrogen response, with nitrogen first taken up by the leaves and then leaf storage affecting tree growth. This conforms more realistically to cellular and physiological processes. If a lag effect in nitrogen response were observed in forests, then this is a location in the model where, by modification of the form of equation (3.12), that lag could be represented. Arithmetically, however, as presently formulated, equation (3.12) simply normalizes the relationships from Aber et al. (1978) so that NF_i ranges from slightly less than zero to a small positive number for each tolerance class within a reasonable range of expected concentrations of available nitrogen.

From Figure 3.7 one can see that nitrogen-tolerant species have a greater response than the other classes except at high nitrogen concentrations (approximately greater than 100 kg/ha/yr). There is a crossover point in the responses of intermediate and tolerant classes, at approximately 100 kg/ha/yr. Below this level, tolerant species have a greater response; above this level, intermediate species have a greater response. Intolerant species have a lower response at all levels of available nitrogen.

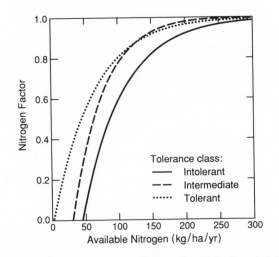

FIGURE 3.7. Nitrogen response function. This is a plot of the function NF_n for the three nitrogen tolerances classes: (1) intolerant of low concentrations of available nitrogen; (2) intermediate in tolerance; and (3) tolerant.

Nitrogen and Site Quality Calculations

The effect of soil nitrogen on site quality is shown in Table 3.4 for the Hubbard Brook site of Table 3.2. Soil texture is a sandy loam intermediate in moisture-holding capacity (150 mm/m), but the nitrogen content in Table 3.4 is 35 kg/ha, less than half the value of the site in Table 3.2 (79 kg/ha). The model is sensitive to changes in soil nitrogen content and is comparatively much less sensitive to changes in soil moisture-holding capacity. When the soil nitrogen is 35 kg/ha (Table 3.4), nitrogen-intolerant species such as white ash can neither grow nor regenerate, in contrast to growth calculated for this species when the soil nitrogen is 79 kg/ha. Even nitrogen-tolerant species, such as white birch, undergo a large decline in nitrogen response when nitrogen drops from 79 to 35 kh/ha. Low soil nitrogen results in a site quality of less than 0.3 for all species. In contrast, under the high nitrogen level of Table 3.2 and the high nitrogen and high soil texture of Table 3.1, site quality ranges up to 0.47. None of these sites is of high quality for any of the species. White birch, a cold-climate, early-successional species tolerant of coarse soils and low nitrogen, maintains the highest site quality under all conditions shown in Tables 3.1, 3.2, and 3.4. Under high nitrogen content, the soil quality for sugar maple and beech approaches that for white birch, and these species can be expected to dominate old-age forests. In the low nitrogen case, it is not clear from inspection of the table whether any late-successional species will persist.

These examples show the relative sensitivity of the model to site conditions, and they emphasize the greater sensitivity of the model to soil nitrogen than to soil texture, which in part determines soil-water-holding capacity.

The calculation of site quality creates a static model of forests somewhat analogous to the biogeographical models implied by the Holdridge Life-zone diagram (Holdridge, 1947), the methods used by Lieth and Whittaker (1975) to project worldwide biomass and production of vegetation, and Box's biogeo-graphical models (Box, 1981). Examination of site quality factors gives insight into the relative dominance of species to be expected, but the dynamic quality of the forest is not represented, and the actual dominance of species can only be discovered by running the model. A map drawn of past, present, and future distributions of vegetation based on the correlations between past distribution of fossil pollen and climate is a model of this kind (Davis and Botkin, 1985); I will refer to this as a *static response surface model*. Analogously, site quality calculations of JABOWA might be used for certain applications, such as deve-lopment of large-scale regional maps of steady-state distribution of vegetation, calculated from environmental conditions.That is, the JABOWA site conditions calculations could be used as a static model for certain biogeographic appli-cations, such as the drawing of global maps of vegetation distribution in relation to environment, an application for which JABOWA has not yet been used.

Soil Basal Area Maximum

One other qualitative factor is added to the growth equation (2.6), the factor $s(BAR)$. $s(BAR)$ is the proportion unoccupied of the maximum potential stem

TABLE 3.4. Effect of low soil nitrogen on site quality $(Q_i)^a$

Month	Temp. (°C)	Precip. (mm)	Water-S (mm)	Snow-S (mm)	Snowmelt (mm)	Snow-D-D (degree-days)	I	A
Jan.	−8.98	101.95	70.48	232.36	0.00	0.00	30.25	0.98
Feb.	−7.01	97.21	70.48	329.57	0.00	0.00	30.25	0.98
March	−2.11	110.20	70.48	331.53	108.23	40.09	30.25	0.98
April	4.14	102.60	70.48	0.00	331.53	226.22	30.25	0.98
May	11.26	123.61	62.30	0.00	0.00	454.56	30.25	0.98
June	16.01	125.28	44.83	0.00	0.00	582.33	30.25	0.98
July	18.71	110.63	35.87	0.00	0.00	685.46	30.25	0.98
Aug.	17.69	119.58	40.64	0.00	0.00	653.65	30.25	0.98
Sept.	13.12	116.60	50.92	0.00	0.00	495.58	30.25	0.98
Oct.	7.17	110.48	68.22	0.00	0.00	327.78	30.25	0.98
Nov.	0.89	125.21	70.48	0.00	0.00	128.81	30.25	0.98
Dec.	−5.83	130.41	70.48	130.41	0.00	0.00	30.25	0.98
Totals		1373.75			439.77			

Month	dw/dt	P (mm)	U (mm)	E_0 (mm)	E (mm)	S (mm)	ρ	DW
Jan.	0.00	0.00	0.00	0.00	0.00	0.00	0.00	0.00
Feb.	0.00	0.00	0.00	0.00	0.00	0.00	0.00	0.00
March	12.46	108.23	0.00	0.00	0.00	95.77	12.46	0.00
April	17.18	434.13	0.00	24.53	24.53	392.43	17.18	0.00
May	−8.18	123.61	0.00	74.06	74.06	57.73	0.00	−8.18
June	−17.47	125.28	0.00	105.47	102.94	39.80	0.00	−17.47
July	−8.96	110.63	0.00	124.84	96.98	22.61	0.00	−8.96
Aug.	4.77	119.58	0.00	109.22	86.95	27.86	0.00	4.77
Sept.	10.29	116.60	0.00	70.35	67.19	39.12	0.00	10.29
Oct.	17.29	110.48	0.00	35.28	35.28	57.90	0.00	17.29
Nov.	14.81	125.21	0.00	3.85	3.85	106.54	12.55	2.26
Dec.	0.00	0.00	0.00	0.00	0.00	0.00	0.00	0.00
Totals	42.19	1373.75	0.00	547.60	491.79	839.77	42.19	0.00

Species	Q_i (Site Quality)	W_iF	WeF_i	$TDEGD_i$	NF_i
Sugar maple	0.060	0.835	1.000	0.789	0.090
Beech	0.060	0.835	1.000	0.793	0.090
Yellow birch	0.055	0.664	1.000	0.913	0.090
White ash	0.000	0.664	1.000	0.320	0.000
Mountain maple	0.056	0.731	1.000	0.844	0.090
Striped maple	0.052	0.731	1.000	0.789	0.090
Pin cherry	0.000	0.859	1.000	0.975	0.000
Choke cherry	0.045	0.859	1.000	0.581	0.090
Balsam fir	0.163	0.664	1.000	0.589	0.417
Red spruce	0.211	0.664	1.000	0.760	0.417
White birch	0.272	0.859	1.000	0.758	0.417
Mountain ash	0.065	0.760	1.000	0.943	0.090
Red maple	0.149	0.900	1.000	0.397	0.417

[a] Variables at site: Elevation: 500.00 m; K: 2.70 mm/day/°C; Snow Temp: −3.40 °C; Degree-days: 1756.79; Soil nitrogen: 35 kg/ha; Adjusted root depth: 0.50 m; Water table: 1.75 m; WFC: 23.49 mm; WK: 16.45 mm.

area on the site, SOILQ.

$$s(\text{BAR}) = 1 - \left(\frac{\text{BAR}}{\text{SOILQ}}\right) \tag{3.13}$$

where BAR is the total basal area on the plot and SOILQ is the maximum allowed basal area. This was added to the model originally so that the model would include the forester's notion of site quality or "site index" (Smith, 1986), that is, an upper limit to the number of trees that could be supported on a site; but the value of SOILQ is set to a very large number and functions only to prevent those rare cases where the abundance of trees would increase uncontrollably. The default setting of SOILQ is 20,000 for $100 \, \text{m}^2$ ($200 \, \text{cm}^2/\text{m}^2$). In actual operation, the basal area does not approach values anywhere near the default level, and this comparatively arbitrary factor has little effect in practice on the dynamics of the model. Users of the model can vary this factor and test its effect. With these factors calculated, the growth, reproduction, and death of each species of tree can then be determined.

A summary of parameters appears in Table 3.5.

Mortality

In 1748, Peter Kalm, a Swedish botanish sent to North America by Linnaeus to collect plants to decorate the gardens of Europe, traveled through much of eastern North America (this introductory story about Kalm is modified from Botkin, 1977, with permission). Arriving in Philadelphia, he visited Benjamin Franklin and saw the first American library, lived with the farmers of Pennsylvania, and then traveled through the wilderness of the Atlantic coastal states and New England on his way to Montreal. He passed through the virgin forests of northern Vermont. "Almost every night," he wrote in his journal, "we heard some trees crack and fall while we lay here in the woods, though the air was so calm that not a leaf stirred" (Kalm, 1770). Knowing no reasons for this death of the trees, he suggested that perhaps the dew loosened the roots of old trees at night or that immense flocks of passenger pigeons settled on the branches unevenly, causing the trees to fall and die. Whatever the reason, he wrote that "they made a dreadful cracking noise." In his travels through the North American wilderness, Peter Kalm discovered that every tree was vulnerable to some chance of death, and death came sometimes in the night.

Modern analyses of mortality rates began with a concern with human death rates, motivated in part by the need of insurance companies to set rates for their policies, and in part by curiosity about fundamental life processes. With the development of the science of ecology in the late nineteenth century, investigations of patterns of mortality focused primarily on animal populations, while early twentieth-century plant ecology emphasized community patterns— the pattern of development of a forest over time in one location, or the patterns of vegetation across a landscape. There was relatively little emphasis on demographic processes that might hold secrets of those patterns. Early analyses of

TABLE 3.5. Basic forest model parameters

(A) Intrinsic growth factors[a]

Species	Maximum Age (years)	Maximum Diameter (cm)	Maximum Height (cm)	B_2	B_3	G	C
Sugar maple	400	170	3350	37.8	0.111	118.7	1.570
Beech	366	160	3660	44.0	0.137	87.7	2.200
Yellow birch	300	100	3050	58.3	0.291	143.6	0.486
White ash	150	150	2440	30.7	0.102	147.5	1.750
Mountain maple	25	14	500	53.8	2.000	72.6	1.130
Striped maple	30	23	1000	76.7	1.700	109.8	1.750
Pin cherry	30	28	1126	70.6	1.260	227.2	2.450
Choke cherry	20	10	500	72.6	3.630	233.3	2.450
Balsam fir	200	86	2290	50.1	0.291	102.7	2.500
Red spruce	400	60	2290	71.8	0.598	50.7	2.500
White birch	140	76	3050	76.6	0.504	190.1	0.486
Mountain ash	30	10	500	72.6	3.630	155.6	1.750
Red maple	150	150	3660	47.0	0.156	213.8	1.570
Scarlet oak	200	30	3050	194.2	3.230	128.7	1.750
Hornbeams	150	30	1520	92.2	1.530	144.4	0.486
Green alder	30	5	300	65.2	6.520	143.3	2.000
Speckled alder	30	8	400	65.8	4.100	196.9	2.000
Chestnut	200	122	2740	42.7	0.175	195.2	1.750
Black ash	70	60	2130	66.4	0.554	96.2	1.750
Butternut	90	91	3050	64.0	0.352	192.2	1.750
White spruce	200	53	3350	121.2	1.140	91.8	2.500
Black spruce	250	46	2740	113.9	1.240	32.0	2.500
Jack pine	185	50	3050	116.5	1.160	142.0	2.000
Red pine	275	91	3050	64.0	0.352	156.4	2.000
White pine	450	101	4570	87.8	0.435	141.2	2.000
Trembling aspen	100	100	3050	58.3	0.291	173.7	0.486
White oak	600	122	3050	47.8	0.198	72.0	1.750
Red oak	400	100	3050	58.3	0.291	107.7	1.750
White cedar	400	100	2440	46.0	0.230	35.7	2.500
Hemlock	600	150	3660	47.0	0.156	86.0	2.000
Silver maple	125	122	3960	62.7	0.257	164.8	1.570
Tamarack	200	85	3050	68.5	0.403	86.3	2.000
Pitch pine	200	91	3050	64.0	0.352	86.5	2.000
Gray birch	50	38	910	40.7	0.535	119.5	0.486
American elm	300	152	3840	48.7	0.160	180.0	1.600
Basswood	140	137	4270	60.3	0.220	169.8	1.600
Bigtooth aspen	70	60	2130	66.4	0.554	176.7	0.486
Balsam poplar	150	100	2440	46.0	0.230	232.5	0.486
Black cherry	258	91	3050	64.0	0.352	166.7	2.450
Red cedar	250	60	1520	46.1	0.384	88.7	2.000

[a] Table 3.5(A) is the same as Table 2.1; it is repeated here for the reader's convenience.

Parameters refer to intrinsic (inherited) properties. B_2 and B_3 are parameters in equations (2.3) and (2.4), and occur also in equation (2.6). C is the parameter in equation (2.2) relating leaf weight to tree diameter. G, which is a product of the constant R times C, is a parameter in the fundamental growth equation (2.6) that determines when the inflection point in the curve will occur (how rapidly a tree growing under optimum conditions reaches one-half its maximum size). Maximum diameter and maximum height are used in equation (2.6). Maximum age is used in equation (3.14) to determine the first probability of mortality.

TABLE 3.5. (*Continued*)

			(B) Functional relationships[b]			
Species	L	N	$DEGD_{max}$	$DEGD_{min}$	DT_{min}	WL_{max}
Sugar maple	3	2	6300	2000	0.567	0.350
Beech	3	2	6000	2100	0.489	0.350
Yellow birch	2	2	5300	2000	0.600	0.245
White ash	2	1	10947	2414	0.400	0.245
Mountain maple	3	2	6300	1800	0.489	0.274
Striped maple	3	2	6300	2000	0.567	0.274
Pin cherry	1	1	6000	1100	0.567	0.378
Choke cherry	1	2	10000	1700	0.567	0.378
Balsam fir	3	3	3700	700	0.211	0.245
Red spruce	3	3	3800	1300	0.489	0.245
White birch	2	3	4000	700	0.544	0.378
Mountain ash	2	2	4000	1800	0.544	0.290
Red maple	2	3	12400	2000	0.322	0.450
Scarlet oak	1	3	8000	3900	0.933	0.450
Hornbeams	1	2	10300	2750	0.933	0.450
Green alder	2	3	3000	540	0.322	0.130
Speckled alder	2	3	5299	2174	0.211	0.050
Chestnut	2	3	8499	3686	0.933	0.450
Black ash	2	1	5300	1700	0.322	0.130
Butternut	1	2	6500	3200	0.933	0.450
White spruce	2	1	3750	600	0.544	0.245
Black spruce	2	3	3800	600	0.156	0.130
Jack pine	1	3	4000	1150	1.250	0.530
Red pine	1	3	4100	2000	1.250	0.500
White pine	2	3	6000	2100	1.000	0.450
Trembling aspen	1	2	5600	600	0.700	0.450
White oak	2	2	10204	2966	0.933	0.450
Red oak	2	2	9600	2400	0.933	0.450
White cedar	2	1	3700	1500	0.100	0.050
Hemlock	3	3	6559	2416	0.489	0.245
Silver maple	2	1	9000	2200	0.400	0.187
Tamarack	1	2	3800	600	0.156	0.050
Pitch pine	1	3	5800	3800	1.250	0.530
Gray birch	1	3	4800	2800	1.000	0.450
American elm	2	1	12000	1900	0.400	0.245
Basswood	3	1	6000	2300	0.567	0.290
Bigtooth aspen	1	2	6000	2100	0.400	0.187
Balsam poplar	1	1	4300	1000	0.400	0.378
Black cherry	2	2	10945	3899	0.567	0.378
Red cedar	1	3	10204	2966	0.700	0.450

[b] Parameters refer to functional response equations (the effect of the specific environment at a site on tree growth and reproduction). L is the shade tolerance class: 1 is intolerant, 2 is intermediate, and 3 is tolerant. Similarly, N is the soil nitrogen tolerance type: 1 is intolerant (good soil nitrogen supplies are required), 2 is intermediate, and 3 is tolerant. $DEGD_{max}$ and $DEGD_{min}$ are the maximum and minimum degree-days under which individuals of the species can grow: DT_{min} is the depth to the water table parameter in meters. WL_{max} is the maximum wilt tolerable by species i.

mortality rates of trees made the simplest assumption: that the chance of death of an individual tree was independent of its age and depended only on intrinsic (genetic) characteristics of its species. This assumption results in a mortality curve for a cohort that has the shape of a negative exponential curve. It is the assumption typically made in lieu of other evidence. For example, for decades it was applied to birds, though their mortality rates were in some ways easier to measure than the curves for trees, and existing evidence, if examined carefully, contradicted the assumption (Miller and Botkin, 1974). At the time that we began JABOWA, comparatively little had been published about the demography of trees. Available information gave us two major insights into patterns of tree death. The first insight was simply the same as Peter Kalm's. Any tree in a forest had a chance of death, from a lightning strike, a hurricane, a fire, or other cause whose locus was random. The second insight was the commonsense understanding of anyone who has nurtured trees: a tree that grows poorly is more likely to die than a tree that grows well. While the same might be said about mammals, it would apply only to young mammals, since mammals have a fixed growth pattern. Trees, in contrast, have indefinite growth in the sense that as long as a tree is alive it continues to increase in biomass. Examining available literature, we found little that would allow a quantitative refinement of a model of tree mortality, to help us set the values of parameters in equations that represented these two mortality processes. Given this situation, we set down the following concepts of mortality processes, expecting them to be rapidly improved on as the model was tested, applied, and used by scientists. To my surprise, those who have taken up JABOWA have left this part of the model essentially unaltered, although there have been some advances in empirical studies, such as Harcombe and Marks (1983), and considerable advances in the study and analysis of vegetation demography, pioneered in part by Harper (1977).

A Model of Mortality Processes for Trees*

In JABOWA, tree death occurs in two ways: (1) *inherent risk of death*: death that might be expected to occur to any healthy tree that is occupying a favorable environment, with or without competition from other trees. This is assumed to be an exponentially distributed event whose probability is a measure of the expected life span of the tree species; and (2) *Competition-induced death*: the death of a tree that is growing poorly.

The first is simply an age-independent function of the maximum observed longevity of individuals of a species, under the assumption that no more than a small fraction of the healthy trees should reach that maximum age. The inherent risk of death is the probability expressed as

$$M_i = 1 \, (1 - \varepsilon_i)^{\mathrm{AGEMX}_i} \tag{3.14}$$

*The following is taken from Botkin, D. B, J. R. Janak, and J. R. Wallis, 1972, Rationale, limitations and assumptions of a northeast forest growth simulator. *IBM Journal of Research and Development* 16: 101–116.

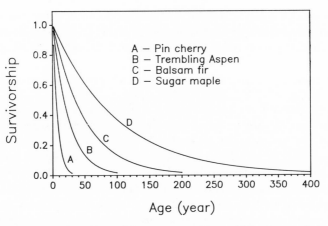

FIGURE 3.8. Tree survivorship without competition. This is a plot of $M_i = 1$ $(1 - \varepsilon_i)\text{AGEMX}_i$ in equation (3.11) for several species.

where M_i is the probability that a tree of species i at age 1 will reach the maximum age, AGEMX_i is the maximum age that an individual of species i should be able to reach, and ε_i is the annual probability of death. The function M_i is plotted for several species in Figure 3.8. In the present version of the model, the fraction of the population that should reach the maximum age is taken to be 2 percent. Setting $M_i = 0.02$,

$$\varepsilon_i = \frac{4.0}{\text{AGEMX}_i} \tag{3.15}$$

For each tree, in each year, a random number is selected; if it is less than ε_i, the tree dies and is removed from the model. The factors M_i and ε_i could have different values for each species, but it has not been necessary to make such distinctions. The model has functioned realistically with a single value for each of these factors.

A tree that does not die by this method is subject to a second kind of mortality which is a function of the current year's growth. Trees that grow poorly are much more likely to die than healthy trees. There are many causes for this higher mortality. Among these causes are: (1) the tree is unable to produce sufficient secondary compounds to resist diseases and insect attacks; and (2) the tree is weaker and more likely to be blown over by wind. The second mortality function is a simple way to approximate the sum of these effects. The program tests whether the last annual growth increment is less than a minimum value, parameter AINC_i. It is assumed that a tree whose diameter increment fell below this minimum for 10 successive years would have only a 1 percent chance of surviving those 10 years. This gives an annual mortality probability of 0.368. Although the minimum growth increment is a species-specific factor, in operation of the model this factor has been set to 0.01 for all species. Each tree that survives the first method, but whose diameter increment is less than parameter AINC_i, is subject to a second risk of mortality. A random number is selected; if it is less than 0.368, the tree dies.

From this discussion it should be clear that three parameters used in the mortality processes were derived from common sense and my own and others' experience in the woods, but are otherwise rather arbitrary. These are: (1) the minimum annual growth increment below which a tree is subject to a higher rate of mortality; (2) the percentage of healthy trees expected to reach the maximum age known for a species; and (3) the percentage of trees expected to survive 10 years in a forest if they are growing poorly. As with other parameters that I have specified, we expected that research on forests would refine these parameters and improve the rationales for their choice. As just mentioned, to my surprise those who have used the model have chosen not to alter this part of JABOWA, even though there have been considerable advances in the study of the population dynamics of vegetation, as summarized in 1977 by Harper, and in the study of the demography of trees. The basis now exists for a reexamination of these aspects of the model.

Reproduction and Regeneration

Population models and population forecasts consider additions to a population as beginning either with birth or with recruitment, which means that individuals reach some minimum age or size at which they are included in the population count. While in principle regeneration in forests could be interpreted as either germination or regeneration, the vast number of seeds produced and distributed widely by trees and the high mortality of these seeds make the reliance on germination computationally unwieldy. Those familiar with forests know that germination of seeds can be extremely patchy. For example, seeds of some species germinate primarily on dead, heavily decayed logs. Such a log may be covered with hundreds of recently sprouted seedlings, most of which survive only a short time, while nearby the open ground or leaf litter has no germinated seeds of that species. The calculation of the addition of a large number of germinated seeds that then immediately (within the same year) suffered a high rate of mortality would add little realism, accuracy, or generality to the model. These considerations lead one to choose recruitment as the logical choice for reproductive processes in a model of forest dynamics.

As I have just noted, trees produce many seeds, most of which do not germinate. Of those that germinate, only a few seedlings survive even a few years, and fewer become saplings. Thus regeneration of mature trees can be divided into several stages: the production of viable seeds by a tree; the dispersal of the seeds; germination; survival of seedlings; and recruitment of saplings (which means the growth of a cohort of seedlings to reach some minimum size class called saplings).

Since trees produce large numbers of seeds, only a few seed-bearing trees are necessary to provide seed to repopulate a large area. A simplifying assumption of version II of the model is that there is an abundance of seed trees in nearby forested areas for all species of trees that can grow under the current climate and soil conditions, and therefore recruitment is independent of the presence of seed-bearing trees within a single plot. In this sense, versions I and II of JABOWA are models of plots *within* a forest. Formally, the simplifying assumption can be stated as: there is an infinite seed pool available. Conceptually, it is simple to make recruitment a function of the presence of mature trees in a set of neighboring plots, but operationally this was unwieldy until the availability of fast microcomputers. Woodby (1991) has produced such a spatially dependent version of the model. In JABOWA-I, recruitment size was defined as saplings with a diameter at breast height of 2 cm. In JABOWA-II, recruitment is of seedlings that have reached at least 137 cm in height. This was a slight modification meant to increase the model's realism. The saplings are recruited into the population with a randomly selected height and a diameter calculated from the relationships governing tree shape discussed in Chapter 2. Only seedlings of those species that can grow under a plot's environmental conditions can be added to a plot in any one year.

Seeds are spread by wind, water, or animals. In general, smaller seeds are wind dispersed and larger seeds are animal dispersed, but the relationship is not universal. Species differ in the length of time that their seeds are viable. Generally, the larger the seed, the longer the viability, but viability varies with the composition of the seed coating and other factors. In northern hardwood forests of North America, seeds of pin cherry appear to retain their viability longer than any other seeds. Pin cherry has large seeds, but they are not the largest (beech produces larger seeds).

The germination requirements of seeds vary widely among species and represent highly specific adaptations to specific environments. For example, the small seeds of birch that are scattered widely in the wind can germinate only if the soil surface is scarified—scraped free of humus and other debris so that the seed makes contact with the mineral soil. Other species, such as hemlock, tend to regenerate only on highly decayed fallen logs. A hectare of forest may contain one small, bare spot where birch seeds sprout, but within that scarified spot hundreds of sprouted seeds might be found during the growing season. Elsewhere in the deep forest shade there may be no hemlock seeds except on the remnants of a single log, but hundreds of sprouted seeds might be found there. The same spatial patterning at a very small scale is true of the early survival of seedlings and subsequent survival of saplings. The spatial patchiness is of too fine a scale to matter for the population dynamics of a forest community, the level of concern of JABOWA.

Because adaptations for seed germination and seedling survival are so specific, few conceptual generalizations can be invoked to model recruitment. To scientists who prefer highly simplified conceptual approaches, this part of the model is more empirical and less satisfying intellectually than the growth and mortality aspects of JABOWA. Realism requires that the model mimic actual responses to site conditions closely.

In the original version of the model, recruitment was influenced only by temperature and soil moisture conditions. A simplifying assumption was made that the effects of environmental factors could be expressed just once, through the effects on growth. In a general perspective on theory, one could argue that the introduction of specific environmental factors in both growth and reproduction might be considered redundant. In reality, environmental factors influence reproduction and regeneration differently, and these separate influences have been added to the current version of the model. This is one of the original simplifying assumptions of the model that has not been retained, because practice has shown that the results are insufficient without more detail.

In JABOWA-I, saplings of a species were added to a plot only if the growing degree-days (variable DEGD) was within the range for growth of that species (between parameters $DEGD_{min(i)}$ and $DEGD_{max(i)}$). In addition, for birch and cherry trees (the early successional species of JABOWA-I), the calculated soil moisture had to be above a species-specific minimum (Botkin et al., 1972a,b). If these two conditions were met, the light available at the forest floor determined which of those species would enter.

For shade-tolerant species, a random choice between zero and two new saplings was selected and was assigned randomly to the species that could grow. For intermediate shade-tolerant species, which in the original version were birches, the number of saplings added varied inversely with leaf weight (directly with light intensity at the forest soil surface). For shade-intolerant species, a large randomized number of saplings was added if leaf weight fell below a species-specific minimum. If the total leaf area was below a first threshold, then between 60 and 75 cherry saplings were added (the exact number in this range selected randomly). If the leaf weight was greater than a second threshold, but less than a third, then between 0 and 13 birch saplings could be added, the exact number selected at random. The number that could enter declined linearly with leaf weight between the upper and lower thresholds.

In the current version of the model, the number of saplings added to a plot in any year is a stochastic function of a maximum (which conceptually represents a genetically determined maximum), multiplied by $f(AL)_L$, the light available at the forest soil surface, and Q_i, the site quality. Remember that Q_i is the product of soil moisture and nitrogen conditions and thermal conditions of the environment (see equation 3.3).

In regard to regeneration, there are three decisions to be made for each species for each year: (1) whether there will be any regeneration for that species; (2) how many saplings will be added; (3) the height (and therefore the diameter as well) of each newly added sapling. In the present version of the model, these decisions are made in the above order, and each is a stochastic function, except where noted. The decisions are slightly different for the three shade tolerance classes (Fig. 3.9).

For shade-intolerant species, saplings can enter if the light intensity at the forest floor is greater than 99 percent of incident sunlight ($AL \geqslant 0.99$) and if the site quality, Q_i, is greater than zero. If these conditions are met, then the number of saplings added of that species i, E_i, is

$$E_i = \zeta * S_i * f(AL)_1 * Q_i \tag{3.16}$$

where ζ is a uniformly chosen random number between zero and one, S_i is the maximum number of saplings of species i that can be added to a plot in any one year, $f(AL)_1$ is the light response function for shade-intolerant and

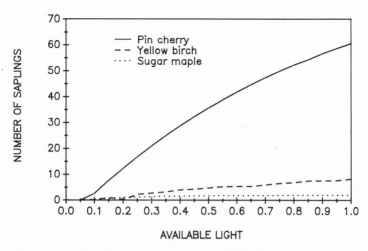

FIGURE 3.9. Relationship of expected saplings to available light for several species (graph of E_i).

intermediate-tolerant species, as defined earlier, and Q_i is the site quality as defined in equation (3.3) (Fig. 3.3a).

Saplings of a shade-intermediate-tolerant species can be added if the light intensity at the soil surface is less than 99 percent of incident sunlight (AL < 0.99), but greater than a species-specific minimum (AL > Γ_i). If light intensity is within this range, then saplings will enter in a given year if

$$\zeta < f(\mathrm{AL})_1 * Q_i \qquad (3.17a)$$

where the terms are defined as above.

This means that the poorer the growing conditions for the species, the less likely that intermediate-tolerant saplings will enter in a given year.

Whether saplings of a shade-tolerant species i can enter is a stochastic function of the environmental conditions. The calculation is exactly as in (3.17a) except that the light-response function for shade-tolerant species is used, so that saplings will be added if

$$\zeta < f(\mathrm{AL})_3 * Q_i \qquad (3.17b)$$

For both shade-intermediate-tolerant and shade-tolerant species, if the inequality defined by equation (3.16) is true (if a random number is less than the product of environmental response functions), then the number of saplings for the species, E_i, that will enter is calculated as follows (Fig. 3.3a):

$$E_i = \zeta * S_i \qquad (3.18)$$

where ζ and S_i are defined as above.

These equations result in the following dynamics of sapling recruitment:

1. Saplings of intolerant species are added only in bright light when the site quality for the species allows for tree growth. If these conditions are met,

TABLE 3.6. Sapling recruitment parameters

Common Name	Scientific Name	L^a	N^a	S_i^a
Sugar maple	*Acer saccharum*	3	2	3
Beech	*Fagus grandifolia*	3	2	3
Yellow birch	*Betula alleghanensis*	2	2	15
White ash	*Fraxinus americana*	2	1	10
Mountain maple	*Acer spicatum*	3	2	2
Striped maple	*Acer pensylvanicum*	3	2	2
Pin cherry	*Prunus pensylvanica*	1	1	60
Choke cherry	*Prunus virginiana*	1	2	60
Balsam fir	*Abies balsamea*	3	3	2
Red spruce	*Picea rubens*	3	3	2
White birch	*Betula papyrifera*	1	3	10
Mountain ash	*Sorbus americana*	2	2	2
Red maple	*Acer rubrum*	2	3	3
Scarlet oak	*Quercus coccinea*	1	3	3
Hornbeams	*Ostrya & Carpinus*	1	2	3
Green alder	*Alnus crispa*	2	3	10
Speckled alder	*Alnus rugosa*	2	3	10
Chestnut	*Castanea dentata*	2	3	0
Black ash	*Fraxinus nigra*	2	1	3
Butternut	*Juglans cinerea*	1	1	3
White spruce	*Picea glauca*	3	1	2
Black spruce	*Picea mariana*	2	3	2
Jack pine	*Pinus banksiana*	1	3	50
Red pine	*Pinus resinosa*	1	3	3
White pine	*Pinus strobus*	2	3	4
Trembling aspen	*Populus tremuloides*	1	2	10
White oak	*Quercus alba*	2	2	10
Northern red oak	*Quercus rubra*	2	2	10
White cedar	*Thuja occidentalis*	2	1	2
Eastern hemlock	*Tsuga canadensis*	3	3	3
Silver maple	*Acer saccharinum*	2	1	2
Eastern larch	*Larix laricina*	1	2	10
Pitch pine	*Pinus rigida*	1	3	2
Gray birch	*Betula populifolia*	1	3	10
American elm	*Ulmus americana*	2	1	3
Basswood	*Tilia americana*	3	1	3
Bigtooth aspen	*Populus grandidentata*	1	2	3
Balsam poplar	*Populus balsamifera*	1	1	3
Black cherry	*Prunus serotina*	2	2	10
East. red cedar	*Juniperus virginiana*	1	3	3

[a] L is the light tolerance type; N is the nitrogen tolerance type; and S_i is the maximum number of saplings that can be added in any single year. For both tolerance types, 1 = intolerant, 2 = intermediate, and 3 = tolerant. Note that S_i is large for type $L = 1$.

a large number of saplings can be added in one year. Whether saplings can be added is deterministic; the number added is a stochastic function.

2. Whether saplings of shade-intermediate-tolerant and shade-tolerant species can enter is a stochastic function of the environmental conditions; the lower the light intensity and the poorer the site quality for a species,

the less likely that saplings will be added. Saplings of intermediate-tolerant species can be added only within a certain light range.

3. Saplings of intolerant and intermediate species cannot enter in the same year, because the minimum light intensity for the entry of intolerants is the maximum for intermediates.

4. For intermediate and tolerant species, only the decision whether saplings will be added is a function of environmental conditions. Once the conditions are met, the number added is independent of light and site quality.

5. The value of S_i is large for shade-intolerant species, intermediate for shade-intermediate-tolerant species, and small for shade-tolerant species (Table 3.6).

6. Saplings of shade-tolerant species can be added in the same year as either of the other types.

In version II, and in the software that is a companion to this book, there is one exception to the rule that the entry of shade-tolerant species lacks specific light intensity boundaries. The exception is for basswood, which cannot enter plots whose total leaf weight is less than 400. Basswood is assumed to germinate and grow in some shade, not in completely open conditions. This is intended to mimic the actual dynamics of that species.

The size of each newly recruited sapling is a stochastic function, meant to reflect variability in the population. Saplings are recruited to a population at a minimum height of 137 cm, which means that they have reached "breast height," the diameter for which dynamics of tree growth are calculated. The actual height of each sapling is randomly selected to be between 137 and 167 cm. Once the height is selected, the diameter is calculated from equation (2.3).

Summary

This chapter describes the influence of the environment on tree growth, mortality, and species reproduction, as these are expressed in the model. Mathematically, the response of a tree to each environmental factor can be expressed as a function that has values from 0 to 1 (or in some cases more than 1), which affects the fundamental growth equations (2.1) and (2.6). Each of these relationships is called a *response function*. A tree responds to the totality of environmental conditions, and a fundamental question is how this response integrates all the individual effects of separate environmental factors. Models can use additive or multiplicative interactions of environmental factors. JABOWA uses multiplicative interactions of light, temperature, drought, soil-water saturation, and soil nitrogen content. Each involves specific assumptions about the relationship between environmental conditions and tree growth and species regeneration. The assumptions, rationales, and mathematical forms of each response function were described. Other assumptions are inferred by the model. These include: (1) only four environment factors are required to explain the dynamics of trees in a forest: light, temperature, soil nitrogen, and soil moisture; (2) spatial position of a tree within a plot does not matter, as long as the plot is small enough so that a tall tree can shade all of its neighbors within the plot during a year; (3) seeds are available from an unlimited pool; reproduction and regeneration are limited by germination and growth conditions, not by the availability of seeds; (4) impact of herbivores can be ignored (or assumed to occur at a constant, time-invariant rate); (5) soil decomposition can be ignored (feedbacks between soil decomposition rates and the availability or nutrients does not affect the dynamics); and (6) growth is deterministic—regeneration and mortality are stochastic functions. Species parameters for response functions are given in Tables 3.5 and 3.6.

With the material provided in Chapters 2 and 3, it is now possible to use the model and explore its implications for the dynamics of forest ecosystems. This is the topic of the next chapter.

Note

1. For readers not familiar with the photosynthesis of woody plants, a note might be useful about the distinction between C_3 and C_4 photosynthetic pathways. Photosynthesis in most plants, and in all but a few woody euphorbias and some shrubs of the Middle East and Soviet Union, have the C_3 pathway, so-called because the initial product is a three-carbon-containing sugar (Kozlowski, Kramer, and Pallardy, 1991). In some plants, mainly monocots including sugarcane and maize, the initial products are four-carbon-containing sugars, hence the name C_4 pathway. Plants with a C_4 pathway are not saturated by light in full sunlight, and the photosynthetic rate is generally higher for these plants at any given light level than for either of the C_3 curves used in the forest model.

In some early and unpublished uses of the model, a C_4 light response function was given to individual species, one at a time. The result was that a species having a C_4 light response function produced much larger individual trees, but the end result, in terms of which species dominated the forest over a long time period, did not differ from previous cases.

4

Implications of the Model
for the Dynamics of Forests

Now that the model has been described, we can begin to explore its implications for forest communities and ecosystems and evaluate how well it mimics known characteristics of forests. In this chapter I discuss implications of the model and explore some linkages between the JABOWA model and other aspects of ecological theory. Later chapters evaluate the model.

It is necessary to separate the process of developing the model as part of doing research from the use of an existing form of the model in research and in applications. In the first, modeling is a *process* involving continual model development. In the second, the model is used in a fixed set of algorithms that serve as a set of assumptions about the dynamics of forests, whose implications can be explored and whose success in mimicking forest dynamics can be evaluated. In this chapter, versions I and II of the model are viewed in this latter sense—each as a set of assumptions about forest ecosystems—so that we can explore the success of the assumptions and the implications of the model. That the equations and algorithms described in Chapters 2 and 3 are sufficient to reproduce observable aspects of forest communities can be taken as an hypothesis that is tested in the rest of this book.

Competition among Trees

The model was originally developed and tested step by step. The first test consisted of evaluating whether the model could successfully reproduce competition among individuals trees. The results of one of the early tests of version I are shown in Figure 4.1. A simulated plot was established representing a small area in a forest of the White Mountains of New Hampshire dominated by one large sugar maple (85 cm in diameter). Forest growth was simulated under a constant climate for 20 years, with the large sugar maple cut at year 9. As long as the large sugar maple survived, all other trees were suppressed. Some suppressed trees died, and even those that survived grew little. When the large sugar maple was cut, the growth rate of the remaining trees increased. Species and size differences in relative growth rates are apparent. Larger trees grew more rapidly than smaller trees. The largest remaining tree, a 28-cm yellow

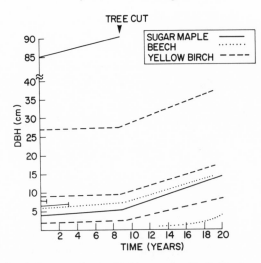

FIGURE 4.1. Simulated diameter (breast high) growth of individual trees on a single plot. Each line represents a single tree and the end of a line signifies that tree's death. At year 9 the large sugar maple was cut. The remaining trees, no longer suppressed by the large maple, show increased growth rates. (Reprinted with permission from D. B. Botkin, J. R. Janak, and J. R. Wallis, 1972, Rationale, limitations and assumptions of a northeast forest growth simulator. *IBM Journal of Research and Development 16*: 101–116.)

birch, grew 10 cm in the next 10 years. But a small (diameter < 5 cm) sugar maple grew faster than a slightly larger beech, almost overtaking it in diameter by the twentieth year. This simple test illustrated that the model successfully reproduced qualitative and quantitative dynamics of competition among trees in a forest, with larger trees suppressing smaller ones, and with size and species differences in growth rates. Many similar tests are possible and can be generated by the reader using the program available as a companion to this book. Thus the model was shown to reproduce realistically local competition among trees in a mixed-species, mixed size-class forest. It was the first model to do so.

The Quantitative Ecological Niche of a Tree

A discussion of competition among trees of different species and the persistence of a set of species brings to mind the concept of the ecological niche and G. E. Hutchinson's famous question (Hutchinson, 1958): Why are there so many species? Another way to view the explicit mathematical relationships presented in Chapters 2 and 3 is that they describe the ecological niche of each tree species. The description of the dynamic realtionships between tree growth and environmental conditions links the conceptual approach of the JABOWA model to the classical ecological concept of the niche. Consider again equations (3.1a), (3.1b), and (3.2). For the general case

$$\frac{\delta D}{\delta t} = \left(\frac{\delta D}{\delta t}\right)_{\text{opt}} * \text{MIN}\{f_1, f_2, \ldots, f_j\} \tag{3.1a}$$

$$\frac{\delta D^*}{\delta t} = \left(\frac{\delta D}{\delta t}\right)_{opt} *f_1 *f_2 * \cdots f_j \qquad (3.1b)$$

where $f_1 \cdots f_j$ are environmental response functions and equation (3.1a) represents the Liebig's law of the minimum and equation (3.1b) represents the multiplicative rule. These environmental response functions are expressed more explicitly as

$$f(\text{environment}) = f_i(\text{AL})*Q_i*s(\text{BAR})_i \qquad (3.2)$$

where the site quality function Q_i, is

$$Q_i = \text{TF}_i*\text{WiF}_i*\text{WeF}_i*\text{NF}_i \qquad (3.3)$$

where TF_i = temperature function
  WiF_i = wilt factor (index of drought conditions that a tree can withstand)
  WeF_i = soil wetness factor (index of amount of water saturation of soil a tree can withstand
  NF_i = index of tree response to nitrogen content of soil

Each equation (3.1a, 3.1b, and 3.2) can be interpreted to be an explict assertion of the niche as defined by Hutchinson (1944) as "a set of points, each one of which defines a possible set of environmental values permitting the species to live" (Hutchinson, 1978). Hutchinson formulated the niche as "an n-dimensional space any point in which is defined by some value of the variables $x', x'', x''', \ldots$ measured along rectangular coordinates" which represents "the conditions for the existence of a species, which requires that the values of x' be between x_1' and x_2' of x'' between x_1'' and x_2'', etc." (Hutchinson, 1978, p. 158.) In equations (3.1a), (3.1b), and (3.2), the environmental limits of a species' persistence are characterized by parameters that define the limits of positive values for each functional relationship for each species.

In equations (3.1a) and (3.1b), the boundaries of the niche are the same, but the interactions among environmental factors are different: as noted before, equation (3.1a) represents Liebig's law-of-the-minimum hypothesis of the interaction among environmental factors, while (3.1b) represents the hypothesis of multiplicative interactions among environmental factors. Thus the equations result in niches that are *qualitatively* the same niche but *quantitatively* different.

The forest model carries the idea of the niche a step further than Hutchinson's definition by creating expressions of the quantitative response of the species to each environmental gradient within the range of values over which a species can persist. Thus through equation (3.2) the forest model defines a *quantitative niche* for each species.

Hutchinson distinguished two niches, the fundamental niche without competitive interactions and the realized niche that includes the effect of competitive interactions (Hutchinson, 1944). The fundamental niche is the set of all environmental conditions within which a species can persist without the influence of competition. The realized niche is the set of all environmental conditions under which a species can persist with competition. In the standard versions of the JABOWA model, equation (3.1b) defines the fundamental niche of a tree, as long as we can assume that all of the parameters have been obtained from studies conducted independently of interspecific competition. In reality, since the parameters are estimated by a variety of methods, this may not be strictly true, and therefore equation (3.2) should be viewed as an estimation of the fundamental niche.

For example, consider the equations that define the temperature response function of species i:

$$TF_i = \max\,(0, TDEGD_i) \tag{3.4}$$

$$TDEGD_i = \frac{4(DEGD - DEGD_{min(i)})(DEGD_{max(i)} - DEGD)}{(DEGD_{max(i)} - DEGD_{min(i)})^2} \tag{3.5}$$

The maximum and minimum degree-days under which a species can grow ($DEGD_{max(i)}$ and $DEGD_{min(i)}$) define the thermal niche dimensions in Hutchinson's sense, while the values that the function TF_i takes on define the quantitative responses of the species i within that niche space. Thus $DEGD_{max(i)}$ and $DEGD_{min(i)}$ define the thermal boundaries of the fundamental niche of the ith species.

The realized niche is found by running the model with a set of interacting (competing) tree species and determining under what conditions each species persists. The results are a projection of the realized niche of each species. Thus *the fundamental niche is formulated as part of the assumptions of the forest model while the realized niche is an outcome of the model.*

By itself, equation (3.2) can be interpreted as a static model of interactions of forest tree species and of forest communities on a static landscape. Mapping the algorithms in equation (3.2) onto geographic information about environmental conditions could produce maps of the distribution of individual species. This use of equation (3.2) is similar to schemes such as the Holdridge life-zone diagram which map vegetation distribution onto temperature and moisture gradients or other environmental gradients (Holdridge, 1947). However, equation (3.2) and the specific equations for each environmental response function produce a more specific and quantitative model than the Holdridge life-zone and similar schemes. From this perspective, one can view the JABOWA forest model as having a foundation in a description of the static, fundamental ecological niche of each tree species. But the JABOWA model also makes clear the difference between those static approaches and dynamic modeling.

The ecological niche expressed in the forest model is consistent with Hutchinson's idea of the niche, but, because it does not include the effects of predators, is somewhat different from the niche as originally formulated by Elton (1927), who wrote that the niche should include "all manner of external factors" that act on a species, and in regard to animals (which were Elton's major focus), the niche should include "its relationship to food and enemies." Interpreted for trees, Elton's concept of the niche would include the effects of insect and mammalian herbivores and fungal and bacterial diseases.

A Complete and an Incomplete Description of an Ecosystem

Another interesting aspect of the forest model is the insight that it gives us about the number of environmental gradients (which we can also call environmental dimensions) required to allow the persistence of a large number of tree species. A little background will make this point clearer. Consider what might be required as a "complete" description of an ecosystem—the position, size, and rate of change of the population of each species and the size and rate of change of environmental pools of resources (Botkin et al., 1979; B. Maguire et al., 1980; Slobodkin et al., 1980). Taken in this individual way, the number of possible connections is huge. For example, suppose we merely focus on one variable and designate the amount of material of one kind (such as total biomass, or the concentration of a specific chemical element such as nitrogen) in the ith compartment (population or nonbiological pool) by X_i and the flow between compartments i and j as f_{ij}. If there are N compartments, then the complete representation of this single-variable system is an N^2 dimensional matrix, where the number of X_i's is N and the numbr of flows (f_{ij}'s) is ($N^2 - N$). Each N and f_{ij} is also a function of time and may be a function of other materials that are not explicitly considered, and the functions are likely to be nonlinear and may be stochastic. The determination of all of these compartments and flows would seem experimentally inaccessible and mathematically intractable. Globally there are approximately 250,000 named species of vascular plants, so that a complete description of all possible interactions among these species would involve a matrix with more than 60 billion ($> 6.25 \times 10^{10}$) elements. Even if we restrict our concern to a large region, the number of possible interactions is impractically large. For example, there are approximately 40 tree species in the boreal and eastern deciduous forests of North America (the forests covered by the version of the model described in this book), so that a complete description of all possible interactions among these species would involve a matrix of 3,540 elements to separate the trees alone.

This seems to lead to a contradiction: the system that we have set out to describe cannot be described completely. Moreover, based on the number of possible interactions among all of the vascular plant species, it might seem that each individual tree faces an impossibly complex task, and that a great many environmental dimensions would be required to allow for the persistence of all of these species. A tree needing information about its environment to survive could not possibly manage all of that information. It appears necessary to simplify the description of such a system, and to select information that would be observed, processed, and used. From the point of view of an individual organism or a population, the sensing of the information about the environment and interactions with the environment—the processes of obtaining necessary resources, avoiding predators, interacting with competitors, and so on—must be simpler than the matrix just described. One answer is, of course, that many species do not interact because historically they have been separated by biogeographical barriers. But even considering the number of tree species actually competing in a forest, the number of dimensions needed to provide separate niches for each species must be much smaller than the $N^2 - N$ possible interactions.

What is the minimum number of environmental dimensions that would allow the persistence of a great many species? Hutchinson has observed that "if we confine our attention to animals, the number of dimensions needed to provide environmentally different niches for a group of organisms small enough to handle, such as a class of vertebrates or an order of insects, is usually likely to be less than four" (Hutchinson, 1978). This is a strong assumption about the character of ecological interactions, but in Hutchinson's analysis, the evidence was based on observations that ranged from qualitative insights, arising from his considerable expertise, to some isolated experiments.

The forest model provides a theoretical foundation to explore this question. If its projections are accurate and realistic, then the model would indicate that, consistent with Hutchinson's statement about the niches of animal species, only four dimensions—light, temperature, water, and nitrogen—are sufficient to describe the niche of tree species both qualitatively (the fundamental niche) and quantitatively (the realized niche). An hypothesis that arises from the model is that these four dimensions are sufficient to account for the dynamic relationships among species of trees and for the dynamic integrated characteristics of a forest. The problem that ecologists have had is to explain the persistence of so many species with so few resources. In the forest model described in this book, compartments and fluxes of just a few resources—light energy, soil water, air temperature, and soil nitrogen—taken within a dynamic temporal framework with stochastic processes, allow persistence of the species in at least some environments. This is a very strong hypothesis; it is tested by evaluating the correspondence between the models projections and our knowledge of the dynamics of real forests. Much of the rest of this book considers aspects of this test.

The Competitive Exclusion Principle as Applied to Forests

Another interesting connection between classical ecological theory and the forest model is with the competitive exclusion principle, which the reader will remember states that two species that have exactly the same requirements cannot co-exist in exactly the same habitat. Instead, one species will always win out over a second, which will become extinct. As Garrett Hardin put it so succinctly, "Complete competitors cannot coexit" (Hardin, 1960). The surviving species will be the one that is more efficient in the use of any, or all, resources, and it will have a higher birth, growth, and/or survival rate than the other.[1]

The competitive exclusion principle requires the following conditions: (1) two noninterbreeding populations, that (2) have the same requirements and do the same things, (3) are *sympatric*, meaning that they occupy the same geographic territory, and (4) one of which grows faster than the other given the same resources (Hardin, 1960).

We can test the forest model to determine whether its projections are consistent with the competitive exclusion principle. In one test, using version II of the model, a "complete" competitor of sugar maple was devised as a hypothetical "beech" species with environmental parameters of sugar maple but with growth parameters of red spruce (red spruce trees grow much more slowly than sugar maple). The maximum number of "beech" saplings able to enter a plot in a year was reduced to one, while sugar maple's maximum was increased to five, so that maple had a clear advantage in regeneration and a slight advantage in growth. The new "beech" species shared exactly the same fundamental niche as sugar maple, but grew more slowly and had a much lower rate of regeneration. Results are shown in Table 4.1 for a hypothetical site in the White Mountains of New Hampshire, with 13 species of the Hubbard Brook Forest able to enter, and a constant climate derived from the Woodstock, New Hampshire, weather records.

Sugar maple and yellow birch dominate the resulting forest. By year 400, sugar maple averages more than 25 individuals/$100 \, m^2$ and more than $25 \, kg/m^2$, while our newly defined "beech" averages about 3 individuals that contribute a negligible biomass of approximately $0.1 \, kg/m^2$. Note that in this experiment the model's standard assumption—that there is a source of hypothetical "beech" seeds produced by mature trees outside the test plot—is maintained. Therefore, beech seedlings cannot be completely eliminated through local competition, and the experiment is not a complete test of the correspondence between JABOWA competitive exclusion and the classical competitive exclusion principle. However, with this small amount of biomass, mature, seed-producing trees of "beech" would be extremely unlikely on the plot, and if seed production were linked to local seed trees, the hypothetical complete competitor of sugar maple would become extinct, since beech seeds would not be viable for 400 years. Thus the model's dynamics appear consistent with the classic competitive exclusion principle, even though the assumptions about the influx of seeds and the stochastic nature of events prevent complete extinction of complete competitors.

When the standard parameters are used, so that beech and sugar maple have different characteristics, these two species share dominance of the forest for most of the 400 years (except during the early stages of succession). Users of the model can test this and other variants of the experiment, using the software available as a companion to this book.

Woodby (1991) developed a spatial version of JABOWA in which regeneration on a plot is determined by the presence of seed-bearing trees on a set of plots (a central plot and adjacent plots) and whose algorithms are otherwise the same as in version II discussed here. If no tree of a species of a seed-bearing age has existed on any of the plots for the last 15 years, regeneration is zero. Under these conditions, a tree species becomes extinct on the set of plots (Woodby, 1991). The two results given here, the test with the hypothetical "beech" and Woodby's spatial version, indicate that JABOWA projects competitive exclusion consistent with the competitive exclusion principle.

TABLE 4.1. Competitive exclusion principle, experiment 1, using hypothetical "beech"[a] and sugar maple. Normal climate (1951–1980 weather record for Woodstock, New Hampshire): CO_2 fertilization switch = off.

Species No.	Tree Species (common name)	Pop. (N/100 m²)	95CI	Basal Area (cm²/m²)	95CI	Biomass (kg²/m²)	95CI	Vol. (ft³/acre)	95CI
				Year 20					
1	Sugar maple	7.8	4.5	0.2	0.2	0.1	0.1	4.3	3.1
3	Yellow birch	9.3	7.1	0.1	0.1	0.0	0.0	1.7	1.2
7	Pin cherry	3.2	1.2	0.0	0.0	0.0	0.0	0.3	0.2
8	Choke cherry	4.9	1.9	0.2	0.1	0.1	0.0	4.9	1.7
Total		25.2	1.8	0.6	0.0	0.2	0.0	11.2	0.9
				Year 100					
1	Sugar maple	30.6	5.0	16.2	5.3	9.1	3.2	670.5	245.2
2	"Beech"	4.9	1.8	0.1	0.1	0.0	0.0	2.5	1.6
3	Yellow birch	17.1	8.0	19.1	6.8	10.9	3.8	1151.8	393.3
4	White ash	1.6	2.5	0.2	0.3	0.1	0.1	3.6	6.2
11	White birch	3.1	2.5	0.2	0.2	0.1	0.0	4.7	3.8
Total		57.3	4.6	35.8	3.3	20.3	1.9	1833.1	177.5
				Year 200					
1	Sugar maple	31.0	3.8	37.2	6.2	25.7	5.2	2248.8	519.7
2	"Beech"	5.3	1.8	0.3	0.2	0.1	0.1	8.2	6.6
3	Yellow birch	4.7	2.5	23.9	10.2	18.2	7.5	2511.5	1018.7
4	White ash	0.1	0.2	0.1	0.1	0.0	0.1	1.4	3.2
Total		41.1	4.6	61.4	6.3	44.1	4.5	4769.9	478.6
				Year 300					
1	Sugar maple	27.4	4.9	38.3	6.7	29.0	6.4	2744.5	707.6
2	"Beech"	3.6	2.0	0.2	0.2	0.1	0.1	7.5	8.7
3	Yellow birch	1.2	0.7	13.9	8.9	12.5	8.1	1949.8	1278.2
Total		32.2	4.3	52.3	6.3	41.7	4.9	4701.8	515.4
				Year 400					
1	Sugar maple	27.0	7.0	36.2	6.8	28.5	6.3	2793.2	715.4
2	"Beech"	3.5	1.9	0.2	0.2	0.1	0.1	9.0	11.2
3	Yellow birch	0.4	0.5	8.0	9.3	8.1	9.5	1343.2	1560.4
Total		30.9	4.3	44.4	5.8	36.8	4.6	4145.4	477.9

[a] Parameters of the hypothetical "beech" are: $D_{max} = 60$; $H_{max} = 2{,}290$; AGEMAX = 400; $B_2 = 71.8$; $B_3 = 0.598$; $G = 50.7$; $C = 2.5$; $DEGD_{min} = 1{,}111.1$; $DEGD_{max} = 3{,}500$; AINC = 0.01; DTMIN = 0.57; WLMAX = 0.350; $W_{max} = 0.0$; $W_{min} = 0.0$; L (shade tolerance type) = 3; N (nitrogen tolerance type) = 2; $S_i = 1$. (Variables are defined as in the text, and in Appendix II.)

Parameters of sugar maple are standard except that $S_i = 5$, to increase the competitive advantage of that species over "beech".

Column headings are: Pop. = density (number of stems per 100 m²); BA = basal area (cm²/m²); Biomass = total biomass (kg/m²); Vol. = merchantable timber volume (f³/acre, the standard United States measure). In all cases, 95CI is the 95 percent confidence interval for the column to the left. All other columns give averages.

Ecological Succession

Ecological succession, the development of an ecosystem over time, has been one of the most studied aspects of plant ecology (Drury and Nisbet, 1973; Horn, 1971; 1974; Odum, 1969; West, Shugart, and Botkin, 1981). Succession has been particular fascinating to ecologists because it is one of the few ecological phenomena that involves clearly recognizable, repeated patterns of change occurring on a short enough time scale to be observed directly or easily inferred. The pattern of secondary succession, familiar in areas of the United States including New England (Hibbs, 1983) and the Great Lakes states, where much land was cleared for farming and logging in the nineteenth century and later abandoned, was among the first processes to be studied by ecologists. For readers not familiar with this process, it is useful to review an example, such as succession that takes place in an abandoned farm field in New England.

In the mid-1970s I toured a farm in East Alstead, New Hampshire (in the southeastern corner of the state) that had been owned by the same family since early in the twentieth century. Its owner knew when each field had been abandoned.[2] We began the tour with a field that had been hayed the year before and had then been abandoned. Seeds of many kinds of "weeds"—annual and perennial herbaceous plants—and seedlings of some trees had sprouted. In nearby fields abandoned 5 to 10 years earlier, saplings had become established and were evident even to the casual observer. Especially abundant were white pine, pin cherry, white birch, and yellow birch, depending on site conditions, species whose seeds are light and readily distributed by wind or are eaten by birds and small mammals and readily transported by them. Individuals of these species are fast growing, particularly in open areas exposed to direct sunlight.

In areas abandoned several decades before our tour, forests of these "pioneer" species were well established and formed dense stands of small trees. Elsewhere in New Hampshire, in stands more than 20 or 30 years old, most of the very short-lived pin cherry had matured, borne fruit, and died (Marks, 1974); these trees cannot grow in the shade of taller trees and do not regenerate in a forest once it has been reestablished.

On the East Alstead farm, after 20 or 30 years since farming, white pine formed even-aged stands that reached 25 to 30 m (50 to 60 ft) high on fields that had been hay. In areas that had been woodlands but on which the trees had been blown down by the 1938 hurricane, birches dominated and formed even-aged stands. In older stands—areas last farmed about 1906, a mixture of species was important, including sugar maple and beech in lower, warmer areas. Elsewhere to the north and at high elevations, red spruce and balsam fir dominated such stands. The New Hampshire farm was too far south and at too low an elevation to have spruce and fir in any numbers, but maple was abundant. Sugar maple, spruce, and fir are slower growing than birch and cherry, but they are shade tolerant and able to grow relatively well in the deep shade of the forest, characteristics that make them well adapted to the later stages in succession. In these older stands the dominant trees are generally taller than those of earlier stages, but the forest is composed of trees of many sizes.

Elsewhere in New Hampshire there are a few locations where even older forests can be found—those last cleared one or two centuries ago. In these, trees are mainly shade-tolerant species—beech and sugar maple in the warmer locations, fir and spruce in the cooler locations.

The abandoned fields in East Alstead, New Hampshire, illustrate the basic features of secondary forest succession in an eastern deciduous forest of North America. There are several aspects of forest succession that one would like to reproduce in a model, including the changes in species composition that I have just described, changes in total community characteristics such as biomass and diversity, and ecosystem characteristics such as the storage of major chemical elements necessary for life.

Succession and the Forest Model

In the model, differentiation of early, middle, and late successional species is accomplished through two factors: the light response functions discussed in Chapter 3 (equations 3.3a, 3b, and 3c; Figs. 3.1, 3.2), and the maximum number of saplings that can enter a stand in any one year. As discussed in Chapter 3, the light response function has a characteristic shape for each light-tolerance class: pioneer species have a greater growth rate at high light intensities than do late successional species, while late successional species have a higher growth rate at lower light levels but their growth rates approach a maximum at a much lower light intensity than the growth of the pioneer species (see Fig. 3.2). There is a crossover point between the two curves at which neither kind of species has a competitive advantage due to light response alone. Thus another strong assumption of the model is that the light response functions of all species can be represented by two or three equations (actually one equation with two or three sets of parameters). Remember that in version II there are two light response functions for growth and three light tolerance classes for regeneration (see Chapter 3). This implies that species fall into one of three groups in regard to their response to light in their environment. As explained in Chapter 3, but is useful to repeat here, a similar general relationship, with tolerant, intermediate, and intolerant categories of species, is used in the model to represent the response of species to available nitrogen in the soil (equation 3.9; Fig. 3.7). These are relationships which will be discussed again later.

Aspects of the implications of JABOWA and other variants of it for forest succession research have been explored by a number of authors (see for example, Barden, 1981; Runkle, 1981; and Kessell, 1981).

*An Example of Succession Projected by the Model**

A simulation of secondary succession for 100 years after clearing using version I at 610-m elevation in the Hubbard Brook Experimental Forest is shown in

Figure 4.2. This area is considerably north of the southern New Hampshire farm that I discussed earlier, but it is part of the same biome, the northern hardwoods forests of North America, grading into the boreal forest at upper elevations and in the extreme north of the state. The model projects that, after an initial high abundance, the density of trees declines rapidly for early successional species (pin cherry, yellow and white birch) and increases steadily for the late succssional beech (Fig. 4.2(A)). No sugar maple or red maple is present at year 15; these two maples and beech increase continuously afterward and, by year 60, become dominant. Fifteen years after clearcut, pin cherry provides most of the basal area, while yellow birch is the only other major contributor to basal area of the plot (Fig. 4.2(B)). All pin cherries disappear by year 30. Afterward shade-tolerant species increase in importance, although yellow birch continues to dominate until year 60. These results are consistent with the natural history observations I have given earlier for succession below 760-m elevation in northern New Hampshire (that is, below the boreal forest zone) (Bormann et al., 1971; Bormann and Likens, 1979; Siccama, 1968). This application, first reported in 1972, demonstrated that the model realistically reproduced ecological succession in northern hardwoods forests. To my knowledge, it was the first model to do so for any forest, or for any vegetation. Version II gives similar results. (Users of the software available for use with this book can try a variety of similar experiments. Because somewhat different weather records are provided with that software, results will not be identical.)

Ecological Stability and Long-term Patterns of Succession

Long-term patterns of succession projected by the model show a peak in abundance in mid-succession and thus differ from the classical ideas that dominated plant ecology in the first half of the twentieth century and continued to play a strong role when the forest model was first developed (Fig. 4.3). In fact these ideas keep recurring in one form or another in discussions of succession (Drury, and Nisbet, 1973; Odum, 1969; Vitousek and Reiners, 1975), ecological stability, and community and ecosystem properties. Succession was believed to end with a fixed climax stage that would reproduce itself over time. This climax stage was believed by many ecologists in the first half of the twentieth century to represent a steady-state that would continue indefinitely unless there was external disturbance (Botkin, 1990). The climax stage was believed to have maximum biomass, maximum species diversity, and maximum storage of chemical elements in organic matter. The climax community was considered homeostatic; the species that dominated this stage were able to produce seeds that could germinate and grow in the shade of the mature trees, so that at any

---

* This section is taken from Botkin, D. B., J. R. Janak, and J. R. Wallis, 1972, Rationale, limitations and assumptions of a northeast forest growth simulator. *IBM Journal of Research and Development* 16: 101–116.

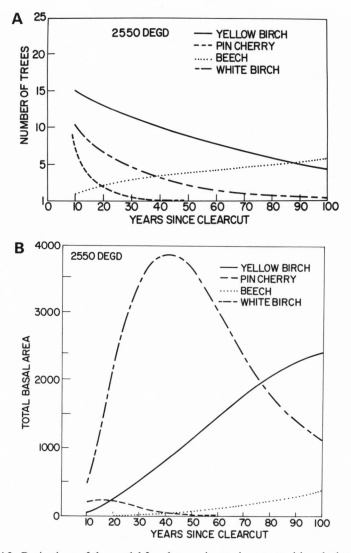

FIGURE 4.2. Projections of the model for changes in species composition during succession: (A) numbers of trees, (B) basal area. Results are from version I for a site at 610-m elevation with 2,550 growing degree-days per year. The forest begins from a clearcut (From D. B. Botkin, J. R. Janak, and J. R. Wallis, 1972, Rationale, limitations and assumptions of a northeast forest growth simulator. *IBM Journal of Research and Development 16*: 101–116, Figs. 12 & 13.)

FIGURE 4.3. Long-term projections of average basal area. These projections are from version I of the model, for the same site as in Figure 4.2. Each line represents the average of 100 plots with identical site conditions, including a deep, well-drained soil and a constant climate (that of Woodstock, New Hampshire, modified to represent the elevation of the site). (A) shows total basal area, and (B) and (C) show basal area of dominant

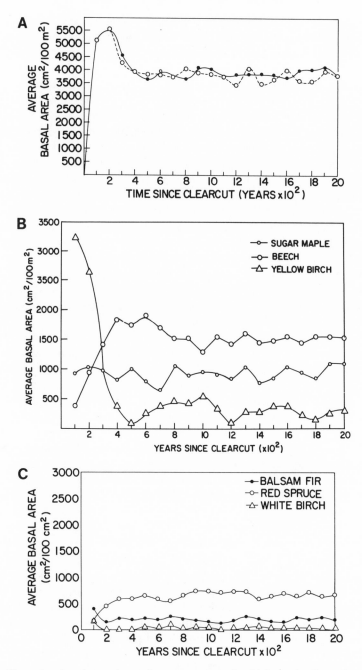

species. Note that basal area is used as a surrogate for biomass in discussions of the model because the two factors tend to vary one-to-one, but basal area is closer to the direct dynamic variables of the model. (Reproduced from D. B. Botkin, J. F. Janak, and J. R. Wallis, 1973, Some ecological consequences of a computer model of forest growth. *Journal of Ecology 60*: 849–872, Figs. 7 & 8, p. 866.)

time there would be a mixture of sizes and age classes of these late successional species in the forest. A snapshot of a cross section of such a forest at one time would look like a snapshot at any other time.

This climax stage was also believed to be characterized by a simple kind of stability that I have referred to elsewhere as "classic static stability" (Botkin and Sobel, 1975). Once subject to a disturbance and then released from that disturbance, the forest would return to the climax stage.

The idea of a permanent and stable climax stage ran deep in the ecological thought of the early twentieth century. As an example, one of the best classic studies of ecological succession was that carried out by William S. Cooper soon after the turn of the century on Isle Royale, Michigan. Cooper studied the northeastern part of the island, which he found to be largely a forest of balsam fir, white spruce, and white birch, with some bogs and areas of maple. The forest, he wrote, "is the final and permanent vegetational stage, toward the establishment of which all the other plant societies are successive steps," and "both observational and experimental studies have shown that the balsam-birch-white spruce forest, *in spite of appearances to the contrary* [my italics], is, taken as a whole, in equilibrium; that no changes of a successional nature are taking place within it," even though "superficial observation would be likely to lead to exactly the opposite conclusion" (Cooper, 1913).

Cooper saw this forest as stable in three ways.* First, it had classic stability: when disturbed, any patch recovered to its original condition, which persisted until disturbed again. Second, any bare surface—be it a barren, dry, rocky ridge top; an open-water bog; or a fertile, well-drained soil—would, if given long enough time without disturbance, develop into the same final balsam-spruce-birch forest. "In other words, those phases of the vegetation that are not uniform in character with the main forest mass are plainly tending toward uniformity." Third, although there were natural disturbances, such as fire and windstorms, their effects were local and balanced out in the larger picture: "The climax forest is a complex of windfall areas of differing ages, the youngest made up of dense clumps of small trees, and the oldest containing a few mature trees with little or no young growth beneath," but "the changes in various parts" balance each other so that "the forest as a whole" remains the same. Every disturbed area was believed to return to the same final condition and to remain constant until disturbed again. Moreover, the forest was believed subject to a uniform rate of small disturbances so that its average composition was always the same. More recently, ecologists refer to this as "landscape-level" stability. Any small area was thus characterized by a stability like that of a clock pendulum; if upset or pushed, it returned to its former condition. The forest as a whole, on a large scale, was composed of a constant number of patches of any one stage at any time, and was therefore uniform. Nature without human influence was constant in both the small and the large. Cooper's study and conclusions were repeated

---

*This paragraph is taken verbatim from D. B. Botkin, 1990, *Discordant Harmonies: A New Ecology for the 21st Century*, Chapter 4.

early in the twentieth century in other areas, and this concept of succession was expounded by major botanists in the first decades of the century.

For example, more recently Whittaker wrote that "through the course of succession, community production, height, and mass, species-diversity, relative stability, and soil depth and differentiation all tend to increase.... The end point of succession is a climax community of relatively stable species composition and steady-state function, adapted to its habitat and essentially permanent in its habitat if undisturbed" (Whittaker, 1970). In a classic paper on succession, Odum wrote in 1969 that ecological succession "culminates in a stabilized ecosystem in which maximum biomass [is] maintained" (Odum, 1969, p. 262). Any decline from this state of maximum biomass and diversity was considered by the earlier plant ecologists to be a "disclimax," analogous to a sickness and not part of the normal sequence of events.

Although ecologists today tend to disclaim a belief in succession that achieves a static climax stage, and within the last few years there has been a rapid expansion in discussions about the lack of equilibrium in ecology, some of which involve discussion of the JABOWA model (DeAngelis and Waterhouse, 1987; Green, 1982), the idea continues to appear in ecological literature. As an example, Pianka wrote in his 1988 textbook, *Evolutionary Ecology*, that an oak-hickory forest "is a stable community in a dynamic equilibrium that replaces itself: such a final stage in succession is termed its *climax*" and that "largely undisturbed areas may be primarily in the climax state.... At climax, production equals respiration and organic materials cease to accumulate" (Pianka, 1988). While there is increasing recognition of the dynamic and stochastic character of early stages in succession (Halpern, 1988), mature forests are still typically characterized as static, as for example in the statement by Waring and Schlesinger (1985) in their excellent summary of forest dynamics that "In old-growth forests, there is little net nutrient accumulation in organic matter, living and dead, and total ecosystem losses should balance inputs," (Waring and Schlesinger, 1985, p. 139) and "Theoretically, nutrient storage in vegetation reaches an asymptotic or cyclic-asymptotic value in old-growth forests that have reached a steady state in the accumulation of biomass" (p. 138). Models of ecological processes such as chemical cycling continue to assume or have as a necessary consequence a fixed steady-state condition as an end point. For example, Sprugel developed a model of changes in biomass during succession for balsam fir patches that by definition achieves a fixed equilibrium state (Sprugel, 1985a, b). Even when there is general acceptance of the idea that forest ecosystems are subject to variation in time, the tendency is to explain the characteristics of mature, old-growth forests in terms of a static condition.

In addition, during the past two decades there has been interest in using Markovian analyses to project future states of forests, ever since the pioneering work of Horn (1971, 1974). However, since the Markov method uses fixed transition probabilities, the final condition will, in general, be a single state (although it is possible to achieve some sets of states that cycle in a regular fashion as the long-term expectation). A Markovian approach therefore has an implicit assumption that an old-age, undisturbed forest achieves constancy.

Although this seems to be contradicted by observation, there continues to be an interest in the use of Markovian analyses. See for example, Acevedo (1981); Lippe, de Smidt, and Glenn-Lewin (1985); McAuliffe (1988); Orloci and Orloci (1988); Sharpe et al. (1985); and Usher (1981).

Such classical ideas of succession are different from the projections of the JABOWA model, an early example of which from version I is shown for a period of 2000 years following clearing in Figure 4.3. The model projects that maximum total basal area occurs in the middle of succession, not at the "end," and that the decrease in basal area and biomass that takes place afterward does not lead to simple, fixed, steady-state values, but to varying magnitudes of basal area and other variables which undergo apparently random fluctuations within certain bounds. Peak total basal area is approximately $5{,}500 \, \text{cm}^2/100 \, \text{m}^2$ ($55 \, \text{cm}^2/\text{m}^2$), which occurs between the first and second century after clearing, after which the basal area declines to a range between approximately 35 and $45 \, \text{cm}^2/\text{m}^2$ (Fig. 4.3(A)).

The reason for the peak in basal area between the first and second centuries after clearing is clarified by Figure 4.4, which shows average basal area of six major species. Yellow birch, a fast-growing, large, early successional tree, accumulates a large basal area (and therefore high biomass) during the first century, producing a more or less even-aged population of that species within the forest. Since seedlings and saplings of that species cannot persist under the deep shade of the forest, regeneration of this species ceases. Meanwhile, the shade-tolerant, late successional species, including sugar maple and beech, (Fig. 4.4(A)) and balsam fir and red spruce (Fig. 4.4(B)) increase in abundance. Peak community basal area occurs during a transition period when yellow birch is still a dominant member of the forest, but individuals of the later successional species also contribute a large amount of biomass and basal area to the forest. The combined contributions of early and late successional species sum to the largest basal area and biomass midway during succession.

When these results were first published (Botkin, Janak, and Wallis, 1972b), they were contrary to the prevailing beliefs about forest succession, but similar findings had been made in a study by Loucks (1970). Loucks analyzed forest stands of a variety of ages in Wisconsin and found that the peak production occurred approximately 200 years after clearing, and that both production and diversity decreased in older stands (Loucks, 1970). Since that time this pattern in forest succession has been more widely accepted (West et al, 1981), but as I mentioned earlier, the old idea persists—there is an implicit assumption in discussions of forest dynamics that succession will lead to a single steady-state unless external disturbance continues. It is important to recognize that JABOWA projects that a forest will undergo variations in biomass and composition indefinitely, within some boundaries, even when the climate is constant. It is also interesting to note that the original 1972 projections amounted to one of the first significant hypotheses generated by the model that was counter-intuitive but since reinforced by observations. Version II projects the same pattern, as do most other adaptations and versions of the model, most applied to other forest types (West et al., 1981).

As I have just explained, the classic idea is that all the desirable characteristics of a forest—the amount of timber, total biomass, fertility of the soil (both in total organic matter and content of chemical elements), and diversity of species—increase during succession to a maximum at the climax stage. Recent evidence from studies of very long successional patterns casts doubt on this idea.

Present knowledge suggests that such a "steady-state" forest will slowly lose some fraction of its stored chemical elements every year by the erosive effects of wind and water (Bormann and Likens, 1979; Bormann et al., 1974; Likens et. al., 1977). For example, in a study on Mount Shasta, California, organic carbon and nitrogen increased in the soil for approximately 200 years and then declined (Dickson and Crocker, 1953). Whether physical erosion or biological aggradation dominates depends on the relative strength of these two processes. For example, in an area of very heavy rainfall, fluvial processes will tend to override the biological effects.

As an example, a sequence of dunes in Eastern Australia provides information for more than 100,000 years of succession (Walker et al., 1981). The pattern of succession at first follows the classic idea: as one walks inland to older and older dunes, the stature of the forest increases as does total soil organic matter and the number of the species. But then these factors decrease and the very oldest dunes are depauperate. Gone are the large trees and the great diversity of species. On the oldest dunes are shrubs of low stature forming a comparatively barren landscape. The soil has been slowly leached of its nutrients, which have been transported so far below the surface that tree roots can no longer reach them. Other evidence suggests that mature forests in North America are also "leaky" and lose chemical elements (Likens et al., 1977). A forest in equilibrium by definition must have a zero net production—no addition of organic matter takes place. Such a forest can lose nutrients through geological processes, but has little means to reaccumulate them (Vitousek and Reiners, 1975). In contrast, an earlier successional forest in which organic matter is accumulating concomitantly increases its storage of chemical elements. Thus the ultimate fate of a never-disturbed forest is to go downhill biologically.

Another example of decline in chemical composition of the soil is succession that occurs following the melting back of glaciers in the southern island of New Zealand. The west coast of New Zealand is famous for its beautiful glaciers and glaciated valleys and harbors. These glaciers are still retreating as part of the long-term trends since the end of the last ice age. The rainfall in this area of New Zealand is very high, as much as 275 in./yr (700 cm/yr). Processes that might take much longer in areas with much lower rainfall occur more rapidly here. Erosion can be observed as one travels from the edge of the glaciers toward the ocean shore. As one walks away from the front edge of the glacier, one passes first over bare rock, then over rock with scattered lichens, then over grasses and lichens, then to a shrub and perennial grass stage, and next to several stages in the development of a rain forest with high species diversity, high biomass, and large trees. This is the state that ecologists earlier in the twentieth century would have called the climax state; they would have assumed that it would persist indefinitely, as long as it was not disturbed. Beyond the

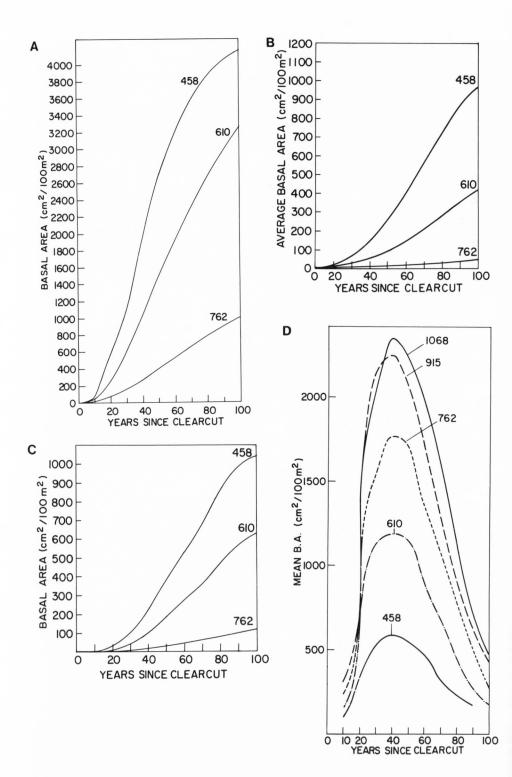

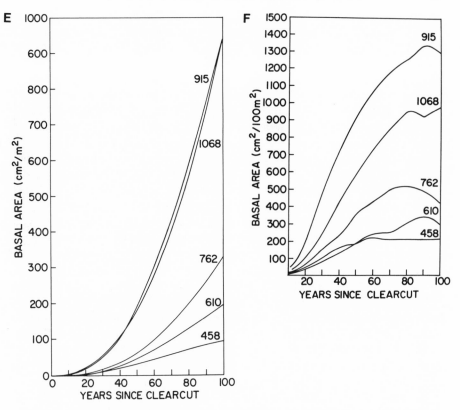

FIGURE 4.4. Average basal area as a function of time since clearing and elevation (in m) for each of six major species of the New England forests. (A) Yellow birch, (B) beech, (C) sugar maple, (D) White birch, (E) red spruce, and (F) balsam fir. Each line represents the average for 100 plots at a single elevation beginning from a clearing with the same initial conditions, including a deep, well-drained soil. (Reproduced from D. B. Botkin, J. F. Janak, and J. R. Wallis, 1973, Some ecological consequences of a computer model of forest growth. *Journal of Ecology* 60: 849–872, Fig. 6, p. 865.)

rain forest farther from the glacier, however, is another stage: one of low shrubs and grasses with low species diversity and low biomass. Inspection of the soil in this final stage shows a well-developed profile with a deeply leached layer. As with the Australian dunes, water has transported most of the important cations below the reach of most plant roots and therefore the rain forests cannot persist.

These ideas were first stated in a general form by Vitousek and Reiners (1975) and by Gorham et al. (1979), who suggested that net ecosystem production is the primary determinant of whether there is a net chemical storage in an ecosystem (Vitousek and White, 1981). When an ecosystem in "steady-state" is disturbed, changes occur in chemical cycling. As a first example, consider a forest that is burned. The fire converts large organic compounds (such as wood) to small inorganic ones (including carbon dioxide, nitrogen oxides, and sulfur oxides). Some of these inorganic compounds are lost to the ecosystem during the fire as particles of ash that are blown away or as gases that move into the atmosphere and are distributed widely. Other compounds are deposited on the soil surface as ash. These are comparatively highly soluble in water and readily available for vegetation uptake. Thus immediately after a fire there is an increase in the local availability of chemical elements in the root zone of the vegetation. This is true even if the ecosystem as a whole has undergone a net loss in total stored chemical elements. If there is sufficient live vegetation remaining after the fire, this "pulse" of newly available elements can be taken up rapidly, especially if there is a moderate amount of rainfall. A very heavy rain following a fire could remove many of these highly soluble elements from the ecosystem, through both of the processes we have discussed earlier: transporting them downward below the root zone and depositing them in the mineral portion of the soil, and transporting them in surface and groundwater downstream and out of the ecosystem. The pulse of inorganic nutrients can then lead to a pulse in growth of vegetation, which in turn can lead to a pulse in the amount of stored chemical elements in the vegetation.

Other disturbances can have similar effects. Severe hurricanes and tornadoes knock down and kill vegetation, which then decays, increasing the concentration of chemical elements in the soil, which are subsequently available for vegetation growth. Storms have another effect on forests: uprooting of trees can bring chemical elements that were near the bottom of the root zone to the surface, where they are more readily usable.

To summarize, if a forest remains undisturbed for a very long period—500 years or more—it will slowly change again. The diversity will decrease—only the most shade-tolerant species will persist—the amount of live organic matter will decrease, and the soil may lose some of its chemical elements. Finally, over an even longer period, there may be a subsequent change. The ecosystem becomes deficient in chemical elements that are lost slowly, and the dominant biota are species that can persist with a low concentration of these elements. In a forest, there is a return to a shrub or herbaceous stage. In primary succession, the earliest stages involve hardy species that can persist on bare rock or inorganic soil. These include small plants such as lichens and mosses. These may aid in retarding erosion and allowing soil to develop, on which trees and shrubs may grow. An important implication of the JABOWA model is that forests are not characterized by classic static stability—by the existence of a single equilibrium state to which the forest always returns following disturbance. Additional related implications of the model are discussed in the last chapter.

The JABOWA model as discussed in this book does not include a feedback between the vegetation and soil in terms of chemical cycling, nor does it include a model of the dynamics of net storage of chemical elements in the soil over time. Some derivations of JABOWA have added a decomposition model and feedbacks between the vegetation and soils, most notably that of Pastor and Post (1985, 1986).

JABOWA projects that a decline in biomass will occur for *internal* reasons—the life history characteristics of early and late successional species which together lead to a maximum biomass in mid-succession—whereas the theories I have just discussed project a decline in biomass for *external* reasons—loss of chemical elements to the ecosystem through fluvial erosion. The two together suggest that there is little reason to expect that the long-term condition of an undisturbed forest will consist solely of the single state that has the highest biomass and diversity. In *Discordant Harmonies: A New Ecology for the 21st Century*, I discuss the connections between the twentieth-century ecological concept of a static climax and prescientific beliefs in a balance of nature (Botkin, 1990). Suffice it to say that there is a direct correspondence between these beliefs.

The Causes of Species Replacement during Ecological Succession

During succession, a well-known process of species replacement occurs, which I have referred to earlier as a change from early successional to late successional species. Over the years the causes of species replacement have been the subject of debate, and Horn has provided an interesting analysis (Horn, 1981). Although the debate will be familiar to many readers of this book, I will repeat it so that we can consider which ideas the forest model reinforces.

There are four explanations about why certain species replace others during ecological succession, as discussed by Connell & Slatyer (1977) and Horn (1981): (1) facilitation, (2) interference, (3) tolerance, and (4) chronic patchiness.

Facilitation means that one species prepares the way for the second, and that without the presence of the first the second could never appear. The occurrence of the first species is a necessary condition for the occurrence of the second. In New England, the appearance of sugar maple in the forest would somehow depend on prior presence of pin cherry. In less extreme cases, facilitation leads to earlier entrance of the second species in the forest, or in greater abundance.

In interference, also called *competitive hierarchy*, individuals of one species prevent individuals of the second from entering a forest stand. When trees of the first species have completed their life cycle and died out, individuals of the second are able to enter. In New England, pin cherry might exude some chemical that retards germination of sugar maple seeds. Removing the first species allows the second species to appear sooner than it would have otherwise.

In the third possibility, tolerance, early successional species neither increase nor decrease the rate of recruitment, growth, or survival of later species. The process is in part simply a difference among life histories: individuals of one species appear first because they get there quicker and grow faster than individuals of the second. Both early and late successional species can grow in the early successional environment, but late successional species are more tolerant of limited resources that occur when competitors are abundant. In New England, sugar maple could grow both in the open and in the deep shade of a forest, whereas birch could grow only in openings. Sugar maple saplings will grow underneath birch or underneath older maples. The sequence of species is simply a matter of time before the late successional species arrive. Removing the first species would not change the time since the start of succession when the second species appears. In New England, elimination of the pin cherry would not have affected the timing at which sugar maple seeds germinated and saplings grew within the forest.

In the fourth possibility, chronic patchiness, succession never really takes place—the species that appears first remains until the next disturbance (Horn, 1981). There may be many species, but species only replace one another after a disturbance.

Which of these four explanations are consistent with the forest model? An examination of the assumptions of the model presented in Chapters 2 and 3 shows that there are no mechanisms for facilitation. Interference could occur only through competition for light, but this competition is across species—each tree is shaded by the sum of all the leaves of taller trees. Species-specific differences in competition for light are manifested in two ways: (1) Some species have a greater leaf weight for a given diameter than others, and in this way one species might have a greater competitive impact on individuals of other species. However, an individual of a high-leaf-weight species would also have a greater competitive interference on other individuals of its own species. (2) Regeneration and growth are functions of available light. Shade-intolerant and intermediate species cannot germinate, and saplings of these species do not grow well, under the shade of other trees. Thus late successional species interfere with the continuation of early successional species. This is a variant of interference that I will refer to as *reverse interference*. Reverse interference is the major mechanism in the forest model that accounts for the projected pattern of species replacement.

There is also some expression of tolerance as a mechanism for species replacement. Early successional species add saplings to a plot when the light intensity is high. The maximum number of saplings that can be added in any single year is much higher for early successional species such as pin cherry, birch, and aspen than for the late successional species such as spruce, fir, and sugar maple. Elimination of the early successional species would allow a more rapid buildup in the population of the late successional species, but this build-up would occur more gradually than occurs following clearcut with a full complement of species.

Chronic patchiness could be forced to occur in single-species stands, but is not characteristic of the model. In conclusion, the forest model is consistent with a combination of reverse interference and tolerance. The accuracy and realism of the projections of the model, as discussed in subsequent chapters, suggest that, at least for eastern deciduous and boreal forests of North America, observed dynamics of forest succession can be accounted for without recourse to facilitation mechanisms, and that a combination of reverse interference and tolerance are sufficient to simulate forest succession in a realistic manner.

Observed Mechanisms of Species Replacement

For many years, ecologists have argued over which of these four possibilities actually occur in nature, in some cases assuming that only one of the four could occur. In fact each occurs. Which possibilities occur in any ecosystem depends in part on the environmental conditions and in part on the characteristics of the pool of species available to take part in succession.

Interference occurs in tropical rain forest areas in Mexico. Following a disturbance, early successional species form dense mats. Species include perennial grasses such as bamboo (*Bambusa spp.*) and *Imperata* species, as well as thick-leaved small trees and shrubs (including *Guazuma* and *Curatella*). These form stands that are suffciently dense so that seeds of later successional species cannot penetrate the thick organic mat and therefore do not germinate. *Imperata* replaces itself, or in some cases is replaced by bamboo which then replaces itself (Gomez-Pompa and Vazquez-Yanes, 1981). Once established, these mat-forming species appear able to persist for a long time. This interference is in contrast to the more common sequence of species that occurs following a disturbance such as a natural fire in the rain forest when early stages do not form as thick a mat, and grasses and early successional shrubs and trees are replaced by other species.

There is some evidence to support the existence of facilitation in tropical rain forests where the appearance of a later species may be affected by nutrient cycling and the local microclimate. For example, in Mexican rain forests, early successional species germinate, become established, grow rapidly, and take up chemical elements that are available in the soil that might otherwise be lost by leaching. As the early successional plants die and become part of the organic matter in the soil, these elements become available to other plants. Without rapid growth by early successional plants, soils would become impoverished of these elements and growth and establishment of the later successional species would be retarded or impaired (Gomez-Pompa and Vazquez-Yanes, 1981).

In some tropical forests, temperature, relative humidity, and light intensity at the soil surface reach conditions very similar to those of mature rain forests after only 14 years (a time by which leaf area index has reached the values of mature forests) (Gomez-Pompa and Vazquez-Yanes, 1981). Once these conditions are established, species adapted to deep forest shade can enter the stand. By growing rapidly and providing shade, early successional species speed up the entrance of later successional species. Hence they seem to facilitate their entrance in succession.

Sand dunes provide another example of facilitation. Along the Great Lakes of North America, sand dunes are formed by the wind. They are unstable and can be breached by storms so that their sand is blown inland. Such instability makes establishment difficult for most species. Dune grass is well adapted to this habitat, spreading rapidly by vegetative runners that have sharp ends and grow readily through the loose sand. A complex underground mat of runners grows outward and grass shoots sprout from these. Grass mats are more stable than the sand, and they tend to stabilize the dunes, slowing down or preventing the movement of the sand.

Without the dune grass, seeds from other plants that fall onto the dune are readily buried to a depth from which they cannot germinate or are blown away by the wind. Once the sand dune is stabilized, seeds from other plants can germinate. Thus once the dune grass is established, species of shrubs and trees, such as white pine (*Pinus strobus*) and white birch (*Betula paperifera*), become established. Dune grass facilitates the occurrence of the later successional species. Once the shrubs and trees are established, dune grass is shaded and does not obtain sufficient light for growth.

Along the shores of Lake Michigan, a series of dunes are established, with the interior ones formed from sand that had previously been blown inland. Because the dune nearest the lake protects the interior ones from the destructive force of the wind, the interior dunes are more stable. A walk inland from the shore takes one through a series of dunes, each representing an older stage in succession. Along the front of the near shore dune is dune grass, while white pine, white birch, and shrubs can be found in the more stable areas of this dune. An increasingly mature forest can be found on interior dunes, dominated by sugar maple, the characteristic species of old stands of this region.

Facilitation also occurs in the early stages of boreal forest bog succession, when sedges that form floating mats provide a substrate that accumulates soil particles and within which seeds of shrubs and wet-adapted trees germinate and grow.

Chronic patchiness occurs in harsh environments such as deserts, where energy or chemical elements required for life are in low supply and disturbances are frequent and large. The physical, degrading environment dominates. For example, in *Larrea–Ambrosia* deserts of California, major species grow in patches of mature individuals along with a few seedlings. These patches persist for long periods until there is a disturbance. Such chronic patchiness could also be expected in environments made harsh by a heavy concentration of toxins in the soil, such as occurs near Sudbury, Ontario, near the large nickel–copper smelters.

In summary, the forest model is consistent with two mechanisms of species replacement during succession: interference (in a reverse form to the way that it is usually stated) and tolerance. The model suggests that facilitation need not be necessary as an explanation of species replacement during succession in the forests to which the model has, so far, been applied. Observations of forests suggest that all four explanations of species replacement can be found in nature, and that there are some rather loose rules as to which can be expected where. In benign environments heavily modified by life, such as occur in a tropical rain forest, facilitation and interference can be important. In harsh environments such as in a desert, nonbiological processes dominate the local environment near the ground and there is little potential for biological mechanisms to play an important role. In temperate and boreal forests, whose environments fall somewhere between these extremes, some forms of interference and tolerance appear sufficient to explain observations.

Changes in Forests with Changes in Climate

Another early result of the forest model was that version I realistically reproduced combined effects of changes in climate and biological conditions during succession. (Version II also simulates these combined effects.) For example, in the mountains of New Hampshire, the model projects that northern hardwoods species—yellow birch, beech, and sugar maple—decline in abundance with elevation, while boreal forest species—white birch, red spruce, and balsam fir—increase in abundance with elevation, consistent with observations (Fig. 4.5).

The model projects that white birch reaches its peak basal area about year 40 at all elevations, with its maximum abundance occurring at the highest elevation considered, 1,067 m. In contrast, yellow birch, a much longer lived and warmer climate species, is projected by the model to increase in basal area at the three lower elevations through the first century, reaching a peak, as has been noted earlier, between the first and second centuries following clearing, and not to grow at all at the upper two elevations.

The elevation 762 m seems to be a crucial one. At this elevation community basal area is at a minimum for all years in comparison with other elevations (Fig. 4.5(A)) and density reaches a maximum in comparison with other elevations between the fourth and six decades after clearing (Fig. 4.5(B)). In reality, this elevation is within a transition in northern New England between the northern hardwoods forest (characteristic of southern, warmer conditions) and the boreal forest (characteristic of northern, colder conditions). The patterns of total basal area and density are second-order characteristics, not as obvious to the casual natural historian as major changes in species composition. Species diversity, measured as the Shannon-Weaver index H', drops above 762 m, because the boreal forest is much less diverse than the northern hardwoods forest (Fig. 4.6) (Pielou, 1969, 1975).

The model's projections of changes with elevation are consistent with prior descriptions of three subdivisions in the forests of the White Mountains of New Hampshire by Harries (1966), who described a lower montane belt dominated by sugar maple and beech up to elevation 732 m; a transition zone that he called the "middle montane forest," which extended from 732 to 1,067 m; and an upper montane forest dominated by the boreal forest species balsam fir and red spruce (Fig. 4.6). Bormann et al. (1971) reported similar findings for Hubbard Brook, with the upper level of sugar maple and beech occurring between 732 and 762 m. The projections are also consistent with observations for the Green Mountains in nearby Vermont, as described by Bormann and Buell (1964).

These early projections of version I of the model are also consistent with observations reported later by Siccama (1974) (who also provided a summary of previous empirical studies), who found that the boreal forest, dominated by balsam fir, red spruce, and white birch, extended down to 792 m elevation in the Green Mountains of Vermont. There were northern hardwood forests at

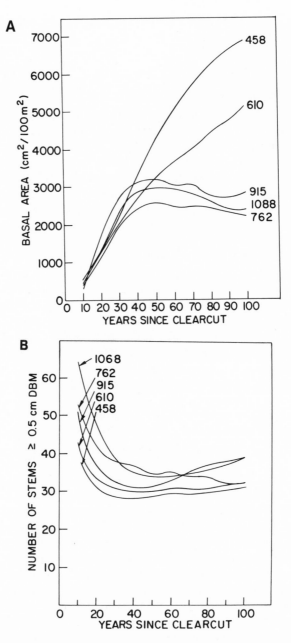

FIGURE 4.5. Projected changes with elevation (in m) of total basal area (A) and total stem density (B). Each line represents the average of 100 plots beginning with the same initial conditions, including a deep, well-drained soil and a constant climate represented by the average monthly temperature and rainfall of Woodstock, New Hampshire. (Reproduced from D. B. Botkin, J. F. Janak, J. R. Wallis, 1972b, Some ecological consequences of a computer model of forest growth. *Journal of Ecology* 60: 849–872, Figs. 4 and 5 p. 866.)

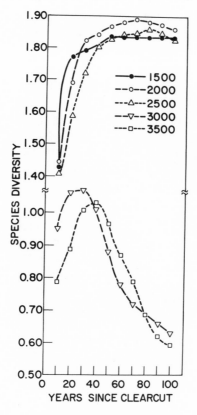

FIGURE 4.6. Species diversity as a function of time since clearcut and climate (represented, in ft, by changes in elevation). Results are shown for JABOWA version I. Each line is the mean of 100 plots beginning with the same initial conditions, including a clearing with a deep, well-drained soil, and a constant climate represented by the Woodstock, New Hampshire, average mean monthly temperature and precipitation records, modified according to standard meteorological lapse rates to account for changes with elevation.

lower elevations, and there was a distinctive tension zone between, where species of both forests occurred. He attributed the transition zone to climatic, not edaphic, factors. However, he gave a specific explanation in terms of climatic processes, which goes beyond the assumptions of the forest model. Siccama stated that the change in vegetation is the result of two factors: (1) a vertical climatic discontinuity (a nonlinear decline) in the length of the frost-free period; and (2) an increase in the frequency of the cloud base at and above 792 m, producing fog drip in the growing season as an additional source of water input for the trees and hoar frost in winter as an additional stress. Version I of the forest model reproduced these changes using standard temperature lapse rates and changes in precipitation with elevation observed for the Hubbard Brook Experimental Forest. (Users of the software available as a companion to this book can compare projected changes in elevation using climatic records from

Bethlehem, New Hampshire, which is farther from the Hubbard Brook site for which the model's projections were made, but for which a longer climate record is available.)

Projected Changes in Species Diversity with Succession and Climate

The model projects a different pattern in diversity at high elevations within the boreal forests than at lower elevations within the northern hardwoods forest. As an example, Figure 4.6 shows the changes in the Shannon index of species diversity H', where

$$H' = -c \sum_i p_i \operatorname{Log} p_i \qquad (4.1)$$

and where p_i is the probability that a random sample will yield an individual of species i and c is a proportionality constant. Here, p_i is calculated as

$$p_i = \frac{n_i}{\Sigma n_i} \qquad (4.2)$$

where n_i is the number of individuals per forest plot of species i. (For a further discussion of this measure, see Pielou [1969, 1975].) At lower elevations, diversity increases through the first 100 years following clearing and appears to approach an asymptote at the lowest elevation (458 m or 1500 ft). A decline in diversity begins at the next two higher elevations. In the boreal forest zone (elevations 915 and 1067 m (3,000 and 3,500 ft), diversity peaks at lower elevations and earlier in succession, at about year 40 to 50. The overall lower diversity at higher elevations is a result of the smaller pool of species in the boreal forest. As a result, just two species dominate older stands, and the transition to this dominance occurs relatively rapidly.

Effects of Soil Conditions on Forest Succession

Biomass, composition, and diversity vary with soil conditions. Here I remind the reader that the response of trees to nitrogen is given in Chapter 3, Tables 3.3 and 3.4 and equations (3.11) and (3.12).

Tables 4.2 and 4.3 compare the effects of soil conditions on forest growth as projected by version II. Table 4.2 shows projected succession in central Michigan on a poor soil, a comparatively coarse sandy loam with very low nitrogen content and poor water availability.

On this site, the biomass remains low, never exceeding $1.1 \, \text{kg/m}^2$, and the population density, which begins with less than one small sapling per square meter, declines, so that a closed canopy forest never develops. Instead the site is so harsh that a dispersed number of small stems, mainly jack pine, grow in an open field. The field would also be occupied by grasses, bracken fern, and small shrubs characteristic of the area such as blueberries and other Ericaceous species. Jack pine remains important throughout, but by the ninetieth year is

TABLE 4.2. Forest succession on a poor soil[a]

Species No.	Tree Species (common name)	Pop. (N/100 m²)	95 CI	Basal Area (cm²/m²)	95CI	Biomass (kg²/m²)	95CI	Vol. (ft³/acre)	95CI
				Year 10					
23	Jack pine	31.1	2.0	0.1	0.0	0.0	0.0	1.5	0.1
24	Red pine	0.7	0.3	0.0	0.0	0.0	0.0	0.3	0.1
Total		31.8	1.4	0.1	0.0	0.0	0.0	1.7	0.1
				Year 20					
23	Jack pine	38.8	2.1	0.6	0.0	0.1	0.0	12.2	0.8
24	Red pine	0.9	0.3	0.1	0.0	0.0	0.0	1.8	0.8
Total		39.8	1.8	0.6	0.0	0.2	0.0	14.0	0.6
				Year 30					
11	White birch	3.2	1.0	0.0	0.0	0.0	0.0	0.3	0.1
13	Red maple	0.9	0.4	0.0	0.0	0.0	0.0	0.1	0.1
23	Jack pine	32.0	2.0	1.3	0.1	0.4	0.0	39.8	2.2
24	Red pine	0.9	0.3	0.2	0.1	0.1	0.0	5.3	2.2
Total		36.9	1.4	1.5	0.1	0.5	0.0	45.7	1.8
				Year 40					
11	White birch	3.5	1.3	0.0	0.0	0.0	0.0	0.6	0.2
13	Red maple	1.1	0.4	0.0	0.0	0.0	0.0	0.5	0.3
23	Jack pine	16.2	1.1	1.4	0.1	0.5	0.0	49.9	3.6
24	Red pine	0.6	0.2	0.2	0.1	0.1	0.0	8.3	3.7
25	White pine	2.8	0.8	0.0	0.0	0.0	0.0	0.7	0.2
Total		24.2	0.7	1.6	0.1	0.6	0.0	60.0	2.2
				Year 50					
11	White birch	3.3	1.0	0.0	0.0	0.0	0.0	1.0	0.3
13	Red maple	1.3	0.4	0.1	0.0	0.0	0.0	1.4	0.6
23	Jack pine	12.3	1.1	1.7	0.1	0.7	0.1	76.8	6.4
24	Red pine	0.4	0.2	0.3	0.1	0.1	0.1	12.8	5.5
25	White pine	4.3	0.9	0.1	0.0	0.0	0.0	2.2	0.7
Total		21.6	0.6	2.2	0.1	0.8	0.0	94.1	3.4
				Year 60					
11	White birch	4.7	1.4	0.1	0.0	0.0	0.0	1.1	0.6
13	Red maple	1.2	0.3	0.1	0.0	0.0	0.0	2.1	0.9
23	Jack pine	9.7	0.9	2.1	0.2	0.9	0.1	108.7	10.3
24	Red pine	0.3	0.2	0.3	0.1	0.2	0.1	16.6	8.3
25	White pine	4.9	0.9	0.1	0.0	0.0	0.0	4.1	1.2
Total		20.8	0.5	2.7	0.1	1.1	0.0	132.7	4.8
				Year 70					
11	White birch	3.6	1.1	0.0	0.0	0.0	0.0	1.1	0.5
13	Red maple	1.2	0.4	0.1	0.0	0.0	0.0	2.5	1.2
23	Jack pine	5.2	0.6	1.6	0.2	0.7	0.1	93.9	11.3
24	Red pine	0.2	0.1	0.2	0.1	0.1	0.1	11.2	8.5
25	White pine	4.3	0.9	0.2	0.1	0.1	0.0	6.4	2.0
Total		14.4	0.3	2.1	0.1	0.9	0.0	115.1	4.1

Footnote [a] on page 130.

129

TABLE 4.2. (*Continued*)

Species No.	Tree Species (common name)	Pop. (N/100 m²)	95 CI	Basal Area (cm²/m²)	95CI	Biomass (kg²/m²)	95CI	Vol. (ft³/acre)	95CI
				Year 80					
11	White birch	3.5	0.9	0.1	0.0	0.0	0.0	1.6	0.6
13	Red maple	1.4	0.5	0.1	0.1	0.1	0.0	3.6	1.9
23	Jack pine	4.1	0.6	1.7	0.2	0.8	0.1	111.3	15.3
24	Red pine	0.1	0.1	0.2	0.2	0.1	0.1	14.0	11.8
25	White pine	4.8	1.0	0.3	0.1	0.1	0.0	11.6	3.4
Total		13.9	0.3	2.4	0.1	1.1	0.0	142.1	4.9
				Year 90					
11	White birch	4.5	1.4	0.1	0.0	0.0	0.0	1.5	0.7
13	Red maple	1.3	0.5	0.2	0.1	0.1	0.0	5.4	2.7
23	Jack pine	3.1	0.5	1.7	0.3	0.8	0.1	122.6	20.1
24	Red pine	0.1	0.1	0.2	0.2	0.1	0.1	12.5	14.0
25	White pine	5.5	0.9	0.4	0.1	0.2	0.0	16.7	4.9
Total		14.5	0.3	2.5	0.1	1.2	0.0	158.7	5.4
				Year 100					
11	White birch	3.7	1.0	0.1	0.0	0.0	0.0	1.8	0.9
13	Red maple	1.3	0.4	0.2	0.1	0.1	0.0	5.1	2.7
23	Jack pine	2.1	0.8	1.0	0.3	0.5	0.1	84.9	22.7
24	Red pine	0.1	0.1	0.1	0.1	0.0	0.1	4.5	8.8
25	White pine	4.4	0.9	0.4	0.1	0.2	0.0	15.3	4.7
Total		11.6	0.3	1.7	0.0	0.8	0.0	111.6	3.7

[a] Results of projections of version II for a site in central Michigan on a shallow (0.2 m depth), dry (water table depth 1 m below surface) clay-loam soil (soil moisture storage capacity 250 mm/m) very low in nitrogen (20 kg/ha). The forest is grown for 100 years from a clearing using, in repeating sequence, the 1951–1980 weather records from Grayling, Michigan. Results are for 40 replicates. Site elevation: 467.9 m; weather station elevation: 242.6 m.

suppressed in density by white birch and white pine. Jack pine is short-lived, and the individuals present at year 10 would not be expected to persist far into the second century. The forest remains open, however, and continues to be occupied by pines, which are characteristic of sandy, dry soils, and by white birch, characteristic of early to mid successional stands on poorer or colder sites.

Table 4.3 shows succession on a fertile and well-watered soil in the same region and with the same weather records as Table 4.2. The forest is dominated by sugar maple and balsam fir from the beginning, with hemlock becoming more important in density by year 50. By year 100 sugar maple contributes most of the biomass (29 ± 3.1 kg/m²), with minor contributions from fir and hemlock. Total biomass is approximately 30 times that of the poor site (Table 4.2) at year 100, but most of this is the result of a single species, sugar maple. Density is high and a closed canopy forest develops.

These two extremes in soil fertility and moisture availability illustrate the range of composition, dominance, diversity, and biomass projected by the model under the same climatic regime.

TABLE 4.3. Forest succession on a rich and well-watered soil[a]

Species No.	Tree Species (common name)	Pop. (N/100 m^2)	95 CI	Basal Area (cm^2/m^2)	95CI	Biomass (kg^2/m^2)	95CI	Vol. (ft^3/acre)	95CI
				Year 10					
1	Sugar maple	13.1	0.6	8.7	0.3	5.4	0.2	435.3	21.5
9	Balsam fir	10.4	0.8	13.3	2.6	7.8	1.6	1318.6	288.2
30	Eastern hemlock	3.5	0.6	0.0	0.0	0.0	0.0	0.6	0.1
36	Basswood	2.8	0.5	0.0	0.0	0.0	0.0	0.4	0.1
Total		29.8	0.8	22.0	0.7	13.2	0.4	1754.9	61.4
				Year 20					
1	Sugar maple	14.5	1.0	11.2	0.7	7.1	0.5	589.5	43.6
5	Mountain maple	1.4	0.4	0.0	0.0	0.0	0.0	0.1	0.1
6	Striped maple	1.1	0.4	0.0	0.0	0.0	0.0	0.2	0.1
9	Balsam fir	10.5	0.8	13.5	2.7	7.8	1.7	1329.0	311.8
30	Eastern hemlock	6.5	1.1	0.2	0.0	0.0	0.0	2.7	0.6
36	Basswood	3.1	0.7	0.1	0.0	0.0	0.0	1.4	0.3
Total		37.1	0.8	24.9	0.7	15.0	0.4	1923.0	61.4
				Year 30					
1	Sugar maple	15.2	1.1	13.5	1.0	8.8	0.7	743.7	64.4
5	Mountain maple	1.0	0.3	0.0	0.0	0.0	0.0	0.1	0.1
6	Striped maple	1.2	0.3	0.0	0.0	0.0	0.0	0.2	0.2
9	Balsam fir	10.9	0.9	12.4	2.6	7.1	1.7	1177.6	314.3
30	Eastern hemlock	7.1	1.4	0.3	0.1	0.1	0.0	4.8	1.2
36	Basswood	3.2	0.7	0.1	0.0	0.0	0.0	2.1	0.7
Total		38.6	0.8	26.2	0.8	16.0	0.5	1928.5	58.7
				Year 40					
1	Sugar maple	16.6	1.2	16.3	1.5	10.9	1.1	934.1	102.8
5	Mountain maple	0.9	0.3	0.0	0.0	0.0	0.0	0.1	0.1
6	Striped maple	1.4	0.4	0.0	0.0	0.0	0.0	0.1	0.1
9	Balsam fir	6.7	0.8	7.2	2.2	4.1	1.4	679.5	263.1
30	Eastern hemlock	8.3	1.4	0.5	0.1	0.2	0.0	10.1	2.7
36	Basswood	3.9	0.8	0.2	0.1	0.1	0.0	4.3	1.4
Total		37.9	0.8	24.1	0.8	15.2	0.5	1628.2	48.7
				Year 50					
1	Sugar maple	17.8	1.4	19.7	1.8	13.5	1.3	1172.1	128.1
6	Striped maple	1.1	0.4	0.0	0.0	0.0	0.0	0.2	0.2
9	Balsam fir	7.1	0.7	6.8	2.2	3.8	1.5	630.9	275.0
30	Eastern hemlock	10.4	1.6	0.8	0.2	0.3	0.1	20.5	5.2
36	Basswood	4.1	0.9	0.2	0.1	0.1	0.0	6.6	2.3
Total		40.5	0.9	27.6	0.9	17.7	0.6	1830.3	57.4
				Year 60					
1	Sugar maple	18.9	1.6	22.8	2.1	16.0	1.6	1414.0	155.1
5	Mountain maple	1.1	0.3	0.0	0.0	0.0	0.0	0.1	0.1
6	Striped maple	0.9	0.3	0.0	0.0	0.0	0.0	0.1	0.1
9	Balsam fir	7.9	0.8	6.5	2.0	3.6	1.4	583.1	254.7
30	Eastern hemlock	10.1	1.5	0.9	0.2	0.4	0.1	24.2	6.2
36	Basswood	4.0	0.8	0.3	0.1	0.1	0.0	7.5	2.4
Total		43.0	1.0	30.6	1.0	20.0	0.7	2029.0	64.6

Footnote [a] on page 132.

TABLE 4.3. (*Continued*)

Species No.	Tree Species (common name)	Pop. (N/100 m²)	95 CI	Basal Area (cm²/m²)	95CI	Biomass (kg²/m²)	95CI	Vol. (ft³/acre)	95CI
				Year 70					
1	Sugar maple	20.0	1.6	27.2	2.3	19.5	1.8	1759.4	177.7
5	Mountain maple	1.6	0.4	0.0	0.0	0.0	0.0	0.2	0.1
6	Striped maple	1.3	0.3	0.0	0.0	0.0	0.0	0.2	0.2
9	Balsam fir	4.9	0.7	3.8	1.7	2.1	1.1	336.5	212.4
30	Eastern hemlock	9.6	1.4	1.1	0.2	0.5	0.1	31.7	7.5
36	Basswood	4.3	0.8	0.3	0.1	0.1	0.0	9.9	3.0
Total		41.7	1.0	32.5	1.2	22.2	0.8	2137.9	76.0
				Year 80					
1	Sugar maple	20.4	1.5	31.1	2.6	22.9	2.1	2111.4	212.8
6	Striped maple	1.1	0.3	0.0	0.0	0.0	0.0	0.1	0.1
9	Balsam fir	5.3	0.8	3.1	1.4	1.6	0.9	238.3	166.1
30	Eastern hemlock	11.2	1.5	1.6	0.3	0.7	0.2	47.8	11.1
36	Basswood	4.2	0.8	0.4	0.1	0.1	0.1	11.7	4.2
Total		42.2	1.0	36.1	1.4	25.3	1.0	2409.3	92.3
				Year 90					
1	Sugar maple	20.9	1.4	34.6	3.0	25.9	2.5	2432.2	257.5
9	Balsam fir	6.7	0.8	3.7	1.5	1.9	1.0	293.9	184.5
30	Eastern hemlock	9.1	1.7	1.5	0.4	0.7	0.2	47.9	14.4
36	Basswood	4.3	0.7	0.4	0.1	0.1	0.1	12.4	4.6
Total		40.9	1.0	40.2	1.5	28.7	1.2	2786.3	108.8
				Year 100					
1	Sugar maple	21.6	1.5	38.0	3.6	29.0	3.1	2757.2	319.3
6	Striped maple	1.0	0.3	0.0	0.0	0.0	0.0	0.1	0.1
9	Balsam fir	4.5	0.7	2.1	1.2	1.1	0.8	164.3	150.2
30	Eastern hemlock	8.1	1.7	1.4	0.5	0.7	0.2	48.4	18.3
36	Basswood	4.6	0.9	0.4	0.1	0.2	0.1	14.8	6.0
Total		39.8	1.0	41.9	1.7	31.0	1.3	2984.7	120.1

[a]Results of projections of version II for a site in central Michigan on a deep (1 m depth), well-watered (water table depth 0.8 m below surface, therefore above soil base) clay loam soil (soil moisture storage capacity 250 mm/m) high in nitrogen (70 kg/ha). The forest is grown for 100 years from a clearing using, in repeating sequence, the 1951–1980 records from Grayling, Michigan. Results are for 40 replicates. Site elevation: 467.9 m; weather station elevation 242.6 m.

The general response of the model's forest community to a gradient in soil nitrogen is illustrated in Figure 4.7(A), which shows biomass at year 100 after a clearing. Except for soil nitrogen, conditions are the same as in Tables 4.2 and 4.3, for a forest stand near Grayling, Michigan, a comparatively dry site, with a soil depth of 0.4 m and a water table below that, so that water is relatively un-available. When soil nitrogen is 0 kg/ha, nothing grows as expected. Biomass at year 100 increases with increasing soil nitrogen content up to a peak at 100 kg/ha nitrogen, then declines slightly, remaining between 5 and 10 kg/m²

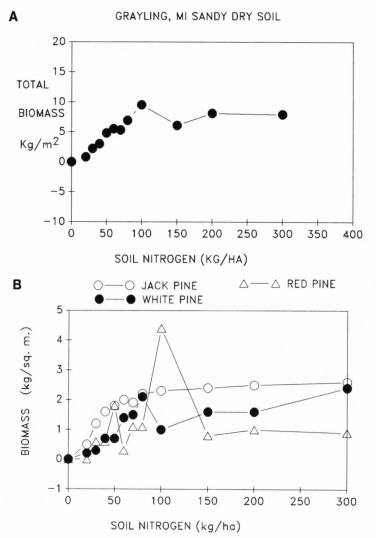

FIGURE 4.7. Response of a forest community to a gradient soil nitrogen content. Site conditions are: elevation, 467.9 m; weather station elevation, 242.6 m; soil moisture capacity, 150.0 mm/m; soil depth, 0.4 m; water table depth, 0.8 m. Weather records are 1950–1980 Grayling, Michigan, used in a repeating sequence. Soil nitrogen content is increased from 0 to 300 kg/ha. (A) Total biomass. (B) Biomass of dominant species.

(Fig. 4.7(A)). Biomass varies with relative success of various species (Fig. 4.7(B)). Total biomass under all nitrogen conditions for this site remains less than one-half that of the biomass of the fertile and well-watered site of Table 4.3, showing the combined limiting effects of water and soil nitrogen which act multiplicatively, as explained in Chapter 3. Jack pine, a species tolerant of low nitrogen as defined by equation (3.8), dominates the poorest nitrogen sites, while

TABLE 4.4. Biogeographical projections. Differences in projected forest conditions 300 years after clearing for an identical site except for geographic location as determined by weather records obtained from a specific site. In each case the 1950–1980 weather records are used in a repeating sequence throughout the 300 years.[a]

Species No.	Tree Species (common name)	Pop. (N/100 m²)	95 CI	Basal Area (cm²/m²)	95CI	Biomass (kg²/m²)	95CI	Vol. (ft³/acre)	95CI
Mount Pleasant, Michigan									
1	Sugar maple	0.6	0.5	0.0	0.0	0.0	0.0	0.3	0.4
8	Choke cherry	1.3	1.3	0.0	0.0	0.0	0.0	0.2	0.2
13	Red maple	0.6	0.6	2.4	3.6	2.6	4.4	365.8	657.4
14	Scarlet oak	0.2	0.2	0.0	0.0	0.0	0.0	0.2	0.4
25	White pine	3.1	2.0	3.0	2.5	2.5	2.6	503.8	606.2
26	Trembling aspen	1.2	1.0	0.3	0.4	0.2	0.3	22.6	40.6
27	White oak	3.5	1.9	1.0	0.6	0.5	0.4	46.1	42.7
28	Northern red oak	3.7	1.9	1.6	0.8	0.9	0.5	88.8	51.7
Total		14.1	0.4	8.3	0.3	6.7	0.3	1027.7	42.8
Munising, Michigan									
9	Balsam fir	2.2	0.9	0.0	0.0	0.0	0.0	0.8	1.0
11	White birch	10.6	5.3	13.4	5.9	10.2	5.1	1711.1	926.8
23	Jack pine	7.1	6.2	4.1	2.5	3.0	2.1	588.6	451.0
26	Trembling aspen	0.6	0.5	1.0	1.1	0.7	0.8	91.3	115.7
Total		20.5	1.0	18.5	1.1	13.9	0.9	2391.9	145.6
Bethlehem, New Hampshire									
1	Sugar maple	15.6	2.8	33.3	7.0	27.7	7.1	2805.6	811.0
3	Yellow birch	0.1	0.1	1.0	2.1	1.0	2.1	166.8	349.0
9	Balsam fir	11.6	1.8	10.0	2.8	5.2	1.7	804.8	305.0
23	Jack pine	0.1	0.1	0.8	1.7	0.8	1.7	174.4	365.1
24	Red pine	0.1	0.1	1.9	3.9	2.3	4.9	383.9	803.4
25	White pine	0.2	0.2	1.9	2.2	1.8	2.2	399.1	479.7
Total		27.5	1.4	48.8	2.6	38.9	2.1	4734.6	217.7
New Brunswick, New Jersey									
1	Sugar maple	2.8	1.1	0.1	0.1	0.0	0.0	2.9	1.8
8	Choke cherry	0.1	0.2	0.0	0.0	0.0	0.0	0.4	0.9
13	Red maple	1.6	0.9	2.8	2.0	2.2	2.0	265.0	270.3
15	Hornbeams	0.2	0.2	0.1	0.1	0.0	0.1	5.5	11.4
25	White pine	3.1	2.0	1.0	0.7	0.5	0.4	81.3	71.9
26	Trembling aspen	0.2	0.4	0.0	0.0	0.0	0.0	0.2	0.3
27	White oak	8.0	4.0	1.3	0.6	0.6	0.3	50.3	27.6
28	Northern red oak	6.2	3.2	2.6	1.1	1.6	0.9	173.2	134.0
39	Black cherry	2.7	1.9	0.1	0.1	0.0	0.0	2.5	2.1
Total		24.9	0.7	8.0	0.3	5.1	0.2	581.3	21.5

[a] Site conditions are: elevation, 100.0 m; weather station elevation, 242.6 m; soil moisture capacity, 150.0 mm/m; soil depth, 0.8 m; water table depth, 1.2 m; soil nitrogen, 60.0 kg/ha.

white pine, a species with intermediate nitrogen tolerance, contributes a slightly smaller amount but reaches the same biomass as jack pine at two nitrogen levels, 80 and 300 kg/ha. In between these values, red pine shares dominance with white pine.

Biogeographic Projections of The Model

The effect of changes in geographic location can be examined by taking one set of site conditions and using these with weather records from a variety of locations. Some examples are shown in Table 4.4 for a sandy, fertile site with a deep soil and moderate soil water availability. When this site is located near the prairie forest boundary, using weather records from Mount Pleasant, Michigan, the dominant species projected by version II are white pine, red maple, and red and white oak. When the site is located near the northern limit of the northern hardwoods forest, using weather records from Munising, Michigan, the dominant species are white birch, jack pine, and trembling aspen. When the site is located in New England but also near the northern limit of the northern hardwoods forest, the dominant species are sugar maple, fir, and white pine. On the northeastern coastal plain, using records from New Brunswick, New Jersey, the dominant species are red maple, red oak, white pine, and white oak. The increase in importance of sugar maple in New England compared with northern Michigan seems to reflect the differences between a coastal and a continental climate; the greater biomass of sugar maple in New Hampshire compared with New Jersey is due to the cooler temperatures of New England. In a coastal climate, rainfall is distributed relatively uniformly throughout the year; in a continental climate, summers are hot and dry. Sugar maple requires more soil moisture and does not occur at the Munising location. Some species, such as sugar maple, are projected to have a broad geographic distribution. Others, such as hornbeams, have a comparatively small geographic distribution.

Total biomass increases from the drier location near the prairie–forest boundary at Mount Pleasant, Michigan, to Munising, Michigan, which has lower evapotranspiration because of its more northerly location. It reaches the greatest amount in New England, where the rainfall is better distributed throughout the year.

These examples show that version II of the model responds realistically to geographic differences in climate. The reader can carry out additional tests of this kind using the model available as a companion to this book and weather records from other locations.

Summary

In this chapter I have considered the model's projections of local competitive events and succession and of the effects of changes in elevation, soil moisture storage, soil fertility, and biogeographic changes in weather patterns. I have also explored some implications of these results for ecological theory. Several conceptual distinctions were made that will be important throughout the rest of this book. These include the distinction between the *process* represented by the continual use and development of the model, as a series of continued tests and improvements, and the *projections* of the model as it is fixed at one stage. *Any fixed stage of the model becomes a set of assumptions about the dynamics of forests, whose implications we can then explore and whose success in mimicking forest dynamics we can then evaluate.* In this chapter I began the evaluation of the model as it has been fixed in versions I and II.

I explored the connection between site quality equations (also called *environmental response functions*) of the forest model and the concept of the ecological niche, suggesting that the site quality equations are a statement of the fundamental niche of a species, and that the operation of the model provides a projection of the realized niche of the species.

The forest model provides a number of hypotheses about forests. Two of these have been brought out in this chapter: the equations and algorithms described in Chapters 2 and 3 have been sufficient to reproduce observations about aspects of forest communities at the temporal and spatial scales discussed in this book; and second, four environmental dimensions—light, temperature, soil moisture, and soil nitrogen content—are, surprisingly, sufficient to account for the dynamic relationships among species of trees and for the dynamic integrated characteristics of a forest. The results presented in this chapter are consistent with these hypotheses, at least as one observes the qualitative natural history of eastern deciduous and boreal forests of North America. Quantitative validation of the model will be discussed in Chapter 5. But to the extent the correspondence between the model and our knowledge of forests has been considered in this chapter, it seems that four environmental dimensions are sufficient to account for many of our observations, including competition among individual trees and changes during succession and the relationship of tree growth and forest composition with climate and soil conditions.

The implications of the model for forest succession were discussed and compared with prior theories of succession. In the model, the differentiation of early, middle, and late successional species is accomplished through two factors: the light response functions discussed in Chapter 3 (equations 3.3a, 3.3b, 3.3c; Fig. 3.1; Fig. 3.2), and the maximum number of saplings that can enter a stand in any one year. These together constitute an hypothesis about the minimum information required to account for successional patterns. Another strong assumption of the model is that the light response functions of all species can be represented by two or three equations (actually one equation with two or three sets of parameters).

The pattern of succession projected by the model differs from earlier theories. JABOWA projects that biomass and diversity peak in mid-succession. Afterward, biomass fluctuates within bounds, but without exact periodicity. The model is an expression of two causes of the species replacement that occurs during the process of succession: reverse interference and tolerance. These seem sufficient to explain observed patterns in temperate and boreal forests.

Succession is projected to vary with the quality of the site. Biomass, composition, and diversity are lower on dry sites, poor in nitrogen, in comparison with well-watered sites with high nitrogen. In this way the model is consistent with natural history observations.

Results presented in this chapter can be repeated and extended by the reader using the programs on the diskette that is available as a companion to this book.

Notes

1. Among the earliest demonstrations of the competitive exclusion principle were the classic experiments conducted by G. F. Gause (*The Struggle for Existence* 1934). Gause grew populations of two paramecia, *Paramecium caudatum* and *P. aurelia*, in separate containers and also mixed in the same container. A constant amount of bacteria was added to each vessel at regular intervals. The paramecia were kept in pure distilled water to which certain inorganic salts were added. The bacteria were cultured separately transferred to a salt solution where they did not multiply, and then added to the vessels containing the paramecium. Through this careful control of conditions, Gause knew that only the bacteria he had added were available to the paramecium. When the two species were grown together, *P. caudatum* eventually became extinct. *P. aurelia* persisted and grew according to a logistic curve, but its carrying capacity was reduced. The effects of the competition were not apparent until the populations became fairly large—until they reached about one-fourth to one-half their maximum levels attained when grown separately. An observer who stopped the experiment after 8 days would not be able to predict the extinction of *P. caudatum* from the population growth curves alone. To the contrary, by day 8 this species appeared to be doing better than the other.

Gause's theoretical analysis was based on the logistic form of competitive interactive equations. Could coexistence ever occur? If the values of the carrying capacities and the interaction terms have exactly the right values, coexistence is *mathematically* possible for these equations. If the saturation values of species 1 is always greater than species 2—for all possible combinations of the population sizes of the two species—then the first species will always win. But if the saturation lines cross, coexistence of the two species is possible. The competitive exclusion principle is formulated excluding this hypothetical possibility.

2. The farmland was owned by Herman and Edith Chase and the history of the farm was provided by Edith Chase.

5

Testing and Comparing the Model
with the Real World

At first glance, testing of a model seems simple. Get some observations, run the model, and see if the projections are the same as the observations. The real problem is much more difficult, subtle, contradictory to our initial prejudices and expectations, and full of ironies. For a system as complex as a forest, the first question is: what variables do we want to test? Even this is far from a simple question, but in most forestry applications, foresters are interested in biomass or merchantable timber. Basic studies in plant ecology often report stem densities and basal area. Since JABOWA's primary variables are stem density and basal area, these primary variables are often of greatest interest when we want to test the model's fundamental properties.

The second problem is finding adequate data. The prevailing belief among ecologists, at least in my own experience, is that ecology is data rich and testing of models fails because of insufficient output from models. On the contrary, as should be clear to readers of the previous chapters of this book, the model generates large quantities of precise data rather easily, and the real problem is obtaining comparable real-world data against which to compare the model's projection.

An example from a classic study in North American plant ecology illustrates the point. In the second decade of the twentieth century, William F. Cooper carefully set out 10- by 10-m plots on Isle Royale, the large wilderness island in Lake Superior that is now a national park. On those plots, Cooper cut down every tree, measured diameters of the stumps, and studied the growth history of the trees (Cooper, 1913). In his thorough studies of the forests of Isle Royale, Cooper published maps of these plots. These provide unique data about the status of the forests of Isle Royale at one time. Although they were among the most detailed data about forest plots taken until that time, these measurements were not repeated elsewhere at a later date and, since the trees were cut down, the plots could not be monitored over time. Cooper did not establish plots where the trees remained and on which he measured diameters, so that change could be monitored and we might use them today to study the changes in Isle Royale's forest. This anecdote is typical of the problems we face when we attempt to compare our ideas about the dynamics of forest ecosystems with real-world observations. Adequate data are usually lacking. This is an especially troubling problem when we attempt to test the accuracy of an ecological model. In my experience, this problem is the opposite of what ecologists expect about models. Ecologists have generally distrusted models in part, because they believe that the projections from models are insufficient to compare with real observations. All too often the opposite is true. Data are insufficient to provide an adequate test of the quantitative output from the model. Once we have constructed a model, the question is how we can best test its projections, and where can we find adequate data for these tests.

A third problem is, having found data, what requirements do we place on them and on the comparisons with the model? For example, one choice is whether we compare the observed diameter of a tree at a specific time with the model's projection of that diameter. Another choice is to compare the observed change in diameter with the model's projection of that change, beginning with some initial observed diameter (in mathematical terms, the question is whether we are testing the model against a measurement of D_{t1} given D at t_0, or against the measurement of ΔD).

Since the change in diameter is a smaller number than the diameter, the second test requires greater accuracy in measurement and in the model. Some have suggested that there is a strong statistical correlation between a diameter of a tree at some starting time and its diameter at any future time. If the fundamental growth equation were true with no environmental limitations (all environmental response functions were set to a value of 1), and if growth, mortality, and regeneration were deterministic, then that would also be an expectation of the model. But, as has been described in Chapter 3, the model adds to that assumption two important dynamic qualities: (1) stochastic processes of mortality for an individual tree and stochastic processes affecting regeneration of a species; and (2) environmental limitations that can vary over time with variations in the environment and, through competition for light, with the population of trees. These qualities occur in a real forest. A simple statistical correlation between diameter at one time and diameter at some future time, sufficiently precise to project growth of all trees, could be expected only in uniform stands, primarily plantations of trees that are well spaced at even intervals, and subject to considerable care, including fertilization, irrigation, and protection from diseases and herbivores. This points to a distinction between what might be a pragmatic method to project growth for a highly managed plantation and what is a reasonable method to project future conditions in a mixed-species, mixed-age-class forest subject to time-varying conditions. It is the latter kind of forest that is the primary concern of this book and of the tests discussed in this chapter.

Goals of This Chapter

Although I would like to provide tests that would be definitive for all behavioral ranges of the model, this is not possible. Only rarely can data adequate to test the model against a specific case history be found. Therefore, the purpose of this chapter is to illustrate the kinds of tests that have been possible and that I have attempted over the years that I have used the model, and to examine what these tests tell us about the accuracy of the model. I urge the reader to study the tests provided here and those reported by others, and then seek sources of data not available to me, and conduct additional tests of the model. Only by testing the model until it bends and breaks can we understand the model's limits and gain the peculiarly interesting insights that arise only from the case-study failure of what seems to be otherwise a realistic and accurate model. Indeed, in spite of the widespread use of the model, there have been comparatively few quantitative evaluations of its projections against on-going observations.

Some Basic Issues Concerning Quantitative Tests of a Model

Before considering the correspondence between JABOWA and observations, it is useful to comment on some terminology, because in ecology the terms *validation* and *verification* have been used with various connotations. Sometimes ecologists use the word *validation* to mean a test of a model or theory against reality, but I prefer to limit the use of this term to its standard meaning in logic and mathematics, which has to do with the logical consequences of a set of premises. A model is valid if it is logically correct; it is verified if it is true. However, because the terms are not yet standardized, I will simply avoid them and discuss quantitative comparisons of the model with observations.[1]

There has been more experience in ecology in tests of statistical validity—the correspondence between two sets of data, or the variance associated with a single set—than there has been in the testing of models. In comparing a model against the real world, we need to shift our point of view somewhat from the perspective of these statistical tests. As I pointed out in Chapter 1 and repeat here, a primary difference in emphasis between standards and criteria for models and those for statistical analysis is the importance of the conceptual basis—explanations in terms of causes and effects—in the criteria used to accept a model. Statistical correlations are sometimes referred to as "statistical models," but typically the correlations are done "blind"—we accept the equation that gives the best fit whether or not this matches some preexisting idea of cause and effect. Sometimes we use the resulting equation as an hypothesis after the fact, a posteriori, and try to develop a cause-and-effect argument for its occurrence.

In contrast, a model such as JABOWA serves as a set of hypotheses. Running the model shows us the implications of those assumptions. We can test both the assumptions and the implications against observations. In the first case we verify the assumptions. In the second case we verify some property, or some range of dynamics, of the model.

As ecologists have learned since the development of computer simulation, verification of hypotheses about ecological systems is a complex process, never completed. The idea of verification seems simple and straightforward: make a prediction, obtain observations, compare observations with predictions. Then go on to test all possible cases for a model. If the observations follow the prediction, the model is verified. If they do not, the model is falsified. The reality is very different, as some earlier papers that confronted these issues made clear (Caswell, 1976; Mankin et al., 1975). Ecological systems are so complex and can occupy so many states that a simple, single test that allows us to accept or reject completely an ecological model does not exist. Analytical models that are open to definitive tests are generally too unrealistic to be useful, and in the history of ecology the discrepancies between observations and theory have been ignored as much as they have been used to reject models. A case in point is the Lotka–Volterra model of predator–prey interactions, which most observations contradict. This has not halted its use over many decades, as I discuss elsewhere (Botkin, 1990).

Ecological systems can occupy a great many states, and we cannot expect to test the model for all possible states. It is more realistic to select a subset of states that are of particular interest and test the model for these. In some cases, our concern is with future states that have never existed, as with forests in a globally warming climate, and for these there is no traditional verification. Our only recourse is to conduct tests that show that the model is accurate and realistic for some existing subset of states that give us faith in the behavior of the model.

Perhaps in the future we will understand the connections between aspects of the dynamics of ecological systems so that we can devise more general tests of ecological theory. But at present all we can do is test the model against available observations and infer from these limited conditions whether the model appears realistic and/or accurate; by doing enough such comparisons, we can determine whether the model seems useful.

With applied problems, verification standards are sometimes clearer. Practical considerations may require that we know whether a specific aspect of an ecological system is understood and predictions are possible, and these practical considerations may provide requirements for accuracy. For example, a 10-percent change in production might be crucial in economic planning, and projections with 10-percent accuracy might be necessary.

An additional distinction is useful between quantitative tests of models used in scientific research and models used in applications. In basic research, the failure of a model can be as useful as its success, because the failure gives us insight into the characteristics of an ecological system. In basic research, our understanding increases both when observations agree and when they disagree with projections of the model. In fact the failure of a prediction may be more interesting and useful to us for a model that is otherwise realistic. The greater the success of a model, the more interesting and useful are its failures. The differences between observations and predictions help us to see the characteristics of the real system, to investigate causes and effects, and to improve our understanding of natural systems. JABOWA provided such insight when, as I mentioned earlier in Chapter 4, its projections of biomass accumulation during succession were different from earlier theories. Where JABOWA is used in scientific research, one of our goals is to find out where the model bends and breaks, to see where there is a difference between observations and predictions. In applications of the model to practical problems, such as those that will be discussed in Chapters 7 and 8, we seek accurate and realistic projections. Tests against quantitative observations increase our faith in the accuracy and realism of the projections.

In testing a model, the first question we ask is what variables to choose as outputs to compare with observations. In the study of forests, several variables are of interest, including density, basal area, biomass, and merchantable volume. However, JABOWA operates from a basis in tree density and individual tree diameter, so that where possible it is useful to test the model using these measures.

The previous chapters show that JABOWA is a realistic model of forest growth, of forest succession, and of changes in forests with climate and soils. The question addressed in this chapter is how quantitatively accurate the model is.

Over the years I have pursued quantitative tests of the forest model and found the location of appropriate data much more difficult than I expected at the outset. Perhaps it will be most instructive to the reader if I provide some of that personal history, as a way to help guide future attempts.

Tests of Accuracy

As I discussed in Chapter 1, forests are highly variable, and it is likely that two forests that appear identical will differ in any measure of interest by 10 or even 20 percent. For example, it is difficult to measure an effect of removal of 10 or 20 percent of the leaves and twigs by herbivores on total tree growth. This provides an expectation for tests. As another example, our recent measurements of forest biomass for large areas have a 95-percent confidence interval that is about 20 percent of the mean (Botkin and Simpson, 1990a,b). If natural systems vary from one another by 10 or 20 percent, we should not demand more of the model: we should not demand that its projections be closer to the observations than 10 or 20 percent. From this line of argument, over the years I have considered acceptable quantitative performance of the JABOWA model to be that the 95-percent confidence interval was within 10 or 20 percent of observations.

The lack of data against which we could test the model became evident in the early 1970s as soon as JABOWA had been developed. At that time I looked for data against which to test its projections and found that there were few forest stands for which an accounting of the diameter of each stem had been published or recorded. In New England, for which the model was originally developed, there were less than a handful for which such counts had been repeated in a regular manner for more than a decade. The U.S. Forest Service maintains a few permanent plots, but for one reason or another none of the data from these plots were adequate. Typically either one remeasurement had been missed, the methods of measurements had changed, or the plot was large and trees were not located within the plot, so that one could not determine how much an individual tree had grown during the time between measurements. In other cases, measurements of tree diameter excluded small trees (typically trees less than 12 cm or 5 inches in diameter), so that the recruitment projected by the model could not be tested.

At the time that JABOWA-I was developed, it was also difficult to use vegetation data gathered directly in the Hubbard Brook Ecosystem Study in validation; at the time there was only one time when the current vegetation conditions had been measured on one watershed, and the history of that forest was at the time not well known.

The forest had apparently been logged twice, once about 1909 when large spruce trees had been removed and again about 1916–1917 when hardwoods and the remaining spruce had been cut.* Analysis of growth rings showed that loggers had left a forest of scattered large hardwoods of beech, birch, and maple. Although the area was probably not completely clear-cut, cutting was sufficient in some areas to allow establishment of cherry and birch (Bormann et al., 1970). In spite of the lack of detailed history, and the possibility of many sources of variation, early comparisons of the first version of JABOWA with Hubbard Brook data were encouraging. Mean values for plots simulated with version I of the model for both years 50 and 60 were within 1 standard deviation of the mean of the 1966 vegetation survey (Bormann et al., 1970). Some 1966 Hubbard Brook vegetation survey plots show characteristics very similar to those of the model. For example, a plot at 553-m elevation with till depth of 1.5 m and 1 percent rock had a total of 18 stems and total basal area of 3658 cm². The average total basal area projected by the model for 100 plots with initial conditions identical to that plot but at 610 m is 3250 cm² at year 50 and 3660 cm² at year 60. The correspondence between projections and observations was poorer for individual species. On the real plot, five yellow birch trees contributed 2376 cm² while the model projected 1400 and 2000 cm² for yellow birch. Six beech contributed 1005 cm² while the model projected 100 and 150 cm² for beech. Two spruce contributed 89 cm², two sugar maple 16 cm², with the remaining contributed by understory species.

Since these tests used version I, the fertility of the soil was not considered. Although there were differences between the projected and observed, the model's projections were sufficiently close to be encouraging. The differences between projected and observed could have been due to logging history or to other unknown historical factors. But perhaps also, any two adjacent plots with the same initial conditions and environmental history could differ by as much as the projected and the observed plots.

My next approach to test the model was to find single point measurements from uncut stands. These data could at least test the model's projections for very old (i.e., long undisturbed) forests. Unfortunately, few uncut stands remained in New England. In New Hampshire only two were known, one called Nancy Pond, which was a high elevation spruce stand in the White Mountains. Another,

---

* This paragraph is taken verbatim from Botkin et al., 1972b.

called The Bowl, was a larger area at lower elevation also in the White Mountains. We obtained some plot measurements for both sites. There were also a few published studies of uncut stands in Vermont (Bormann and Buell, 1964).

Version I of the model predicted that white birch would be present at elevations below 762 m during early stages of succession and red spruce would be an important member of old-age stands below this elevation.* Although spruce, fir, and white birch did not occur in the Hubbard Brook vegetation plots in the 1966 survey below 709 m, these species grow at lower elevations in nearby locations, and were apparently more important in the past. An early report by Chittenden (1905) on conditions of undisturbed forests in northern New Hampshire stated that white birch grew at all elevations but was much more important at higher ones, and that red spruce was "the characteristic tree" of the northern part of this area regardless of elevation. Forests between 550 and 1070 m were "Characterized by the prevalence of spruce in mixture with balsam and yellow and paper birch" (Chittenden, 1905). This report attributed an increase in hardwoods to the effect of lumbering. Thus the long-term projections of version I are consistent with what is known about the original forest cover of New Hampshire, although they were contrary to natural history observations in the mid-twentieth century.

An interesting difference between projections of version I and reports of real forests is the distribution of stems by size class in stands 50 to 60 years old. Three reports, one of an old-age stand in Vermont and one of a 25-year-old stand in New Hampshire (Marquis, 1969), and the report discussed previously by Chittenden, agree on the following: stems of size 2 to 10 cm are more numerous than stems of 10 to 20 cm or greater than 20 cm, and the number of 10 to 20-cm stems is not greatly different from the number of larger ones. Version I of the model projects that the number of stems 10 to 20 cm at year 50 and 60 would be four to five times the number of larger stems. However, projections for stands older than 70 years agree with the observations just mentioned. In this case, the model's projections appear to lag the observations by a decade.

In the intervening years I have continued to search for acceptable data sets. Papers by Siccama (1988) and Siccama and Vogelmann (1987) have been helpful in locating existing plot data, but few of these provide adequate information for a test of the model.

Tests of the Model against Reconstructions of Short-term Forest History

Given the limited published data on forest growth, Janak, Wallis, and I decided that we should attempt to obtain our own data to test the model. To make the test broader than the Hubbard Brook Experimental Forest, I decided to obtain plot samples in representative stands throughout the White Mountains of New

---

* This paragraph and the next are taken verbatim from Botkin et al., 1972b.

Hampshire. General locations were selected to represent the range of sites. At each site, the corner of a plot was chosen at random to eliminate local bias. The plan was to measure the current condition of the forest on 10-by-10 m plots and also core every tree on the plot to recover the last 10 years' growth record and to age the stand (based on the age of the oldest tree). From this growth record the plot could be set back to its condition 10 years earlier and, assuming no mortality of trees greater than 2 cm in diameter had occurred in that time interval, the model could project the growth of that plot forward to the present. The test would be a comparison of the present conditions of the plot as measured and as projected by the model from the estimated condition 10 years earlier.

During 1973 I sampled more than 50 plots. There are several limitations to these data. The plots were located in back country far from weather stations, so that weather records were extrapolated from those in Woodstock, New Hampshire, using only elevational corrections. These follow standard lapse rates as discussed in Chapter 2. The test was incomplete because trees that had died and fallen within the last 10 years were not accounted for, and on many plots some trees were rotten or diseased and it was impossible to obtain cores from which growth rings could be established. While plots with these data gaps were useful for other purposes, the intended comparisons of the model with observations were not possible for such plots.

Table 5.1 gives results for six of the plots (those with the most complete data). To test the projections of the model for the growth of the trees, regeneration and mortality were turned off and only the deterministic growth routine was used.

In most cases, the estimates of basal area are within 10 percent of the observed. The projections were least accurate for balsam fir, more accurate for sugar maple, yellow birch, pin cherry, spruce, and white birch. These results indicated that version I was accurate within the crieteria I had previously established for total basal area and for most species. The results suggested that parameters for balsam fir in JABOWA-I were incorrect or the least accurate, and they have since been corrected.

Projections of individual tree diameters fall generally within 10 to 20 percent of the observed (Table 5.1B). One expects that projections for individual tree diameters would have a greater error than projections for basal area by species, since the overprojection and underprojection for individual stems would tend to average out.

The comparisons in Table 5.1 are between the final size of a tree or the final basal area of a species. Another comparison is between projected and observed changes in diameter. However, over a 10-year period, diameter growth is small and, lacking detailed information about climate, a stringent demand is placed on the model. On average, the model projects diameter growth (change in final diameter) within 100 percent of observations.

Another limitation of these tests is the short time scale involved. The idea behind this test was to use a short enough time interval so that we could assume that no trees present on the plot 10 years ago had died, decayed, and disappeared and could no longer be accounted for in our survey. That would give us a

TABLE 5.1. JABOWA validation run (deterministic): Projected and observed change in basal area and in tree diameters, White Mountain forest plots. Predictions of the model are compared with measured conditions on forested plots in the White Mountains of New Hampshire. Plots were chosen to be representative of a variety of conditions, but within a chosen locale the corner of a plot was located at random. On each plot the species, diameter and height of each tree greater than 2 cm were recorded. An increment core was taken from each tree and the most recent 10 years' growth rings were measured. The size of each tree 10 years earlier was calculated by subtracting the last 10 years' growth from the current radius. Using a deterministic version of JABOWA-I, the growth of the trees from 10 years earlier to the time of measurement was projected. The first column gives the calculated basal area of each species on a plot in year 1. The second column gives the projected change in basal area for each species on that plot at year 11. The third column gives the observed basal area in year 11 (the year of actual measurement).

Plot descriptions

A. *Plot 6*: Northeast aspect and slope of 18%. Hardwoods and softwoods (particularly sugar maple) were regenerating. Main species red spruce, balsam fir, sugar maple, white birch; other species scattered throughout plot. Average soil depth 20 cm; maximum soil depth 70 cm.

B. *Plot 8*: Southwest aspect and slope of 35%. Dominated by sugar maple, yellow birch, and red spruce. Two beech were located just south of plot with thick regeneration beneath. Balsam fir, white birch, mountain maple, and striped maple also present. Average soil depth 20 cm; maximum soil depth 70 cm.

C. *Plot 10*: South-facing slope of 35% with numerous large trees (mainly spruce and yellow birch) and lots of seedlings and small saplings (spruce, beech, and sugar maple). No yellow birch regeneration seen. Fir was uncommon on plot. Soil very rocky, with depth ranging from 10 to 50 cm.

D. *Plot 20*: Slope of 25% with southern aspect at top of a knoll. Red spruce and white birch were dominant species; several others were moderately common. Regeneration of fir and striped maple common; that of birch, fairly common; and that of spruce, beech, and mountain maple uncommon. Soil depth ranged from 15 cm to 1 m, with few rocks.

E. *Plot 21*: Plot with slope of 5%, dominated by red spruce, with balsam fir also present in fair numbers and occasional white birch. Canopy fairly open (many dead stems on ground, apparently blown over after having succumbed to suppression). Regeneration was abundant—mostly fir, but also spruce, white birch. Red maple seedlings common. Soil depth ranged from 20 to 80 cm; no rocks.

F. *Plot 31*: Plot on small level knoll with southern aspect. Few rocks; soil depth ranged from 5 to 30 cm. Beech dominant and also only species that showed good regeneration. Spruce saplings common, but all large spruce were dead. Pin cherry and yellow birch appeared to be dying out. Some yellow birch regeneration (small seedlings) on rotting spruce log.

Plot environmental conditions

Plot number	Elevation (m)	Soil Depth (m)	Rock (%)	Growing degree-days	Evapotranspiration index
6	546	0.2	5	3188.2	235.4
8	838	0.2	4	2593.9	232.6
10	506	0.3	15	3268.6	303.1
20	664	0.3	2	2946.8	336.3
21	625	0.4	0	3027.2	370.9
31	686	0.1	1	2903.4	180.4

TABLE 5.1. (*Continued*)

(A) Basal area

Species	Total basal area year 1 (cm^2)	Observed basal area year 11 (cm^2)	Projected basal area year 11 (cm^2)
Plot 6			
White birch	873.0	1075.8	1136.1
Yellow birch	135.1	169.7	167.5
Balsam fir	957.9	1044.6	1466.8
Mountain maple	27.4	31.4	53.4
Sugar maple	712.6	939.3	1082.3
Red spruce	649.2	776.5	718.4
Totals	3355.6	4037.3	4624.7
Plot 8			
Yellow birch	679.6	820.3	818.9
Balsam fir	2.5	2.5	45.5
Mountain maple	10.0	15.2	16.2
Sugar maple	870.4	1151.8	1130.3
Red spruce	153.6	188.6	228.0
Totals	1716.3	2178.5	2239.1
Plot 10			
Beech	116.3	169.4	215.2
Yellow birch	829.7	1008.9	1013.4
Balsam fir	26.4	31.1	66.9
Sugar maple	485.4	606.9	675.5
Red spruce	1889.6	2063.8	2025.2
Totals	3347.3	3380.3	3996.4
Plot 20			
White birch	801.9	1106.3	1304.3
Yellow birch	321.7	466.2	387.3
Pin cherry	419.5	647.8	595.1
Balsam fir	447.3	597.4	904.4
Red spruce	1197.9	1574.2	1465.7
Totals	3188.5	4392.0	4657.0
Plot 21			
White birch	214.4	257.8	387.9
Balsam fir	356.8	740.1	686.5
Red spruce	3057.6	3726.4	3551.0
Totals	3628.9	4724.4	4625.5
Plot 31			
Beech	1907.1	2827.4	2505.7
Yellow birch	82.5	133.4	118.6
Pin cherry	108.2	193.9	173.7
Balsam fir	4.9	7.5	39.6
Striped maple	13.4	51.5	23.9
Red spruce	104.1	149.9	146.8
Totals	2220.6	3363.6	3008.6

TABLE 5.1. (*Continued*)

(B) Individual tree diameters by species (cm) for three plots

Species	Observed year 1	Observed year 11	Projected year 11
		Plot 6	
White birch	23.2	26.4	26.2
	10.7	13.9	13.0
	20.3	21.9	23.3
Yellow birch	13.1	14.7	15.1
Balsam fir	3.4	4.9	7.3
	29.7	30.4	33.7
Mountain maple	2.8	3.6	4.8
Sugar maple	18.2	19.7	22.4
	18.2	20.8	22.4
	5.2	8.7	8.2
	14.0	15.5	17.8
	4.3	7.7	7.1
Red spruce	26.7	29.1	27.9
	2.5	4.1	3.3
	2.9	2.9	3.6
	10.1	10.8	11.1
		Plot 8	
Yellow birch	22.2	23.2	24.3
	19.3	22.5	21.2
Balsam fir	1.8	7.8	7.6
Mountain maple	3.5	4.4	4.5
Sugar maple	11.7	14.1	13.4
	3.7	5.1	4.9
	12.4	14.1	14.1
	1.8	2.9	2.8
	10.0	12.8	11.6
	8.2	8.6	9.7
	4.5	5.7	5.8
	18.5	20.1	20.3
	0.6	1.7	1.2
	12.4	14.0	14.1
	10.5	12.7	12.1
Red spruce	3.5	4.8	5.1
	2.3	3.5	3.7
	5.3	6.2	7.0
	12.1	12.9	14.1
		Plot 20	
White birch	16.0	20.4	21.2
	14.3	18.5	19.0
	23.6	25.5	29.1
Yellow birch	13.3	16.2	14.6
	15.2	18.2	16.6
Pin cherry	17.0	20.4	20.0
	8.2	11.5	10.2
	13.2	16.5	15.8
Balsam fir	9.7	11.1	14.7
	11.3	12.3	17.1
	18.6	22.1	25.2

TABLE 5.1. (*Continued*)

Species	Observed year 1	Observed year 11	Projected year 11
Red spruce	5.2	6.3	6.0
	12.0	14.2	13.1
	6.4	7.7	7.3
	5.5	7.7	6.4
	8.3	10.5	9.3
	2.5	6.9	3.2
	2.3	3.1	3.0
	7.0	8.8	8.0
	18.3	19.4	19.9
	12.4	15.0	13.7

From Table 5.1, pp. 46–47, and Table 5.2, pp. 50–51 in Botkin, D. B., 1981, Causality and Succession. In West, D., H. H. Shugart, and D. B. Botkin, eds., *Forest Succession: Concepts and Applications*. Springer-Verlag, New York, reprinted with permission.

complete census of a plot at the beginning and end of 10 years. We were concerned that using a longer time period would simply increase unknown values (trees that had died and decayed to the point that they were no longer counted in our survey). This seemed logical at the time, but in retrospect it is clear that a decade is a short time for tree growth, and that diameter increment for most trees over that short a time will fit a straight line fairly well. The nonlinear, stochastic, dynamic properties of the model, those features that were most useful and most interesting, were not really tested by this comparison.

In the intervening years I have continued to search for studies in the boreal and eastern deciduous forests of North America where the diameters of individual trees have been measured for at least three time periods at least one decade apart. Searching for such a data set has been like the Spanish explorers searching for El Dorado—people continually tell me that they know where such a data set exists, although they don't have it themselves, and that if I only call or write a certain person, I will find that data set. But the series of contacts that has resulted from each incident has eventually petered out or, if there is such a set, it suffers from the problems I mentioned earlier, such as that the methods used in the first measurement were not continued in the later measurements, or that individual trees were not located so that only plot totals were given and we could not determine how an individual tree grew. Unfortunately, this has been the case for permanent plots on federal and state forest lands as well those at private research stations. All this points to the need for an important, simple to do, but rarely accomplished task: the creation of permanent forest plots that are remeasured at regular intervals using a standard measurement methodology. One motivation for writing this book is the hope that some readers and users of the model will be motivated to seek, find, or develop such plots.

In the past few years, somewhat longer records have become available. These are the subject of the next section.

Hubbard Brook Experimental Forest 22-Year Record

The Hubbard Brook Experimental Forest provides some of the best quantitative information to test the model. A vegetation sample was made of the primary control watershed (known as watershed 6) in 1965, with 208 10-by 10-m plots located on a regular grid throughout the watershed (Bormann et al., 1971). On each of these plots, the diameter and species of all stems measuring 2 cm or more were recorded. Remeasurements were made three times, in 1977, 1982, and 1987. These data seemed, at first glance, to offer a fairly complete quantitative test of the model. Unfortunately, two characteristics of these data have limited their utility. Since the original survey, the exact position of the original 1965 plots could not be relocated. Subsequent remeasurements were made on larger plots that were believed (but not known exactly) to include the original plots. These data also lacked plot by plot measurements of soil nitrogen, soil depth, and soil texture. For these reasons, the data do not provide a complete, independent test of the model. These data provide a method to *calibrate* the model and, as we shall see, to provide insights into the realism and accuracy of the model.

These data were collected by, and are used here with permission of, Thomas J. Siccama (Botkin et al., in prep.). Each of the 208 plots was taken as an initial condition, and 10 iterations were run for each of 22 years. The 208 plots measured in watershed 6 of Hubbard Brook can be viewed as providing 208 replicates of an initial condition or, when divided into 10 elevational bands, as a set of replicates of approximately 20 replicates each. Tests of the model show that 10 replicates of each of these plots are sufficient to reduce the variance to an asymptotic percentage of the mean, and therefore 10 replicates were run for each plot in the tests reported here. The model projected the growth of 10 replicates of each plot, with output provided for each year for 22 years. A time series of temperature and rainfall was obtained from the Bethlehem, New Hampshire, weather station, modified by standard climatic lapse rates. Because the field observations did not include plot by plot measurements of soil nitrogen content, a series of tests was run for a range of soil nitrogen from 30 to 90 kg/ha in increments of 1 kg/ha. Soil texture (in terms of millimeters of soil moisture holding capacity per millimeter depth of water) was assumed to be 150 mm/m. Soil depth was measured for each plot. Once the projections were made, the percent difference between projected and observed was calculated for each case as $[(P - \text{Obs})/\text{Obs}] * 100$.

For the 22-year (1987) remeasurement, the overall best fit for all elevations for biomass occurs at nitrogen values of 63 to 64 kg/ha, for which the projected and observed differed by ± 0.9 percent (Table 5.2). The overall best fit for density occurred at nitrogen values of 47 to 48 kg/ha. Projections using nitrogen values between 58 and 70 kg/ha were within 5 percent of observed biomass, while values between 43 and 51 kg/ha were 5 percent of observed density. Projections using nitrogen values between 55 and 75 kg/ha were within 10 percent of the observed biomass, while projections using nitrogen values between 40 and 55 kg/ha were within 10 percent of the observed density. These ranges are large enough to suggest that the model is only moderately sensitive to soil nitrogen content (Table 5.2). However, the results are inconsistent with respect to density and biomass. Using the results as hypotheses, the model projects a lower expected value of soil nitrogen from stem density than from biomass.

Accuracy varied considerably with elevation zone. The best fit for density, basal area, and biomass occurred for decreasing values of nitrogen with increasing elevation, although there is some variability (Table 5.3). The most accurate projections are for density, the least accurate for biomass. This is not surprising, because biomass is a derived factor, calculated from tree diameters using dimensions analysis procedures that have their own errors and add to the total error of prediction.

Table 5.4 shows the difference between projected and observed densities for each of the three remeasurements, 1977, 1982, and 1987. For all elevations, the best fit occurs at a nitrogen value between 47 and 54 kg/ha, which is in good agreement. We can conclude that the model projects that 50 kg/ha is a reasonable average value for nitrogen in the watershed based on density measurements, and that the model can project densities within 1 percent for specific nitrogen values.

Given my experience in the field, I was surprised at the accuracy of these projections. As I said earlier in this book, forests appear highly variable, and a difference of 10 percent in biomass between two plots that appeared identical would not surprise me. Even though this is a test in which the projections are fit to the observations by changing the nitrogen values, it is surprising to me that any projections are within 1 percent of the observed. Some readers might reply that any linear fit over a range of values is bound to pass through the data at some point, but it is in the abstract just as likely that all values would miss all the data.

A nitrogen value of 63 kg/ha *underprojected* the biomass at the lower four elevations (and was 15% or more in error for the lower three elevational zones), and *overprojected* the biomass for four of the five upper zones. This tendency is general throughout the watershed. The best fit for the lower three zones is near the maximum value tested, 88 to 89 kg/ha, while the best fit for the highest zone occurs at nitrogen level 44 kg/ha. (Table 5.3C). The best fit for zone 8 is at a nitrogen level between 51 and 53 kg/ha.

Thus the comparison of the model with the 22-year Hubbard Brook data yields several testable hypotheses:

H 1. Nitrogen content of the soil decreases with elevation.
H 2. The average available nitrogen content of the soil is approximately 63 kg/ha, with a range of 58 to 70 considered a reasonable expectation.
H 3. The available nitrogen content of the upper elevations is approximately 44 to 53 kg/ha, while that of the lowest elevations is approximately 88 kg/ha or greater.

These hypotheses can be tested, and they offer a new method to test the model. The first hypothesis is, in general, consistent with natural history

TABLE 5.2. Comparison of projected and observed values, all plots, Hubbard Brook watershed 6, 1987 data. Results of a comparison between projections of JABOWA-II and observed forest growth. The original vegetation survey was conducted on 208 plots in 1965. Resurveys did not locate the original trees, but included the original area, so the test is not complete. This table gives observations and projections after 22 years. Since the original survey did not include measures of soil nitrogen, projections of the forest model were conducted using a series of nitrogen values from 30 to 90. For each nitrogen level, replicates of the model were run for each of the 208 plots (each plot used as an initial condition) and the values averaged. Each column gives the percent difference between projected and observed measures as: (observed − projected)∗100/observed. The right-hand column uses the sum of density and basal area.

The observations were divided into 10 elevation zones, but in this table all the observations were used. Other comparisons were done between the model's projections and specific elevational bands.

Available Nitrogen	Density % Diff.	Basal Area % Diff.	Biomass % Diff.	Available Nitrogen	Density % Diff.	Basal Area % Diff.	Biomass % Diff.
30	− 86.7	− 93.9	− 95.5	60	13.8	− 3.1	− 3.4
31	− 86.7	− 93.7	− 95.4	61	14.3	− 3.1	− 3.2
32	− 49.3	− 60.3	− 61.7	62	16.3	− 1.6	− 1.8
33	− 30.3	− 42.6	− 43.1	63	17.7	− 0.7	− 1.1
34	− 23.3	− 38.2	− 39.1	64	18.3	0.7	0.9
35	− 17.4	− 33.6	− 34.6	65	18.2	− 0.4	− 0.9
36	− 16.6	− 33.4	− 34.5	66	19.9	2.6	2.6
37	− 14.7	− 31.6	− 32.8	67	21.4	4.3	4.2
38	− 12.8	− 30.3	− 31.3	68	23.6	4.4	4.3
39	− 11.4	− 28.6	− 29.5	69	23.1	3.1	2.6
40	− 9.1	− 26.2	− 27.2	70	23.1	4.4	4.5
41	− 9.0	− 26.9	− 27.9	71	25.0	7.4	7.3
42	− 7.4	− 24.7	− 25.6	72	24.6	6.8	6.9
43	− 4.6	− 22.9	− 23.7	73	27.8	9.2	9.3
44	− 4.0	− 21.2	− 21.8	74	26.3	9.4	9.9
45	− 2.4	− 20.0	− 20.8	75	28.1	10.0	10.0
46	− 1.4	− 19.2	− 20.0	76	29.6	10.8	11.2
47	− 0.7	− 17.8	− 18.4	77	29.8	11.2	11.4
48	0.9	− 17.8	− 18.7	78	30.3	11.5	11.7
49	2.0	− 16.1	− 16.9	79	31.4	12.3	12.6
50	3.5	− 14.0	− 14.5	80	31.8	13.1	13.8
51	4.5	− 13.7	− 14.3	81	33.3	15.1	15.7
52	6.1	− 10.6	− 10.7	82	32.6	14.9	15.5
53	7.5	− 11.2	− 11.6	83	33.4	15.3	15.6
54	7.0	− 9.9	− 10.5	84	34.6	17.0	17.7
55	8.9	− 8.8	− 9.3	85	34.3	17.1	17.8
56	10.2	− 7.2	− 7.4	86	34.7	16.9	17.7
57	10.0	− 7.7	− 8.1	87	35.7	18.4	19.0
58	11.0	− 5.7	− 5.7	88	37.2	19.4	20.1
59	13.1	− 4.7	− 4.8	89	36.1	19.2	19.9
				90	38.6	20.0	20.6

information. The highest elevations are dominated by boreal forest conifers that produce litter with a low nitrogen concentration. Comparatively cold and wet conditions of the upper elevations might not be conducive to nitrogen fixation by free-living soil bacteria. In contrast, northern hardwoods that dominant the lowest elevations produce litter with a high nitrogen content, and the somewhat warmer and drier conditions might be more conducive to nitrogen fixation in the soil. Thus this prediction is consistent with natural history ideas about a

TABLE 5.3. Comparison of projected and observed values by elevation zone, Hubbard Brook Watershed 6, 1987 data. Results of tests described in Table 5.2 for nitrogen values that give best fit of model to data for each of 10 elevational zones in watershed 6.

Elevation Zone	Available Nitrogen (kg/ha)	Density % Diff.	Basal Area % Diff.	Biomass % Diff.
		Best Fit for Density		
1	65	0.0	− 13.8	− 17.1
2	90	− 10.4	− 17.5	− 14.5
3	70	0.4	− 6.4	− 5.3
4	43	− 0.1	− 27.2	− 28.3
5	36	− 1.6	− 18.9	− 22.2
6	47	0.0	− 14.1	− 13.9
7	41	1.4	− 21.5	− 24.7
8	33	2.8	− 31.9	− 36.8
9	35	0.6	− 31.9	− 33.6
10	33	− 6.2	− 33.1	− 29.8
		Best Fit for Basal Area		
1	76 & 80	11.8	− 0.9	− 2.9
2	90	− 10.4	− 17.5	− 14.5
3	88	10.4	0.0	1.3
4	71	32.4	− 0.1	− 2.2
5	51	11.0	− 0.9	− 2.8
6	59	17.0	− 0.2	0.2
7	55	29.6	0.0	− 3.0
8	52	35.0	2.7	− 0.4
9	58	52.0	− 0.5	− 3.8
10	51	30.3	− 0.2	2.8
		Best Fit for Biomass		
1	85	12.7	2.0	0.0
2	90	− 10.4	− 17.5	− 14.5
3	81	5.0	− 2.5	− 1.2
4	68	34.6	1.2	− 0.3
5	53	15.8	1.9	− 0.2
6	59	17.0	− 0.2	0.2
7	58	34.3	2.4	− 0.7
8	52	35.0	2.7	− 0.4
9	61	53.2	3.8	0.9
10	44	28.0	− 2.8	0.5

forest in New Hampshire, but the model was not constrained a priori to reach such a conclusion.

Interestingly, Huntington, Ryan, and Hamburg (1988) carried out a nitrogen inventory in a Hubbard Brook watershed next to the one used in the tests just

TABLE 5.4. Comparison of projected and observed values, Hubbard Brook Watershed 6, best fit for density, 1977, 1982, 1987. Results of tests described in Table 5.2 for nitrogen values that give best fit of model to data for each of 10 elevational zones in watershed 6.

| | Best Fit for Density | | | |
Elevation Zone	Available Nitrogen (kg/ha)	Density % Diff.	Basal Area % Diff.	Biomass % Diff.
		1977		
All	54	0.3	− 12.4	− 12.8
1	80 & 83	0.3	− 13.6	− 17.8
2	90	− 7.5	− 17.6	− 18.0
3	73	0.1	− 11.9	− 14.2
4	44	− 0.2	− 23.9	− 25.7
5	50	− 0.1	− 5.5	− 7.2
6	53	0.3	− 11.8	− 11.6
7	45	0.2	− 11.7	− 13.3
8	33	3.9	− 23.7	− 27.3
9	39	− 0.3	− 25.3	− 25.3
10	33	2.7	− 19.0	− 15.6
		1982		
All	50	0.1	− 14.9	− 15.4
1	78	0.0	− 14.2	− 17.6
2	89	− 8.6	− 17.5	− 17.8
3	67	− 1.3	− 14.0	− 15.0
4	42	0.4	− 25.8	− 27.2
5	51	0.2	− 6.7	− 8.5
6	49	− 0.6	− 14.0	− 14.0
7	39	− 0.2	− 17.8	− 20.4
8	33	9.5	− 24.5	− 29.0
9	38	0.6	− 32.9	− 34.1
10	33	3.5	− 22.2	− 19.8
		1987		
All	47	− 0.7	− 17.8	− 18.4
1	65	0.0	− 13.8	− 17.1
2	90	− 10.4	− 17.5	− 14.5
3	70	0.4	− 6.4	− 5.3
4	43	− 0.1	− 27.2	− 28.3
5	36	− 1.6	− 18.9	− 22.2
6	47	0.0	− 14.1	− 13.9
7	41	1.4	− 21.5	− 24.7
8	33	2.8	− 31.9	− 36.8
9	35	0.6	− 31.9	− 33.6
10	33	− 6.2	− 33.1	− 29.8

TABLE 5.5. Nitrogen concentration (kg/ha) in a Hubbard Brook watershed[a]

Elevation	Nitrogen Concentration		
(m)	Forest Floor	Mineral Soil	Total Solum
510–560	1200	6100	7300
560–610	1500	6000	7500
610–660	900	5200	6100
660–690	1000	5600	6400
690–730	1600	5600	7800
730–750	1500	6200	7600

[a] Samples were obtained from 71 × 71 cm blocks in watershed 6.

From Table 6 in Huntington, T. G., D. F. Ryan, and S. P. Hamburg, 1988, Estimating soil nitrogen and carbon pools in a northern hardwood forest ecosystem. *Soil Science Society of America Journal* 52: 1162–1167.

described. I did not examine their results until after tests were completed and the interpretation just given had been written down. Contrary to the prediction of the model, the observed inventory shows that total nitrogen is higher in the soil on the ridge than lower down (Table 5.5). Minimum total soil nitrogen content occurred in the 610- to 660-m elevational band, while the values in the 560- to 610-m and 730- to 750-m (highest elevation) are similar.

However, trees respond not to total soil nitrogen content reported in that paper, but to available nitrogen, which is represented more accurately by amounts added in fertilization experiments (Mitchell and Chandler, 1939) and by mineralization rates, whose range approximates the 30 to 90 kg/ha used in the tests conducted here (Pastor, Gardner, and Post, 1987).

In addition, it might also be that differences in land- use can alter the natural soil nitrogen availability. The projections from the forest model assume that all elevations were cleared at the same time and have experienced the same human land use since then. Some previous studies suggest that as much as half the forest floor nitrogen can be lost in the first 5 to 15 years following logging (Covington, 1981; Federer, 1984). Heavier or more recent logging at the lower elevations than at the higher might explain the observed differences in a way that would be consistent with the projections of the model. The observed pattern is not yet explained, and the difference between the projection of the model and these observations is of interest. Here projections of the model serve a useful purpose in helping to stimulate an interest in additional research as a way of explaining the discrepancy between observed and predicted.

Another point that can be raised is that one could vary other variables in the model, such as soil-water content (or the factors that determine it, such as evapotranspiration), and possibly obtain an equivalent fit from an independent, alternative hypothesis. Those tests have not yet been carried out, but could be done by users of the software that is available as a companion to this book. This example serves to illustrate the difficulties in what seem at the outset to be a comparatively straightforward test of a computer simulation of a complex system.

Other investigators have sought similar tests of other versions of the model against observations. Those are not reviewed here because, as I explained in the preface, the purpose of this book is to provide the reader with a specific discussion of the versions of JABOWA as I have used them. In part the motivation for this is to keep the discussion within reasonable bounds in terms of the number of equations and within a reasonable length for the entire book. In part the motivation is to focus the discussion in a way that will be most useful to those who want to apply the software that contains version II and is available as a companion to this book.

Some readers may be interested to know that, in addition to attempts to achieve direct quantitative tests such as those described in this chapter, and in addition to the other comparisons reviewed in Chapter 4, over the years there have been many informal comparisons between the natural history expertise of individual scientists and the model. These are anecdotal, and form an entirely different category of tests of the model than the primary focus of this chapter.

Version I provided output, if requested, for the age and size of every tree at death. The original reason we added this output was to develop information on the maximum size projected to be reached by trees under specific climates. The model was designed to be general, so that parameters for a species would apply throughout its range. A question that Janak, Wallis, and I expected to be asked was whether the model would realistically reproduce the maximum size reached by trees in suboptimal parts of the range. For example, the largest sugar maples (and many other eastern deciduous forest species of North America) are found in the southern Appalachians, including the Great Smoky Mountains. The maximum diameter parameter for sugar maple applies to a tree from that region.

Given that potential size, what actual size would sugar maple reach in the harsher environments of New Hampshire? The answer we found was that the model projected that sugar maple reached smaller maximum sizes at death under Woodstock, New Hampshire, weather records than allowed for by the D_{max} parameter, and that these maximum sizes were consistent with those observed in the Hubbard Brook Experimental Forest. Although we expected to be asked the question about the maximum size reached by trees, it was never asked, and the results were never published. Because of the lack of interest in such output, and because such output rapidly fills up disks, this information is not available in the current software. It is a simple matter, however, to add the code required to provide such output to the model, if the need were to arise.

Summary

In Chapter 4 we found that the model is realistic, especially in the sense that it reproduces changes in forest composition over time and during ecological succession that are consistent with natural history observation, and in the sense that we showed soon after the model was developed that its projections were consistent with available published data about existing forests. In this chapter we have found that tests that go beyond that level, and seek quantitative comparison between projections of the model and observations, are severely limited by the lack of empirical observations, especially long-term monitoring of forest stands. Because of these data limitations, the forest model has not been tested quantitatively to the degree that I expected given the extensive use that it has had. One of the most important needs in the future for modeling of forest growth is, ironically, simple but repeated monitoring of forests.

Where data are available for quantitative tests, the model appears reasonably accurate. Ironically, tests of the model against the real world reveal limitations of data and lack of long-term measurements of forests more than they reveal lack of capability of the model. The model provides output that is more precise than observations. In all cases for which some tests have been possible, the data required for a complete test of the model are not available. In the end, only partial tests of the model are possible at present, but these give confidence that the dynamics of the model are realistic and accurate, generally within 10 percent of observed diameter and basal area.

The lack of continuing programs in monitoring of forest conditions is a serious problem for the science of ecology and for the management of biological resources. While there are obvious difficulties in creating such programs, they could play a fundamental role in our attempts to improve our understanding of ecosystems. The irony is that the model provides abundant hypothetical data, little of which can be tested against observations. Perhaps there are additional useful data in the records of private industries that could be applied to this task. One of my hopes in writing this book is that some readers will be motivated to seek or create such empirical studies.

Note

1. Some ecologists active in modeling use validation in a specific sense to mean tests of the model against independent data. Verification is sometimes used to mean ensuring that calculations are done properly, in contrast to the more traditional use meaning the test of the truth of something.

6

Sensitivity of the Model to Errors
in Observations and in Concepts

The Purpose of Sensitivity Tests

In earlier chapters I have shown that the response of the model changes with values of parameters and environmental conditions. How sensitive the model is to these changes is important information. Superficially, it might seem that the best model would be the least sensitive. But a model that was completely insensitive to parametric values could not distinguish among species—all species would respond the same way. On the other hand, a model that was very sensitive to variations in the value of a parameter would be impractical, because it is difficult to estimate the real quantity of the parameters, and a model that was too sensitive would give widely varying results with small changes in parameter estimations. As I have said before, but which bears repeating here, in my field experience, even under the best of conditions, estimations of parameters from field measurements are likely to be in error by 10 percent or more. It is therefore useful to test the sensitivity of projections against errors at 10 percent. If we cannot measure parameters more accurately than within 10 percent, then we should demand no more from the model.

This discussion suggests that the most useful model would have an intermediate sensitivity. The problem is somewhat analogous to the design of aircraft. Early in the twentieth century when the first aircraft were designed, it was unclear how much sensitivity to external forces made aircraft easiest to fly. Superficially, one might think that the most stable (but therefore the least sensitive) aircraft would be the best, but such an aircraft would fly straight ahead regardless of the change in forces on it, including the controlling forces exerted by the pilot. A completely insensitive (completely stable) aircraft could not be guided, while a too-insensitive (too-stable) aircraft would require extremely strong forces to turn. On the other hand, a very sensitive (unstable) aircraft would change direction at the slightest change in the winds or in the setting of the control surfaces, and would require constant adjustment and attention. In the end, it was found from experience that airplanes flew most easily if they had intermediate sensitivity to external forces. In my experience with ecological models, the same seems to be the case. We need models that respond realistically to changes in the values of parameters and in environmental conditions, but that are not so sensitive that we cannot obtain realistic projections without absolutely precise measures of parameters.

Consider a specific parameter: the temperature response, $DEGD_{max_i}$. Suppose the model gave the same results no matter what the value of this parameter. The model would be unrealistic, unable to represent the way that forest trees respond to environment. On the other hand, suppose the model were extremely sensitive, so that a change in one degree-day in $DEGD_{max_i}$ resulted in a 10-percent change in the temperature response. Since the estimates of $DEGD_{max_i}$ of a species, obtained from range maps as explained in Chapter 2, could easily vary by 10 degree-days, the model could produce projections differing by 100 percent from two estimates of the same parameter.

The ideal model would have the same sensitivity to parameter values as real trees. To know what this real sensitivity is, however, would require not only accurate estimation of parameters but analysis of the sensitivity of real trees in real forests to changes in parameters, research that is yet to be done. This discussion suggests that the sensitivity of a model to parameter estimation must lie somewhere between extreme sensitivity and extreme insensitivity.

There are several reasons to study the sensitivity of the model. If we find that the model is not overly sensitive to errors in the estimation of parameters, then our faith in the realism of the model's projections will increase. In addition, sensitivity tests can give us additional insight into the dynamics of forest communities.

In this chapter I explore some aspects of the sensitivity of the forest model. Formally, *a sensitivity test is a test that determines how great a change occurs in the value of an output variable with a change in the value of either an input variable or a parameter intrinsic to the model.* A very large number of sensitivity tests could be conducted for a model with the complexity of JABOWA. Inclusion of all possible sensitivity tests would be impractical in a single volume. The availability of the model on diskettes for operation on a personal computer allows a user to conduct additional sensitivity tests. The purpose of this chapter is to explore the general issues underlying sensitivity tests and demonstrate the general level of sensitivity of the model. It is not to provide an unwieldy and unreadable volume of tables of all possible tests. Sensitivity tests can be grouped into two types: *intrinsic sensitivity,* which is sensitivity to parameter estimation, and *external sensitivity,* which is sensitivity to input or external variables, such as air temperature and precipitation.

Intrinsic Sensitivity: Tests of Sensitivity to Errors in Parameter Estimation

The forest model can generate volumes of output, especially for sensitivity tests. My purpose here is to illustrate types of sensitivity tests, not conduct all of them. In this section I consider the sensitivity of the model to seven intrinsic model parameters.* These are tested in a specific situation for two species. Limiting consideration to two species keeps the results to an amount I can write about in a single book and the reader can handle reasonably. The species I have chosen are two dominants of the Superior National Forest of Minnesota, balsam fir and sugar maple. Balsam fir is a dominant of mature boreal forest stands on wet soils, whereas sugar maple dominates moist, fertile, well-drained soils in the northern hardwood forest. The site I have chosen represents a major biogeographic transition—the transition from boreal to northern hardwood forests. In such a transition, the two species can occur together on moist sites. One expects, however, that each species, as well as the entire forest community, will be more sensitive to changes in parameters at such a transitional site than in the center of a species range (that is, in the midst of a major biome). In the transition, species are near the edges of their ranges and near the limits of the temperature response curve, as defined in Chapter 3 (equation 3.6). In the center of a biome, the dominant species are near the optimum of their temperature response curves. Inspection of equation (3.6) suggests that a species will be relatively insensitive to changes in parameter estimations there. The temperature response curve is steep as the function approaches its limits, and this occurs in ecologically transitional sites.

In most tests presented here, changes are made in the parameters for balsam fir, and the responses of both fir and maple are examined. The response of fir represents the response of a species to changes in its intrinsic characteristics; the response of sugar maple represents the response of a competitor to these changes. The parameters are:

1. S_i, the species-specific maximum number of saplings that can be added to a plot in any one year (affecting regeneration)
2. Maximum longevity of a species in years ($AGEMAX_i$) (affecting mortality probabilities)
3. Minimum diameter increment required per year for an individual tree to avoid being subjected to a higher probability of mortality ($AINC_i$) (affecting mortality probabilities)
4. Maximum and minimum number of degree-days under which the species could grow ($DEGD_{max_i}$ and $DEGD_{min_i}$, respectively) (affecting individual tree growth)
5. An index of soil-water saturation, which is the minimum depth to the water table that a species can withstand ($DTMIN_i$) (affecting individual tree growth)
6. An index of drought, the minimum soil moisture conditions a species can withstand ($WLMAX_i$) (affecting individual tree growth)

Intrinsic model parameters were varied by 10 percent above and below the standard values.

All tests were made beginning with a standard plot representing a typical

---

*This section is adapted from an article in the journal *Climatic Change*.

old-age, undisturbed stand of balsam fir in the Superior National Forest of Minnesota.

There are two steps in the sensitivity tests: (1) choosing or creating initial conditions; and (2) running the model to perform the sensitivity tests. One could use observed conditions from a real plot as initial conditions, but then there would likely be a transient response of the model to these conditions, and the sensitivity would have to be interpreted against this response. It is simpler and clearer to create hypothetical old-age stands generated by the model and use these as initial conditions, and this is the approach I have taken. The initial tree population for this hypothetical but typical stand was generated by running the model for 400 years from a clearcut with 50 replicates and using the 1951–1980 Virginia, Minnesota, weather records (the nearest weather station to the southern limit of the Superior National Forest and its designated wilderness, the Boundary Waters Canoe Area) in a repeating sequence. The initial plot conditions are: 0.081 ± 0.002 (95% C.I.) stems/m^2 of balsam fir occupying 27.9 ± 1.5 cm^2/m^2 basal area, soil depth 1 m, depth to the water table 0.8 m, soil moisture-holding capacity of 250 mm water/m depth of soil, and a moderate soil fertility of 50 kg/ha. The resulting initial stand is dominated by balsam fir, but contains enough sugar maple to allow a rapid change in dominance to this species, if the parameter change results in sensitivity.

As in the creation of initial conditions, the "normal" climate for each test described below is a recurring sequence of the 1951–1980 weather records from Virginia, Minnesota. In each test, the model was run for 90 years with 50 replicates for each trial. A trial consisted of increasing or decreasing one parameter by 10 percent. The values of the parameters used in these tests are given in Table 6.1.

Results of the first tests are given in Tables 6.1 and 6.2. Basal area (Table 6.1) and density (Table 6.2) are not sensitive to 10-percent error in estimation of most parameters. Errors of 10 percent in the estimate of the maximum longevity of either species and minimum diameter increment below which mortality rates increase do not affect the abundances of either species. Nor do 10-percent errors in the estimation of soil moisture parameters affect the results significantly.

In this case, sensitivity at the 10-percent level exists only for the parameters that control degree-day limits of the species. Decreasing the maximum number of degree-days for balsam fir by 10 percent leads to 50 percent less basal area by the year 2010 than when the standard parameter value is used, and a 66-percent lower value by the year 2070 (Table 6.1). A change in the minimum degree-day parameter has no effect on the abundance of balsam fir. Sensitivity to the maximum and not to the minimum degree-day parameter is due to the location of the Superior National Forest relative to the range of balsam fir, which as I explained earlier (and which should be apparent from discussions in previous chapters), is near the southern boundary of the geographic distribution of balsam fir. Changing the northern limit by 10 percent has a very small effect on the shape of the temperature response curve (equation 3.6) for balsam fir growing near Virginia, Minnesota, while changing the value of the southern limit has a strong effect on balsam fir growing in that vicinity.

TABLE 6.1. Effects of forest model sensitivity to parameter values for balsam fir basal area, using Virginia, Minnesota, 1951–1980 weather records[a]

Parameter[b]	Values		Basal Area (cm^2/m^2 ± 95% CI)			
	Control	Test	Year 10	Year 30	Year 60	Year 90
Balsam Fir						
Control (degree-days: 2250–2950)			27 ± 1	18 ± 3	14 ± 2	18 ± 2
AGEMAX	200	220	29 ± 2	19 ± 3	15 ± 3	20 ± 3
AGEMAX	200	180	28 ± 2	19 ± 3	16 ± 3	16 ± 3
AINC	0.01	0.02	28 ± 1	18 ± 3	14 ± 2	18 ± 2
AINC	0.01	0.005	28 ± 1	18 ± 3	14 ± 2	18 ± 2
SAPL	2	2.2	27 ± 2	20 ± 3	15 ± 2	18 ± 3
SAPL	2	1.8	28 ± 1	17 ± 3	14 ± 2	17 ± 3
DEGD$_{min_i}$	700	770	28 ± 2	17 ± 3	16 ± 3	18 ± 3
DEGD$_{min_i}$	700	630	29 ± 3	20 ± 3	14 ± 2	17 ± 2
DEGD$_{max_i}$	3700	4070	29 ± 1	21 ± 3	20 ± 3	21 ± 4
DEGD$_{max_i}$	3700	3300	25 ± 1	10 ± 2	6 ± 1	6 ± 1
DTMIN	0.21	0.23	28 ± 2	17 ± 3	18 ± 3	20 ± 3
DTMIN	0.21	0.19	27 ± 2	18 ± 3	14 ± 3	20 ± 3
WLMAX	0.245	0.27	28 ± 1	18 ± 3	14 ± 2	18 ± 2
WLMAX	0.245	0.22	28 ± 2	17 ± 3	14 ± 3	21 ± 3
S_i	2	4	29 ± 2	20 ± 2	27 ± 2	33 ± 3
Sugar Maple						
Control (degree-days: 3135–3899)			3 ± 0.2	2 ± 1	3 ± 1	2 ± 1
AGEMAX	400	440	3 ± 0.2	3 ± 0.6	3 ± 0.9	3 ± 0.8
AGEMAX	400	360	3 ± 0.2	2 ± 0.6	3 ± 1	3 ± 1
AINC	0.01	0.02	3 ± 0.2	2 ± 0.5	2 ± 1	3 ± 1
AINC	0.01	0.005	3 ± 0.3	2 ± 0.6	3 ± 0.9	2 ± 0.8
SAPL	3	3.3	3 ± 0.2	2 ± 0.5	3 ± 0.8	3 ± 1
SAPL	3	2.7	3 ± 0.2	2 ± 0.5	2 ± 0.8	2 ± 0.9
DEGD$_{min_i}$	2000	2200	3 ± 0.2	0.5 ± 0.3	0.2 ± 0.1	2 ± 0.1
DEGD$_{min_i}$	2000	1800	3 ± 0.2	6 ± 0.8	17 ± 2	30 ± 3
DEGD$_{max_i}$	6300	6930	3 ± 0.2	2 ± 0.5	2 ± 0.8	1 ± 0.6
DEGD$_{max_i}$	6300	5670	3 ± 0.2	3 ± 0.6	3 ± 0.7	3 ± 0.7
DTMIN	0.57	0.62	2 ± 0.3	2 ± 0.5	2 ± 0.7	2 ± 0.8
DTMIN	0.57	0.51	3 ± 0.3	2 ± 0.6	3 ± 0.9	4 ± 1
WLMAX	0.35	0.39	3 ± 0.2	2 ± 0.6	3 ± 0.8	3 ± 0.9
WLMAX	0.35	0.32	3 ± 0.2	2 ± 0.5	2 ± 0.7	2 ± 0.6
S_i	3	6	3 ± 0.2	1 ± 0.5	2 ± 1	3 ± 1

[a] Initial plot conditions are: 0.081 ± 0.002 (95% CI) stems/m^2 of balsam fir occuping 27.9 ± 1.5 cm^2/m^2 basal area; soil depth 1 m; depth to the water table 0.8 m; soil moisture-holding capacity 250 mm water/m depth of soil; available nitrogen 50 kg/ha. The population in the initial stand is given in Appendix IV, Table A-IV.1.

[b] AGEMAX = maximum age of the species in years; AINC = minimum diameter increment per year (cm) to avoid higher mortality probability; SAPL = maximum number of saplings of a species that can enter a plot/year; DEGD$_{max_i}$ = maximum number of degree-days for growth; DEGD$_{min_i}$ = minimum number of degree-days for growth; DTMIN = minimum tolerable depth to the water table; WLMAX = maximum evapotranspiration rate permitted for tree growth (which can also be thought of as the maximum wilt tolerable by species i). S_i is the maximum number of saplings of species i that can be added to a plot in any single year.

TABLE 6.2. Effects of forest model sensitivity to parameter values for balsam fir stems/100 m^2 $\pm 95\%$ CI, using Virginia, Minnesota, 1951–1980 weather records[a]

	Values			Stems/100 m^2 $\pm 95\%$ CI		
Parameter[b]	Control	Test	Year 10	Year 30	Year 60	Year 90
			Balsam Fir			
Results with default values			9 ± 0.5	8 ± 0.7	9 ± 0.9	10 ± 0.7
AGEMAX	200	220	9 ± 0.5	7 ± 0.6	9 ± 0.9	10 ± 1
AGEMAX	200	180	8 ± 0.6	7 ± 0.6	8 ± 0.7	9 ± 1
AINC	0.01	0.02	9 ± 0.5	8 ± 0.7	9 ± 0.9	10 ± 0.7
AINC	0.01	0.005	9 ± 0.5	8 ± 0.7	9 ± 0.9	10 ± 0.7
SAPL	2	2.2	9 ± 0.6	9 ± 0.7	10 ± 0.0	11 ± 1
SAPL	2	1.8	8 ± 0.5	6 ± 0.6	7 ± 0.7	9 ± 0.7
SAPL	2	4	15 ± 1	19 ± 1	26 ± 2	31 ± 2
DEGD$_{min_i}$	700	770	9 ± 0.5	8 ± 0.6	8 ± 0.8	10 ± 1
DEGD$_{min_i}$	700	630	9 ± 0.4	7 ± 0.6	8 ± 0.8	10 ± 0.9
DEGD$_{max_i}$	3700	4070	9 ± 0.5	8 ± 0.6	9 ± 0.9	11 ± 1
DEGD$_{max_i}$	3700	3300	8 ± 0.5	4 ± 0.6	5 ± 0.5	5 ± 0.6
DTMIN	0.21	0.23	9 ± 0.4	8 ± 0.7	9 ± 0.8	10 ± 1
DTMIN	0.21	0.19	9 ± 0.6	7 ± 0.6	8 ± 0.8	11 ± 1
WLMAX	0.245	0.27	9 ± 0.5	8 ± 0.7	9 ± 0.9	10 ± 0.7
WLMAX	0.245	0.22	9 ± 0.6	7 ± 0.6	8 ± 0.7	11 ± 0.9
			Sugar Maple			
Results with default values			4 ± 0.3	4 ± 0.5	5 ± 0.5	4 ± 0.5
AGEMAX	400	440	4 ± 0.2	4 ± 0.6	4 ± 0.7	4 ± 0.6
AGEMAX	400	360	5 ± 0.3	5 ± 0.5	5 ± 0.5	4 ± 0.6
AINC	0.01	0.02	4 ± 0.3	4 ± 0.5	4 ± 0.5	4 ± 0.6
AINC	0.01	0.005	4 ± 0.3	4 ± 0.4	4 ± 0.5	4 ± 0.6
SAPL	3	3.3	5 ± 0.3	5 ± 0.4	5 ± 0.6	5 ± 0.6
SAPL	3	2.7	5 ± 0.2	3 ± 0.4	3 ± 0.5	3 ± 0.5
SAPL	3	6	9 ± 0.6	10 ± 0.9	11 ± 0.8	11 ± 0.9
DEGD$_{min_i}$	2000	2200	4 ± 0.3	2 ± 0.4	2 ± 0.4	1 ± 0.3
DEGD$_{min_i}$	2000	1800	5 ± 0.4	9 ± 0.6	12 ± 1	16 ± 1
DEGD$_{max_i}$	6300	6930	4 ± 0.2	3 ± 0.5	3 ± 0.5	5 ± 0.5
DEGD$_{max_i}$	6300	5670	5 ± 0.3	5 ± 0.5	5 ± 0.6	5 ± 0.5
DTMIN	0.56	0.62	4 ± 0.3	4 ± 0.6	4 ± 0.4	4 ± 0.5
DTMIN	0.56	0.51	5 ± 0.3	4 ± 0.6	5 ± 0.5	5 ± 0.6
WLMAX	0.35	0.39	5 ± 0.3	4 ± 0.5	4 ± 0.5	4 ± 0.6
WLMAX	0.35	0.31	4 ± 0.3	4 ± 0.4	4 ± 0.6	4 ± 0.5

[a] See Table 6.1 for initial plot conditions.
[b] See footnotes Table 6.1 for definition of parameters.

The response of sugar maple is the mirror image of the response of balsam fir. An increase in the minimum degree-day parameter for sugar maple effectively allows the species to grow farther north and has a large effect on the abundance of that species in the Superior National Forest, which is near its northern boundaries. On the other hand, 10-percent alteration of the maximum degree-day parameter, which determines the southern distribution of sugar maple, has no effect on the results (Table 6.1).

Need for Better Temperature Parameter
Estimation Procedures

The parameters $DEGD_{max_i}$ and $DEGD_{min_i}$ were determined by comparing maps of a species' distribution with maps of temperature isotherms. As I explained before, this method was selected simply for convenience when the model was first developed, and I expected that later users of the model would find more accurate methods. In general, they have not. In part this seems to be a result of the success of the model and its only moderate sensitivity to estimation of the temperature response parameters. However, the user should know the assumptions of this method, assumptions that could lead to errors in the accuracy of projections and the reader might at this point want to refer back to relevant sections of Chapters 2 and 3. First, this method assumes that the distribution of a species is in steady-state with the current climate. This is not likely to be the case, since over the millennia all tree species have migrated continually across the landscape in response to climate change (COHMAP 1988; Davis, 1983; Ritchie and Yarrantow, 1978; Swain, 1978; Wright, 1976). That the method has worked is probably the result of the long time required for trees to migrate, so that in terms of the estimation of this parameter, there is a reasonable correspondence between modern distributions and modern climate. In addition, the method assumes that all individuals of a species have the same temperature response—that there are no ecotypic variations. It is possible that ecotypes (local populations with distinctive genetic characteristics) near the boundary of the distribution of a species have evolved to be better adapted to limiting temperature conditions than the average for the species, and therefore they may be able to persist longer than expected from the standard estimates of the parameters under rapid climatic change. That this is possible is suggested by the observation of local differences—on the same hillside—within a species as a function of elevation. (Ledig and Korbobo, 1983).

As I discussed in Chapter 3, but bears repeating here, the general shape of a species' response curve to temperature level is based on general concepts of physiological response to temperature (Kozlowski, Kramer and Pallardy, 1991; Kramer and Kozlowski, 1979), and the maximum and minimum temperatures expressed in terms of yearly degree-days reflect the actual geographic range limits of the species (Botkin et al., 1972a, b). The resulting parabolic function could be studied and verified, particularly at the extremes. The actual shape for some species may be a truncated normal curve (similar to a beta distribution) expressing the effect of differential responses of ecotypes, such as we used in an algal community model (Lehman et al., 1975a, b).

Bend until It Breaks

The tests against a 10-percent change in parameter estimation are encouraging because they suggest that the model is not overly sensitive to factors whose measurement cannot be expected to be done with an accuracy better than 10 percent. However, it is useful to carry the tests a step further and find out what percent change in a parameter will produce a significant change in the model's output. Tests of this kind were conducted for the same seven parameters and same site and initial conditions as in the previous tests. The output, given in terms of biomass, appears in Tables 6.3, 6.4, and 6.5 and in Figures 6.1, 6.2, and 6.3.

In the following discussion, it is useful to note that balsam fir reaches a biomass at year 90 of $8 \, kg/m^2$ with default parameters. The model is not very sensitive to changes in the maximum number of saplings of balsam fir that can be added each year; the biomass of balsam fir ranges from 8 to 15 at year 90

TABLE 6.3. Parameter error required to produce significant change in biomass after 90 years of growth[a]

	SM biomass (kg/m^2)	BF biomass (kg/m^2)
(A) S_i: Maximum number of balsam fir saplings that can enter plot during one year		
S_i		
1	4.8 ± 0.8	10.8 ± 2.2
2 (default)	5.6 ± 0.9	8.8 ± 1.8
3	5.6 ± 0.9	8.6 ± 1.4
6	5.9 ± 1.0	11.7 ± 1.9 (signif. differ.)
12	5.5 ± 0.8	12.7 ± 1.6
24	5.0 ± 0.8	14.5 ± 1.0
(B) AGEMAX$_i$: Maximum age of balsam fir		
AGE$_i$ (yr)		
50	9.1 ± 0.9	0.1 ± 0.1
100	7.6 ± 0.9	3.5 ± 1.6 (signif. differ.)
200 (default)	$5.6 + 0.9$	8.8 ± 1.8
250	5.7 ± 0.8	$11.4 + 1.8$
300	4.7 ± 0.8	13.3 ± 1.9 (signif. differ.)
400	4.9 ± 0.6	17.2 ± 2.3
500	4.2 ± 0.7	20.6 ± 1.8
(C) AINC$_i$: Minimum diameter increment of balsam fir tree to avoid being subject to higher probability of mortality		
AINC$_i$		
0.005	5.5 ± 0.9	8.5 ± 1.8
0.01 (default)	5.6 ± 0.9	8.8 ± 1.8
0.02	5.5 ± 0.9	8.5 ± 1.8
0.04	6.2 ± 0.9	8.5 ± 1.8
0.06	6.1 ± 1.0	6.3 ± 1.6
0.08	8.0 ± 1.0	3.8 ± 1.2 (signif. differ.)
0.10	8.5 ± 1.1	2.0 ± 0.5
0.20	8.5 ± 1.1	0.0

Footnote [a] on page 166.

TABLE 6.3. (*Continued*)

	SM biomass (kg/m^2)	BF biomass (kg/m^2)

(D) DEGD$_{max_i}$: Maximum number of degree-days (above 40°F) per year under which balsam fir tree could grow

DEGD$_{max_i}$		
1028	8.8 ± 1.1	0.0
1400	9.4 ± 1.0	0.0
1600	9.4 ± 1.0	0.0
1800	8.7 ± 0.9	0.0
1850	7.7 ± 1.0	0.0
1900	7.6 ± 1.1	2.2 ± 0.9 (signif. differ.)
1950	6.3 ± 1.1	7.1 ± 1.7
2000	5.4 ± 0.8	8.9 ± 1.8
2056 (default)	5.6 ± 0.9	8.8 ± 1.8
3083	5.7 ± 0.9	11.6 + 1.7 (signif. differ.)

(E) DEGD$_{min_i}$: Minimum number of degree-days (above 40°F) per year under which balsam fir tree could grow

DEGD$_{min_i}$		
98	6.1 ± 0.9	7.3 ± 1.6
195	5.0 ± 0.8	8.1 ± 1.9
390 (default)	5.6 ± 0.9	8.8 ± 1.8
585	5.1 ± 0.9	10.0 ± 1.7
790	5.8 ± 0.9	10.2 ± 1.9
1170	5.0 ± 0.8	12.5 ± 2.3 (signif. differ.)

(F) DTMIN$_i$: Minimum depth to water table that balsam fir tree can stand

DTMIN$_i$		
0.01	5.6 ± 0.9	8.8 ± 1.8
0.05	5.4 ± 0.8	10.5 ± 2.0
0.10	6.0 ± 0.8	8.3 ± 1.8
0.15	5.8 ± 1.0	9.4 ± 1.8
0.21 (default)	5.6 ± 0.9	8.8 ± 1.8
0.25	6.7 ± 1.0	7.6 ± 1.5
0.40	6.1 ± 0.9	7.1 ± 1.9
0.50	6.3 ± 0.9	6.4 ± 1.7
0.60	6.9 ± 0.9	4.8 ± 1.2 (signif. differ.)

(G) WLMAX$_i$: Minimum soil moisture conditions that balsam fir tree can tolerate

WLMAX$_i$[b]		
0.10	5.7 ± 0.8	7.8 ± 1.6
0.20	5.7 ± 0.9	9.0 ± 1.8
0.25 (default)	5.6 ± 0.9	8.8 ± 1.8
0.40	5.7 ± 0.9	9.7 ± 1.9

[a]Tests are for biomass of sugar maple (SM) and balsam fir (BF) 90 years after parameter modification for the site described in Table 6.1. Initial forest is an old-age balsam fir stand described in the text and used in Table 6.1. Values are means ± 95% confidence intervals.
[b]Default value WLMAX for balsam fir is 0.245.

TABLE 6.4. Biomass of sugar maple and balsam fir at year 90 as a function of drought tolerance parameter, $WLMAX_i$, for balsam fir[a]

$WLMAX_i$	SM Biomass (kg/m^2)	BF Biomass (kg/m^2)
0.4	8.7 ± 1.1	9.9 ± 2.3
0.5	7.7 ± 1.1	9.4 ± 1.6
0.6	8.6 ± 1.2	9.8 ± 1.9
0.7	7.0 ± 1.2	9.9 ± 1.6
0.8	7.4 ± 1.0	10.6 ± 1.8
0.9	7.0 ± 1.0	9.3 ± 1.9
1.0	6.8 ± 1.1	11.3 ± 2.2
1.1	7.1 ± 1.2	11.4 ± 2.3
1.2	7.9 ± 1.2	9.2 ± 1.5
1.3	7.7 ± 1.1	10.0 ± 2.0
1.4	7.3 ± 1.0	9.8 ± 2.0
1.5	7.1 ± 1.1	10.5 ± 2.0

[a] Soil depth is 1.0 m and depth to the water table is 1.0 m

TABLE 6.5. Biomass of jack pine at year 90 as a function of drought tolerance parameter, $WLMAX_i$[a]

$WLMAX_i$	Jack Pine Biomass (kg/m^2)
0.1	0.0
0.4	0.0
0.45	0.3 ± 0.1
0.46	0.4 ± 0.2
0.48	1.0 ± 0.3
0.49	0.9 ± 0.3
0.5	1.0 ± 0.3
0.53 (default)	0.9 ± 0.3
0.9	1.7 ± 0.5
1.0	1.9 ± 0.4
1.5	1.6 ± 0.5

[a] Site is a coarse, sandy soil near Grayling, Michigan. Weather records are from Grayling, Michigan. Soil depth is 1.0 m and depth to the water table is 1.2 m. Elevation: 900 ft; soil texture 50.0 mm/m; percent rock 0.0; available nitrogen 55.0 kg/ha. Biomass is at year 90 beginning with an established jack pine stand whose population composition is given in Appendix IV, Table A-IV.2.

when the maximum number of saplings ranges from 1 to 24 (Fig. 6.1(A)). For unexplained reasons, biomass of balsam fir reaches a minimum at year 90 when the maximum number of saplings is three or four; the biomass increases slightly when the maximum number of saplings is reduced to one. The increase in biomass with increasing recruitment is expected, but the model is less sensitive to changes in this parameter than I had expected.

The model is more sensitive to changes in the maximum age of balsam fir than to the maximum number of saplings that can be added in any one year. Balsam fir becomes extinct when the maximum age is set at 50 years (Fig. 6.1(B)). Biomass of this species at year 90 increases almost linearly from zero at a maximum age of 50 to 20 kg/m^2 at a maximum age of 500 years. Thus increasing the maximum age by 10 times increases balsam fir biomass by 20 kg/m^2, while a 10-fold increase in the maximum number of saplings increases balsam fir biomass by 7 kg/m^2. The connection between extinction and maximum longevity of balsam fir is of interest for questions about biological conservation, as explored by Woodby (1991).

The model is also sensitive to changes in the second mortality parameter, the one that affects trees that grow poorly. The default value (0.02) is large enough to have little effect on the response of the model. Halving and doubling this value, as indicated in Figure 6.1(C), has no effect on the biomass of balsam fir. The model responds to changes in this parameter only when the value is tripled to 0.06, which means that a tree growing 0.06 cm/year or less is subject to the higher probability of mortality. Balsam fir biomass at year 90 drops rapidly with increases in this parameter beyond 0.06, and the species becomes extinct when the parameter reaches 0.2 cm/yr. These results suggest that trees must be able to withstand considerable competition from shading and low growth rates in order to persist in a multi-species forest.

Figure 6.2 shows how a competitor is affected by changes in parameters of a species. In this case the response of sugar maple is shown for changes in balsam fir parameters. Biomass of sugar maple at year 90 declines slightly as the maximum age of balsam fir increases from the default of 200 to 500 years, and the decline approaches an asymptote of 4 kg/m^2 (Fig. 6.2(A)). Sugar maple biomass at year 90 increases when the maximum age of balsam fir decreases below the default, but the increase is relatively modest. At the balsam fir maximum age of 50 years, at which balsam fir becomes extinct, sugar maple biomass at year 90 reaches 9 kg/m^2. The relatively small rise in sugar maple biomass may be due to competition with other species.

⟶

FIGURE 6.1. Sensitivity of balsam fir biomass to changes in fundamental population dynamics parameters of balsam fir. Results are shown after 90 years of simulated growth for a site in northern Minnesota representing the Superior National Forest and its designated wilderness, the Boundary Waters Canoe Area. Site conditions are given in Table 6.1. (A) Changes in the regeneration parameter, S_i, the species-specific maximum number of saplings that can be added to a plot in any one year. (B) Changes in maximum age of balsam fir. (C) Mortality parameter 2, $AINC_i$, the minimum diameter increment required per year for an individual tree to avoid being subjected to a higher probability of mortality.

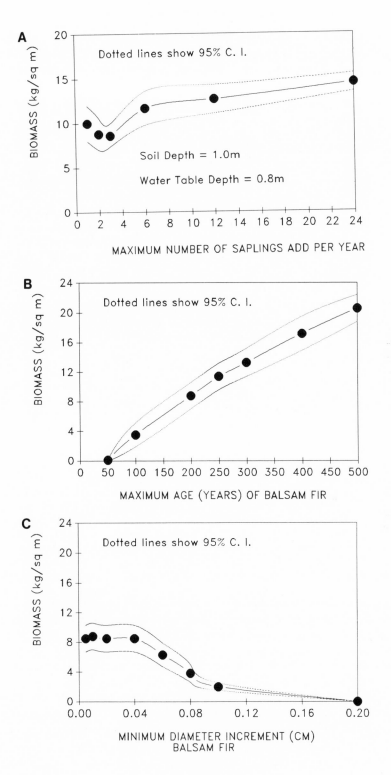

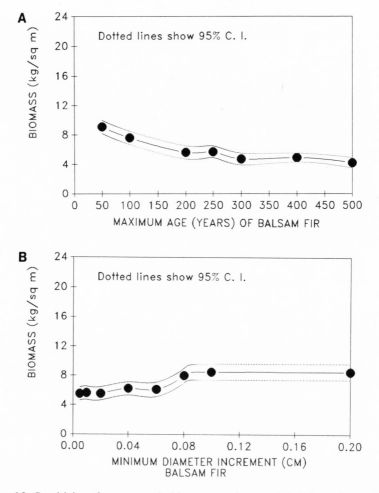

FIGURE 6.2. Sensitivity of sugar maple biomass to changes. (A) Balsam fir maximum age. (B) Balsam fir mortality parameter 2 (AINC). (C) Balsam fir maximum degree-day parameter. (D) Minimum degree-day parameter. Results are shown for year 90 for a site in northern Minnesota representing the Superior National Forest and its designated wilderness, the Boundary Waters Canoe Area. Site conditions are given in Table 6.1.

Sugar maple is less sensitive to changes in balsam fir parameters than is balsam fir, as might be expected. For example, sugar maple biomass at year 90 increases from $5 \, \text{kg/m}^2$ to $8 \, \text{kg/m}^2$ (Fig. 6.2(B)), while balsam fir biomass becomes extinct over this same range.

The sensitivity of balsam fir to large changes in degree-day parameters (beyond the 10% value of the first sensitivity experiment) is shown in Figure 6.3. Remember that the maximum degree-day parameter controls the warmer (southern) limit of the species, while the minimum degree-day parameter controls the colder (northern) limit of the species. Decreasing the maximum degree-day parameter moves the southern boundary of the species northward. When that

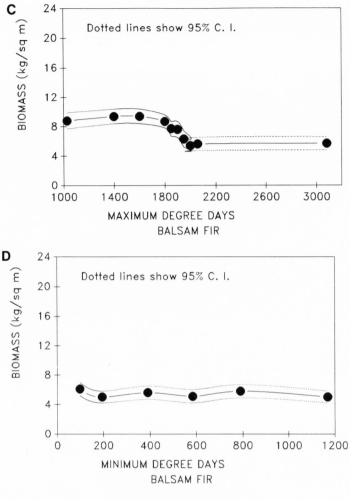

FIGURE 6.2. (*Continued*)

is done, balsam fir biomass declines rapidly at the test site. The default value, 2,056, is near the value that occurs at the test site, which means that this site is near the southern boundary of balsam fir's range. As I explained earlier, the site was chosen purposefully with this in mind. The model projects that a species whose southern limit is 1,800 degree-days or less cannot grow at this site (Fig. 6.3(A)). Extending the southern limit (by increasing the maximum degree-day parameter) beyond 1,800 allows the species to persist, but increases beyond this value have only a small effect on balsam fir biomass.

Figure 6.2(C) shows the effect of a change in the warm (southern) limit of a species on a competitor. Sugar maple biomass reaches an asymptote of $8\,kg/m^2$ when balsam fir is eliminated. One might expect that sugar maple would continue to increase as balsam fir becomes less and less abundant, but

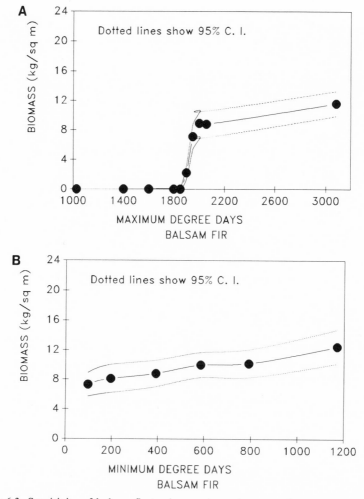

FIGURE 6.3. Sensitivity of balsam fir to changes in environmental response parameters of balsam fir. (A) Maximum degree-day parameter; (B) Minimum degree-day parameter. Results are shown after 90 years of simulated growth for a site in northern Minnesota representing the Superior National Forest and its designated wilderness, the Boundary Waters Canoe Area. Site conditions are given in Table 6.1.

in a multispecies forest others species in addition to sugar maple increase as balsam fir declines. Biomass of sugar maple declines to approximately $6 \, kg/m^2$ under a balsam fir maximum degree-day parameter of 1,900 to 3,000. At this site sugar maple appears to have two dominant states in regard to balsam fir: a high abundance of $8 \, kg/m^2$ and a low abundance of approximately $6 \, kg/m^2$.

At the test site, balsam fir is much less sensitive to changes in the colder (northern) limit defined by the minimum degree-day parameter (Fig. 6.3(B)). As the northern limit is moved to the south (the value of the parameter increases), balsam fir biomass increases. This is caused by a steepening of the temperature response function—the parabola that characterizes this response covers a smaller range so that the function increases more sharply at the extremes. Moving the cold limit northward stretches out the curve and lowers the biomass at the Virginia, Minnesota, test site; moving the cold limit to the south compresses the curve and increases balsam fir biomass at this site. Sugar maple shows little response to changes in the cold degree-day limit for balsam fir (Fig. 6.3(D)), which is not surprising in light of the small effect that changes in this parameter have on balsam fir itself.

To summarize the results so far, a significant increase in biomass of balsam

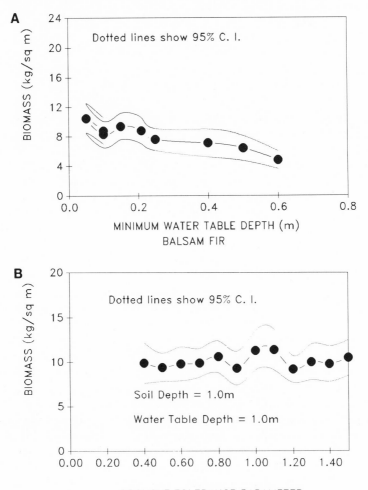

FIGURE 6.4. Sensitivity of balsam fir to (A) minimum water table depth and (B) drought tolerance parameter. Conditions are as in Table 6.1. Results are shown for year 90 after start of model run.

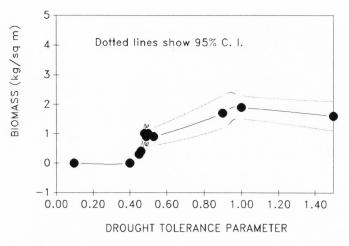

FIGURE 6.5. Biomass of jack pine at year 90 as a function of the drought tolerance parameter, $WLMAX_i$. Site conditions are given in Table 6.5.

fir does not occur until the maximum number of saplings that can enter in any one year is tripled, from two to six (Table 6.3(A); Fig. 6.1(A)). Increasing or decreasing the maximum age by 100 years significantly changes the biomass of balsam fir, but an increase of 50 years (a 25% increase) does not lead to a significant change (Table 6.3(B)). Biomass increases almost linearly over the range from a maximum age of 50 to 500 with only a slight tendency toward an asymptote. Biomass is zero at a maximum age of 50. Balsam fir with longevity restricted to 50 years cannot persist in the test site. An increase in maximum age of balsam fir leads to a decrease in the biomass of sugar maple, but the slope is gentler than the increase in biomass of balsam fir, and a significant change occurs only at the extremes (Fig. 6.2(A)).

The sensitivity of the model to $AINC_i$, the second mortality parameter, is shown in Figures 6.1(C) and 6.2(B). Remember that this parameter sets a threshold. A tree with diameter growth below that threshold for 10 years has only 1 chance in 100 to survive. The default value, 0.02 cm/yr, gives the same results as when the parameter is set at zero. This indicates that the default value does not affect the dynamics at the test site. Doubling the value to 0.04 cm/yr results in no change, but an increase beyond that level leads to a rapid drop-off in balsam fir biomass until the species becomes extinct at a value of 0.2 cm/yr. This experiment suggests that any factor or set of factors that produces a high chance of mortality in balsam fir, for all sizes and ages of individuals with diameter increments equal to or less than 0.2 cm/yr, will result in extinction of this species. This could be tested empirically by creating a stress in a test stand that reduces diameter increments below this level and observing the fate of balsam fir for 10 years.

As a wetland species, balsam fir should be able to withstand high water tables, and that is the result indicated in Table 6.3(F) and Figure 6.4(A). For the Virginia, Minnesota site, balsam fir is insensitive to changes in the soil-water saturation parameter; it can withstand a soil saturated to 10 cm below the surface, and its biomass does not show a decline until the water table depth is lowered to 60 cm. Balsam fir is also insensitive to changes in the drought tolerance parameter for a slightly drier site, shown in Table 6.4 and Figure 6.4(B), where the soil depth and the water table depth is 1 m. Examining only the response of balsam fir at this one site, one might conclude that the minimum water table depth does not affect the dynamics of the model and might be incorrectly formulated.

A more correct impression of the sensitivity of the model is obtained when we examine the response of an additional species, one characteristic of a dry site. Jack pine is such a species, and its response is shown in Figure 6.5 and Table 6.5 on a typical site for that species, a coarse sandy soil of low fertility in central Michigan. Jack pine is sensitive to changes in the drought tolerance parameter. Remember that

$$\text{WiF}_i = \max\left\{0, 1 - \left(\frac{\text{WILT}}{\text{WLMAX}_i}\right)^2\right\} \qquad (3.7)$$

where

$$\text{WILT} = \frac{(E_0 - E)}{E_0} \qquad (3.8a)$$

and E_0 is the potential evapotranspiration and E is the actual evapotranspiration.

Equation (3.8) produces a steep-shouldered curve, suggested by the steep rise in jack pine biomass at the Grayling, Michigan, site as the drought tolerance parameter increases from 0.4 to 0.6.

Another important kind of sensitivity test is the response of a model to extrinsic factors. In the case of the forest model, an important test of this kind is the sensitivity to choice of weather records. Tests of this kind have been carried out for projections of the effects of global warming on forests, which are discussed in the next chapter.

Summary

As a generalization, the forest model is moderately sensitive to errors in estimation of intrinsic parameters, which affect the fundamental population dynamics (parameters that control the maximum number of saplings that can be added in any year and the increase in mortality at low growth rates). As revealed by the examples discussed here, the location of a site relative to the range of a species affects the relative sensitivity of the output to errors in parameter estimates. This is as one would hope and expect; it indicates that the model properly responds to changes in site conditions.

The sensitivity is moderate, which is encouraging. As I discussed at the beginning of this chapter, moderate sensitivity means that the output of the model is likely to be robust against accuracy of field measurements, yet sensitive enough to show changes with an expected range of environmental conditions.

The realism of the projections of the model would seem to be in part a result of its moderate sensitivity.

To the extent that one accepts JABOWA as a realistic model of forest dynamics, the sensitivity tests yield some insights into these dynamics. These can be used as hypotheses about populations of trees within forest ecosystems. Among these insights are: there seems to be a reasonable correspondence between the present geographic ranges of trees and present climate, at least sufficient to approximate the temperature boundaries of the trees, in spite of continual long-term migration of tree species across the landscape. This reinforces the use of the correlation between pollen analysis and past climates to project future geographic ranges of tree species.

Trees must be able to withstand highly suboptimal conditions and the consequent slow rate of growth, if they are to persist within a forest. Survival of existing trees is more important for persistence of a species in a forest than the quantity of individuals recruited each year to the population; thus longevity and the ability of mature trees to survive years of very poor conditions are more important than recruitment. This is based on the result that the model is more sensitive to changes in mortality probabilities than to the maximum number of saplings that can be added in any one year. Changes in genetically determined thermal boundaries for a species can change the growth of an individual tree within these boundaries except at the very center of the range. This is based on the shape of the temperature response curve. Narrowing the thermal range steepens the temperature response function, and increases the growth at any point except at the exact center. The influence is quantitatively greater at the edges of the range than near the center. These examples of interpretation of sensitivity tests show that such tests are not only valuable as part of the technical evaluation of the model, but also as a method to help us learn about the dynamics of forested ecosystems.

There are many other factors for which sensitivity tests could be made. In addition to the tests of other parameters, there are tests that concern larger assumptions of the model, including tests of the effects of plot size on the model, tests of combinations of species characteristics not present with existing species (such as great longevity in a shade-intolerant species); tests of how the model would respond if all species had the same light response function for growth but not for regeneration, and vice versa; tests in which regeneration were made deterministic while mortality remained stochastic, and vice versa. The reader can no doubt think of others. Users of the software available as a companion to this book can conduct many of these tests.

With the accuracy of the model demonstrated in Chapter 6, and the moderate sensitivity demonstrated in this chapter, we can have greater confidence in the application of the model to applied problems and to its extension as a method in fundamental research. The next chapters explore the application of the model to applied problems.

7

Use of the Model for
Global Environmental Problems

The Biosphere

In the 1980s, concern grew about global environmental problems. A popular belief developed that the uptake and storage of carbon dioxide by forests could help remove carbon dioxide added to the atmosphere by human activities, and thereby slow global warming. Interest in global environmental and large-scale, regional environmental issues ranged from a renewal of a holistic belief in the Earth as a life-support system, perhaps best publicized through the Gaia hypothesis of Lovelock and Margulis (Lovelock, 1979; Lovelock and Margulis, 1974), to technical analyses of the fate of specific compounds that might have global effects. Concern with global environmental problems has led to the call for new, international cooperative research programs, such as that proposed by the NASA Earth Systems Science Committee (Bretherton et al., 1986).

There are two kinds of global environmental problems: system-level problems and multiple point-source problems. In the first, human activities lead to a change in some general characteristic of the biosphere, and this in turn has global effects. Human-induced global climate change is an example of this kind of problem. In multiple point-source problems, local or regional human effects occur at so many places around the Earth that global changes are possible. Acid rain is an example of such a regional multiple point-source effect (Evans, 1982). In this chapter I will consider one example of each kind of problem. The examples are global warming and acid rain.

The idea that life and the environment are connected at a planetary level began to develop early in the twentieth century, with such works as Henderson's 1913 *Fitness of the Environment* and Vernadsky's 1926 *The Biosphere*, with the famous 1956 conference whose proceedings were published as *Man's Role in Changing The Face of The Earth* (Thomas, 1956), and with pioneering analyses by Hutchinson of biogeochemical cycles (Hutchinson, 1950; 1954).

Much of what we understand about global environmental problems derives from models. Earlier analyses of global environmental problems, conducted from the 1940s to the 1970s, depended heavily on simple compartment models of the kind discussed in Chapter 1. This was the case for the earliest analyses of possible anthropogenic influences on the global carbon cycle, and the possibility of a human-induced global warming (see, for example, Bolin and Eriksson, 1959). In these early models, all vegetation of the Earth was treated as belonging to a single compartment or to two compartments. For example, Ekdahl and Keeling (1973) used a model in which there were two vegetation compartments, long-term and short-term carbon storage. While unrealistic, these models served a useful initial purpose, providing a way to begin to analyze the global carbon budget under steady-state conditions.

In recent years, understanding of global environmental problems has improved. We have developed a new perspective and new concepts of global ecology (Botkin, 1980, 1984, 1990; Lovelock, 1979, 1988; Lovelock and Margulis, 1974; Rambler, Margulis, and Fester, 1989), in part the result of increasing knowledge of specific phenomena. We now understand that life influences the environment at a global level (Botkin, 1980, 1984; Lovelock, 1988), and that life has changed the environment at a global level for more than 3 billion years (Botkin et al., 1989a; Rambler, Margulis and Fester, 1989). This new understanding has led to an interest in the development of a science of the biosphere, meaning a science dealing with the planetary life-support system that includes and sustains life, a system made up of the lower atmosphere, all of the oceans, soils, all of life, and those solid sediments in active interchange with the biota (Schlesinger, 1991).

We now recognize that terrestrial vegetation plays an important role in the biosphere (Broecker, et al., 1979; Woodwell and Pecan, 1973; Woodwell et al., 1977), affecting the global cycle of carbon and other elements, and affecting climate in four ways: (1) through the carbon cycle by affecting the concentration of greenhouse gases; (2) by altering the surface roughness (and therefore affecting exchange of energy between the atmosphere and the surface through friction; (3) through reflection and absorption of electromagnetic spectrum; and (4) through evaporation of water (which affects the transfer of energy from the Earth's surface to the atmosphere).

Forests play an important role at a global level in biogeochemical cycles and in climate dynamics. Because of the longevity of trees and the high organic content of many forest soils, forest ecosystems are the major storehouse of organic carbon. The annual influence of terrestrial vegetation—primarily forest vegetation—can be seen in the annual cycles of the CO_2 concentration in the atmosphere measured since 1957 at Mauna Loa Observatory by Keeling and colleagues (1989), which is caused by the uptake of carbon by terrestrial vegetation during the summer and its release through respiration in the winter.

Standard estimates suggest that forest trees and soils together store more carbon than is found in the atmosphere (Botkin, 1977; Lieth & Whittaker, 1975). Direct measures that provide the first statistically valid estimates of biomass and carbon storage for a large region of the Earth indicate that above-ground vegetation in boreal forests of North America contain 22 ± 5 billion metric tons of organic dry matter and 9.7 ± 2 billion metric tons of carbon (Botkin and Simpson, 1990a, b). If the boreal forests of the rest of the world have equivalent concentrations of biomass and carbon, then above-ground vegetation in all boreal forests could contain 60 billion metric tons of biomass and 27 billion metric tons of carbon (Botkin and Simpson, 1990a). This is approximately five times the carbon estimated to be added annually from the burning of fossil fuels (Woodwell et al., 1977).

With our new understanding of the biosphere and with growing concerns about global environmental problems, there is a need to improve our ability to project the response of forests to global environmental change, and to understand the role of forests in these changes. The issues divide into two major categories: those that concern climate dynamics (and are fundamentally concerned with energy exchange processes of the biosphere) (Schneider, 1989a, b) and those that concern global chemical cycles (and deal with global budgets of chemical elements) (Schlesinger, 1991). There is a divergence in modeling approaches between these two categories. Models of climate dynamics have advanced rapidly to complex computer simulations of nonlinear systems (Hansen et al., 1983, 1988; Schlesinger and Mitchell, 1987; Schlesinger and Zhao, 1988; Schneider, 1989), while models of global biogeochemical cycles generally remain simple, steady-state compartment models soluble analytically or on small calculators or microcomputers (Botkin, 1982, 1984).

How theory of global ecology is best approached remains a little studied subject, about which we have many questions but few answers (Botkin, 1982). How can we develop a general theory for the biosphere? The complexity of the biosphere is overwhelming. There are huge amounts of information about biota at a local scale, and about intricate connections among the major components of the biosphere (Botkin et al., 1979). What has become known as the "scaling issue" adds to the difficulty. The scaling issue concerns the question: How can we deal with disparate phenomena with very different time and space scales and rates of change? O'Neill et al. (1986) made an important contribution to our understanding of how we might approach ecological phenomena as a hierarchy of interconnected systems. We need to build on these insights to develop theory and models for the biosphere system that is composed of the atmosphere, oceans, soils, solid surfaces, and the biota, including vegetation, each component with its own characteristic temporal and spatial rates of change. At present, there is greater understanding of how to model the response of vegetation to global change than there is about how to model the effect of vegetation on the rest of the biosphere.

The Possible Effects of Global Warming on Forests

As an example of how the effects of global environmental change on forest vegetation might be projected, I will describe research that several of us have done on the possible effects of global warming on forests. I hope the progress we have made will be helpful to others in developing more general theory for the biosphere.

Projections of possible climate change that could occur with a continuing increase in the concentration of greenhouse gases in the atmosphere have been made primarily through the use of computer models of the Earth's climate, as mentioned earlier. These are known as general circulation models, or GCMs

(Hansen et al., 1983, 1988; Schlesinger and Mitchell, 1987; Schlesinger and Zhao, 1988; Schneider, 1989). By the late 1980s, the major general circulation models projected that mid-latitudes would experience pronounced warming and drying out of soils, suggesting that there might be major changes in mid-latitude forests (Smith and Tirpak, 1989). Simple inspection of projected temperature and precipitation regimes suggests major disruptions in the distribution of vegetation. Studies correlating past distributions of pollen deposits with past climatic regimes have been used to project a redistribution of vegetation under global warming. The redistribution is drastic (Davis and Zabinski, in press). However, these projections assume a steady-state climate and a steady-state relationships between climate and vegetation, neither of which is true. However, results of the use of the JABOWA model discussed in Chapter 4 suggest that this assumption is not a bad approximation for the estimation of present relationships between air temperature and the geographic range of a tree species. This gives added support to the plausibility of the projections made from pollen analysis. (A statistical correlation is calculated between the distribution of fossil pollen at some time in the past and the reconstructed climatic conditions. A necessary assumption of an extrapolation from these is that the pollen distribution was in steady-state with the climate, and that the climate was in steady-state or else changing so slowly as to not affect the distribution of pollen as measured.) In addition to these correlations, another kind of analysis is needed to provide insight into the non-steady-state, transient responses of vegetation to a non-steady-state climate, so that the rate of change of forests can be estimated. The forest model offers a method to achieve this analysis.

Since the late 1980s, other scientists as well as my colleagues and I have used JABOWA and its variants to gain insight into the possible effects of global warming on forests (Botkin, Nisbet, and Reynales, 1989; Botkin, Woodby, and Nisbet 1991; Pastor and Post, 1988; Solomon, 1986). The first issue one faces is how to combine projections from global circulation models, which operate on a coarse spatial scale in which one unit of the Earth is several degrees of latitude and longitude, with the forest model, which operates only on a small spatial scale, in our case 10- by 10-m plots. Our solution is to assume that projections of the climate models are uniform over a single unit. Within a grid unit, we select a first-order weather station that will provide recent climatic records. Then either a set of forest sites near this weather station are located, or hypothetical sites are defined, with certain site characteristics, and the population on these sites is generated by the forest model. Our initial work has concentrated on forests of the midwestern states of the United States, including an area of importance to commercial forestry, a designated wilderness, the habitat of an endangered species, and an area that is now primarily in agriculture, with small farm woodlots, but that was originally near the prairie–forest border. The prairie–forest border site is near Mount Pleasant, Michigan, a heavily settled area where commercial forests are still an important economic resource. Forests are transitional between northern hardwoods and oak-dominated forests. To the north, a site near Virginia, Minnesota (encountered in Chapter 6), represents heavily forested lands in and near the Superior National Forest, where commercial forestry has been important, and that Forest's legally designated wilderness, the Boundary Waters Canoe Area (BWCA), important for recreation and biological conservation. A third site, near Grayling, Michigan, was chosen to represent the managed habitat area of the Kirtland's warbler, a site that will be discussed in Chapter 8.

Projected climatic changes were derived from several global climate dynamics models, whose output was modified especially for these studies by Dr. R. Jenne of the National Center for Atmospheric Research. The climate models we used are the NASA GISS "normal" climate; GISS "twice-CO_2" climate; GISS transient A and transient B; Princeton University's GFDL "twice-CO_2" climate; and the Oregon State University's (OSU) "twice-CO_2" climate. The most interesting and important of these scenarios is the transient provided by the NASA GISS model, because it projects changes that might occur from 1980 into the future for 90 years. The other scenarios are for a steady-state condition where the greenhouse effect is equivalent to that imposed by a carbon dioxide concentration twice that present in 1980. While this is interesting theoretically, it is a condition that would exist only temporarily, and its use does not allow analysis of the transient response of forests to global warming.

Output from the general circulation models are used to modify real weather records that are obtained in computer form for stations in the Great Lake states and made available by R. Jenne of the National Center for Atmospheric Research. At each weather station, 30-year records (1951–1980) were prepared for control (normal) conditions and treatments (weather as modified by the output from climatic dynamic models). In control runs, the JABOWA-II model used the actual 30-year weather record, with the record repeated for simulations of more than 30 years in a direct sequence, so that the weather for year 31 is the same as that for year 1, the weather for year 32 is the same as that for year 2, and so on.

In each treatment, projected weather records were prepared as follows: ratios between the model's "normal" steady-state output and the CO_2-enhanced climate were calculated by R. Jenne and provided in computer format. Ratios were calculated for mean monthly temperature and mean monthly rainfall. The actual mean monthly temperature and precipitation were then multiplied by the appropriate ratios to generate "treatment" climates.

To model forest growth in the Boundary Waters Canoe Area, where there are no local weather stations closer than the one at Virginia, Minnesota, weather records were modified to represent an area to the north, following an approach we have used for other sites (Botkin et al., 1972, 1977). We assumed that an area to the north of a weather station would experience weather equivalent to that at an elevation higher than that of the weather station. Therefore we used standard temperature and precipitation lapse rates and selected a site at a higher, imaginary elevation above the Virginia, Minnesota, weather station. In the work reported here, 60 replicates were obtained for each trial. Each replicate began with 1951 data and followed the same weather sequence.

Projected Response of Midlatitude Forests
to a Rapidly Changing Climate

We conducted a number of experiments using both steady-state twice-CO_2 climate projections and the projected transition in climate from 1980 to 2070 that was developed by NASA Goddard (Hansen et al., 1988). The latter is the most useful and interesting, and I will concentrate on these results. Readers interested in more details about the twice-CO_2 steady-state experiments can find these discussed in Botkin et al., (1989b).

The next problem that one faces is how to select a few representative forest sites out of the huge number that could be chosen. We chose to concentrate on four kinds of sites, each with a different soil. We used all four sites in some experiments, but not all sites for all kinds of forest stands. The four sites are: a deep, comparatively dry, coarse soil; a deep, relatively wet, fine soil; a shallow, wetland, heavy-clay soil; and a shallow, dry, coarse soil. These provide a broad range of forest conditions, from bogs dominated by cedar and old balsam fir to early successional stands on coarse, thin soils dominated by aspen and white birch.

Global climate models project very great changes in temperature and rainfall with global warming. These models are known by their acronyms; the NASA model of Hansen et al. (1988) is known as the GISS model; that of Schlesinger and Zhao (1988), developed at Oregon State University, is known as the OSU model; and a model developed at Princeton University is known as the GFDL model. The GISS, GFDL, and OSU twice-CO_2 scenarios convert Mount Pleasant from a place where winters commonly have months with mean temperature below freezing to a place where there are series of years in which no month has an average temperature below freezing. The GISS and GFDL models increase January temperatures approximately 10°F (5°C). While the mean January temperature never exceeded the mid-20°sF (never reaching 0°C) from 1950 to 1980, the projected interval for Mount Pleasant ranges from a high in the 30°sF (0° to 10°C range) to a low in the mid-20°sF (below 0°C). The 1951–1980 maximum January mean value becomes the minimum value in the projected climatic warming. The OSU model projects January means that are only slightly colder.

Similar increases occur for July mean temperatures. The GISS model projects average temperatures from the mid-70°sF to above 80°F (25°C), while the 1951–1980 average temperatures range from the high 60°sF (about 20°C) to the mid-70°sF (about 25°C). The GFDL model projects even more severe summer temperatures, with July means exceeding 90°F (33°C) and never descending below 82°F (27°C). The OSU model projects somewhat cooler July mean temperatures than the two other models, with most values remaining in the low to mid-70s°F (near 20°C).

Several projections have been made using the NASA GISS model. One, known as the "transient-A scenario," assumes a "business-as-usual" burning of fossil fuels in the future, so that there is an increasing rate of emission of carbon dioxide as per capita industrial output is maintained at 1980 levels into the future and population increases; another projection, known as the "transient-B scenario," assumes some energy conservation. Although we have made projections for both scenarios, for the sake of simplicity I will concentrate on the projections from the business-as-usual transient-A scenario from the NASA climate model.

The GISS model transient A projects an increase in July mean temperatures above the 1951–1980 average after 1990. Transient B, whose projections continue only 60 years after 1980, shows increases on the order of several degrees Fahrenheit (about 1 to 2°C) by the fifth decade. Such summer increases might greatly increase evaporation of water from soils and trees, and lead to a much drier environment.

Applying these temperature regimes as input to the forest model, we find that global warming would lead to major changes in forest composition (Figs. 7.1 and 7.2). Near Virginia, Minnesota, northern hardwood forests, now characteristic of areas to the south, replace boreal forests. Which species dominate varies with soil type and soil-water conditions. On the site in the Superior National Forest, where balsam fir now dominates, sugar maple would become the dominant: in general, northern hardwood species would replace boreal forest species (Fig. 7.1). Upland sites where white birch or quaking aspen are now dominant are also taken over by sugar maple. White cedar bogs become treeless, because there is no ecological analog to northern white cedar immediately to the south in the midwestern states.

To the south, near Mount Pleasant, Michigan, hardwood forests that are today transitional between sugar maple-dominated hardwoods and oak forests are projected to change to oak woodlands or savannahs, found to the south under current conditions, or even to treeless prairies found farther west (Fig. 7.2).

Wood production and the accumulation of biomass can be greatly affected, but the effect depends on soil and soil-water conditions. Forests on dry, sandy soils at the Mount Pleasant site change to prairie and savannah, so that wood production declines to an insignificant amount. Soils with abundant water may continue to support trees, but with a somewhat lower wood production and biomass. At the Superior National Forest site, wood production ceases on cedar bogs converted to treeless shrub bogs. In contrast, upland sites that are presently too saturated with water for maximum tree production experience an increase in wood production and biomass, as the saturated soils dry out and the sites improve for tree growth.

The most impressive aspect of these projections is the rate at which changes occur (Figs. 7.3–7.8). Significant changes in the forests are projected to occur as early as the year 2010, and no later than year 2040. The transient climatic regimes lead to surprisingly rapid changes, but the rate and kind of changes are very sensitive to soil moisture conditions. By 2010, a 400-year-old stand dominated by balsam fir on deep, fertile, moist soil changes to a stand with one-third of the balsam fir basal area under the 1951–1980 weather regime

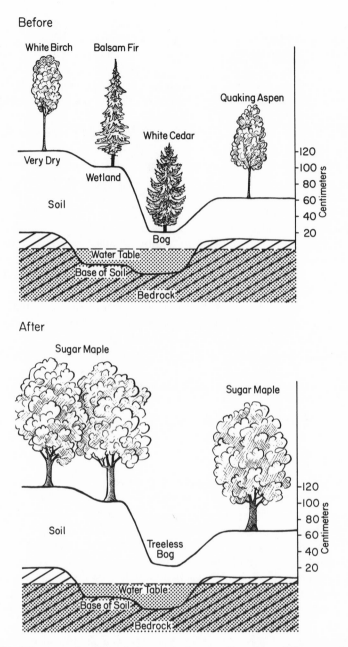

FIGURE 7.1. Diagram of projected change in forests of northern Minnesota beginning in 1980 (Before) and continuing for 90 years (After) under the 1950–1980 climate record.

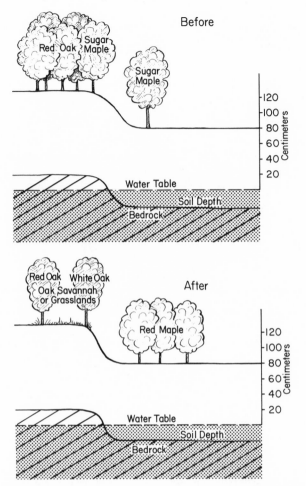

FIGURE 7.2. Diagram of projected change in forests of southern Michigan beginning in 1980 (Before) and continuing for 90 years (After) under the GISS transient-A global warming climate.

(Fig. 7.3(A)). Sugar maple replaces fir as dominant and its total biomass nearly triples (Fig. 7.4(B)). In a wetland, a 400-year-old white cedar forest declines to a treeless bog with total biomass less than 0.1 kg/m^2 (Fig. 7.4).

On a drier but fertile sandy upland soil, white birch, dominant under normal climate, declines to about 10 percent of its initial level in about 40 years, and it too is replaced by sugar maple (Fig. 7.5(A, B)). After 90 years, total biomass declines to about one-half of normal (Fig. 7.6(A, B)). Thus the response of the forests is very sensitive to soil conditions. This is because the climatic changes lead to a great increase in evapotranspiration; although rainfall increases, the evaporative losses increase more and water becomes limiting. This explains why upland sites can be converted to savannahs while wetlands can maintain substantial, if reduced, forest growth of species presently dominant farther south.

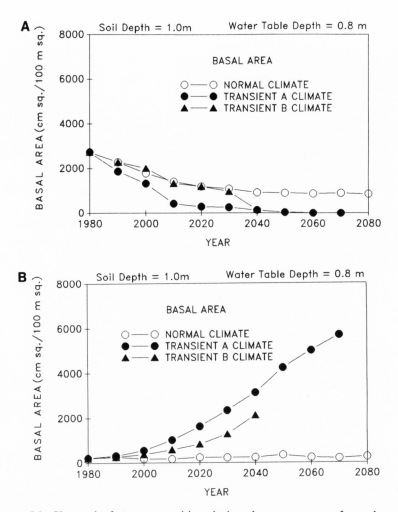

FIGURE 7.3. Change in forest composition during the next century for a deep, wet, sandy soil in the Boundary Waters Canoe Area. Projections are for a 400-year-old balsam fir stand that is characteristic of northern Minnesota and the northern portion of the Great Lakes states. (A) Changes in balsam fir, the initial dominant species. (B) Changes in sugar maple, which is present but of minor importance under the 1950–1980 climate, but increases under global warming.

FIGURE 7.4 (*top*). Changes in forest composition during the next century for a cedar bog in the Boundary Waters Canoe Area. Projections are for a 400-year-old white cedar bog that is characteristic of certain water-saturated soils in northern Minnesota and throughout the northern portion of the Great Lakes states.

FIGURE 7.5 (*bottom*). Changes in white cedar basal area during the next century for a deep, dry, sandy soil in the Boundary Waters Canoe Area. Projections are for a 400-year-old forest dominated by white birch in 1980. Such stands are typical in northern Minnesota and throughout the northern portion of the Great Lakes states. (A) Changes in white birch. (B) Changes in sugar maple.

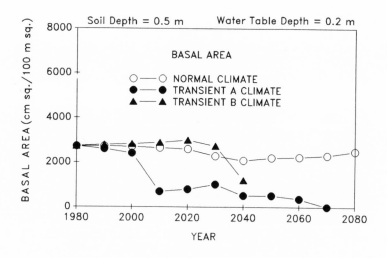

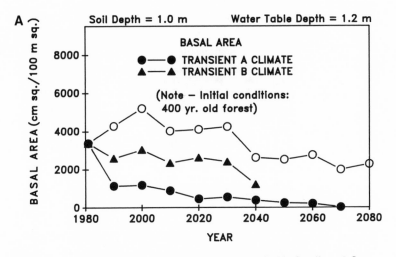

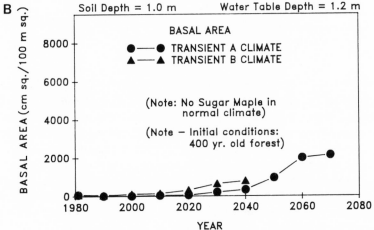

187

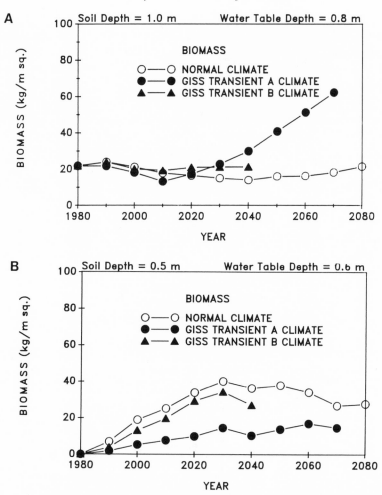

FIGURE 7.6. Changes in balsam fir biomass for (A) the balsam fir stand shown in Figure 7.3 and (B) a clearcut stand on a thinner, drier soil representative of the Boundary Waters Canoe Area and northern Minnesota.

Upland well-drained sites that begin from a clearing develop under the 1951–1980 climate regime as typical young boreal forests dominated by trembling aspen. Under the twice-CO_2 climates, succession from a clearing follows patterns typical of northern hardwoods forests. Upland sites are soon dominated by sugar maple, while wetlands (which under the normal climate develop into larch) regrow under global warming climates to red maple. Thus both old-age forests and successional stands developing from a clearcut change from boreal to nothern hardwoods. And in these modified northern hardwoods, sugar maple seems to increase in importance and dominate a greater variety of sites. By inference, there seems to be a decline in vegetation diversity.

Results are similar for the site to the south near Mount Pleasant, Michigan, where we projected growth following a clearcut. Under the normal climate, white oak dominates, sugar maple is second in importance, and red oak and white pine are also important (Fig. 7.7(A)). Under global warming, sugar maple and white pine are unimportant throughout the period, the climate having become too warm or dry for those species (Fig. 7.7(B)). Red oak replaces white oak as dominant, but red oak declines rapidly toward the end of the period.

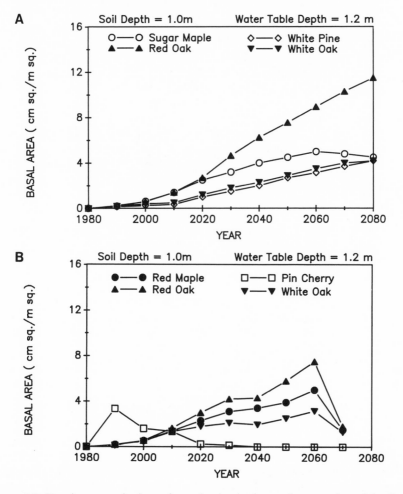

FIGURE 7.7. Development of a forest from clearing in the southern portion of the Great Lakes states. The forest represented here is near Mount Pleasant, Michigan. (A) A deep, dry, well-drained, sandy soil and "normal" (1951–1980 Mount Pleasant) weather records. (B) The same site as in (A) with weather records modified by the GISS transient-A climate, representing the transition from current conditions to climate under increasing CO_2 concentrations. (C) Conditions as in (A) except that the water table is higher and the site wetter. (D) Conditions as in (B) except that the water table is higher (as in C) and the site wetter.

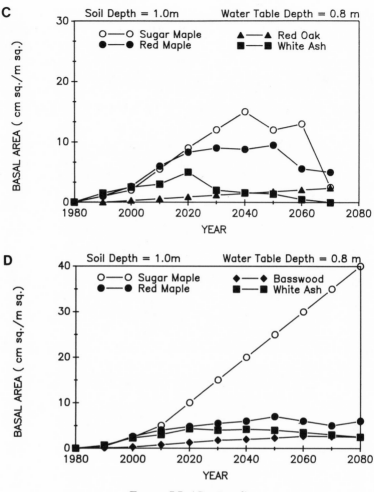

FIGURE 7.7. (*Continued*)

Red maple and white oak at first increase, but then decline rapidly by the end of the period. By 2070 only a few small trees remain.

On a somewhat wetter site, sugar maple attains a high abundance under the normal climate (Fig. 7.7(C)) but disappears by the end of the global warming period (Fig. 7.7(D). On both drier and wetter sites, total tree biomass drops precipitously and is low by the end of the period (Fig. 7.8(A,B)). Although the response is slow, the effects are drastic at this site. If grasses could grow under the global warming conditions, the forests would be converted to savannahs or prairie, which did occur in small pockets in presettlement times. (In fact, during a warm period between the end of the last continental glaciation and the twentieth century, thousands of years ago, a prairie "tongue" extended from the Mississippi River through what is now Iowa all the way to Long Island, in what is now New York State, accounting for the present occurrence on the east

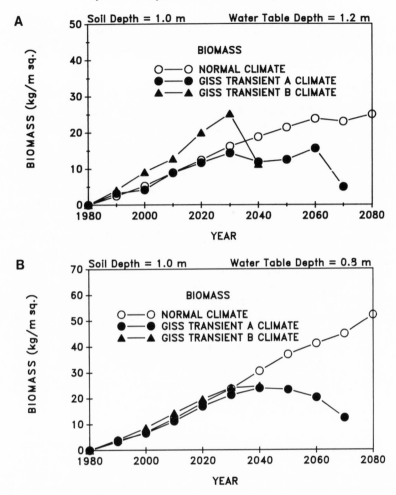

FIGURE 7.8. Changes in biomass during the development of a forest from clearing in the southern portion of the Great Lakes states. The forest represented here is near Mount Pleasant, Michigan. (A) Deep, dry, well-drained, sandy soil (normal 1951–1980 Mount Pleasant records). (B) Wet, sandy soil (weather records as modified by the GISS transient-A climate, representing the transition from current conditions to climate under increasing CO_2 concentrations).

coast of the United States of prairie grasses such as little bluestem, *Andropogon scoparius*.

The effects of global warming on the forest would extend beyond species dominance. Chemical cycling, storage of organic matter, and rates of decomposition differ between conifer-dominated boreal forests and northern hardwood forests. Hence, major changes in vegetation could lead to large changes in the ecosystems. The flux of chemical elements from forests to streams could change.

The habitat for wildlife would be altered, and dominant species of wildlife could change. For example, boreal forests suitable today for moose might become favorable for white-tailed deer. It is possible that the entire character of the Boundary Waters Canoe Area would be altered.

Comparison of Projected Effects on Forests of Transient Climates with Global Warming Steady-State Climates

The transient results are consistent with the projections for twice-CO_2 steady-state climates under all three models, as illustrated here for Mount Pleasant weather records. For twice-CO_2 steady-state conditions, the three climate models differ in the magnitude of their projections. The GFDL model projects the most severe climate changes; the OSU model projects the mildest changes. On drier sites under normal climate, total biomass accumulation after 100 years of growth exceeds $20 \, kg/m^2$, while under the GISS and OSU twice-CO_2 climates, biomass accumulation after 100 years is less than $10 \, kg/m^2$, and under the GFDL twice-CO_2 climate, tree biomass is negligible.

Biomass accumulation is greater on wetter sites under all steady-state climates, but the GFDL model projections lead to much less biomass, approximately $10 \, kg/m^2$ compared with the normal forest biomass of approximately $50 \, kg/m^2$ at year 100. On dry soils, major changes occur in dominant species. With the GISS model twice-CO_2 steady-state climate, the forest model projects an open savannah of low biomass (biomass $< 2 \, kg/m^2$, compared with an average of $25 \, kg/m^2$ under normal climate). The global warming forest is dominated by oaks, stems of which are sparse and small even after 100 years. While the normal forest would have commercially useful hardwoods, the forest under the steady-state global warming climate would not produce commercially useful hardwoods even after 100 years. Under such a shift, wildlife species would change from those adapted to the more northerly closed forests to those adapted to grassland—savannah.

The GFDL model gives an even more extreme result: at year 100 many very small red and white oak and red maple trees remain, but these contribute almost no biomass. For example, at year 100 there is an average of only $13.6 \, cm^2/100 \, m^2$ red oak, equivalent to a single sapling with a diameter of $4 \, cm$ at breast height. Assuming grasses could survive under these conditions, one expects that the forest would be converted to a savannah or a grassland sparsely populated with very small trees. Although the OSU model projects the mildest climatic change, its twice-CO_2 climate leads to a sparse, open forest with $6 \pm 3 \, kg/m^2$ biomass, which would appear to vary from a savannah to an open woodland. Consistent among the projections for the three steady-state climates is a considerable decline in biomass accumulation that occurs after the third decade on drier sites and after the seventh decade on the wetter site. The GISS and GFDL climates result in negligible biomass accumulation on the drier site for all time periods. This suggests that upland areas would be converted to savannahs or open woodlands.

On deep, well-waterd sites with sandy soils, under normal conditions the forest is transitional, dominated by sugar maple and red maple along with other species characteristic of wetlands and flood plains, including white ash and hemlock. The GISS model twice-CO_2 climate results in a forest dominated heavily by red maple with some oaks. None of the northern hardwood forest species occur. Red maple increases in basal area under the three steady-state twice-CO_2 climate regimes. Sugar maple continues to increase throughout the 100-year simulation under the normal climate, but no sugar maple grows in any of the steady-state twice-CO_2 climates.

Limitations of the Projected Responses of Forests to Global Warming

The general pattern that emerges is that the CO_2-induced climatic change leads not only to much warmer but also to much drier conditions than occur at present. Although rainfall increases, total evapotranspiration increases more. As a result, less soil moisture remains for tree growth. Wetter sites become warmer and somewhat drier and biomass generally declines. Sandy wetter sites are able to support forests, but resulting forests are characteristic of areas to the south. The dominant species shift from those with commercial value to those of little commercial value. On drier sites the climatic shift is severe enough to convert substantial forests to open woodlands, savannahs, or grasslands with small scattered trees.

It should be emphasized that the results given here are for good sites for forest growth. Soils are deep. Even the relatively sandy soils are moderately fertile. Wetter sites have minimally saturated soil. Thus the projections may be more optimistic than what could occur in general. Additional experiments with the models on a greater variety of soils would be useful to improve the realism of the projections.

It is important to consider limitations and possible errors associated with these projections. We can categorize these concerns into: unaccounted for factors that might reduce the rate of climate change and unaccounted for factors that might reduce the response of trees. Some of these are legitimate concerns, but others, often brought up in discussions of global warming, have relatively little basis in fact or in the dynamics of forested ecosystems.

Among the most important concerns are factors that might change the projections of the general circulation models. The models that provided the climate projections used here were developed in the 1980s. They lacked a dynamic model of ocean currents and of land vegetation and made very simple assumptions about soils. It is possible that changes in ocean circulation could increase the transfer of heat energy from a warmed atmosphere to the ocean waters, thereby delaying or decreasing the global warming response. It is also possible that land vegetation, which affects the climate through the uptake of carbon dioxide, reflection of light, evaporation of water, and changes in the surface roughness, could alter the boundary conditions for the atmosphere and thereby change the climate projections. There are some possibilities of biota–climate positive feedbacks that might enhance global warming. For example, if permafrost melts over large areas, large amounts of organic matter might decay, releasing methane as well as carbon dioxide and adding to the greenhouse effect. If forests change to open woodlands or savannahs, then the flux of carbon from the atmosphere into the trees would decrease, and the loss of forest soils would, at least temporarily, add to the transfer of carbon dioxide to the atmosphere and increase the greenhouse effect. The concerns with these possible effects are legitimate. What has been presented in this chapter is the projection of forest responses to presently available projections of climate change.

There are also some climatic changes to which trees might respond, but which the present forest model does not take into account. For example, in some boreal forest areas at high latitudes, projected temperature changes occur primarily in the winter, raising the temperature in part of Siberia from $-40°C$ to $-35°C$. There might be an effect on the frost hardiness of trees, but there is little empirical information about such an effect, and the model does not include effects on trees from changes in temperature below $0°C$, except as these affect snowmelt, as explained in Chapter 2. Applications of the model discussed in Chapter 4 suggest that the model is realistic without such details, but it is useful for the reader to be aware of those factors not accounted for in the model.

Some plant physiologists have claimed that global warming will not lead to a decline in forest growth, and might actually increase forest growth, on the grounds that an increase in the carbon dioxide concentration in the atmosphere has several potential benefits, including acting directly as a fertilizer and acting indirectly to reduce water use. This conclusion is based on experience with plants grown in greenhouses or in otherwise controlled conditions and supplied with as much water and fertilizers as they need (Bingham, Rogers, and Heck, 1981; Bryan and Wright, 1976; Canham and McCavish, 1981; Downton, Björkman, and Pike, 1980; du Cloux and Vivoli, 1984; Funsch, Mattson, and Mowry, 1970; Hollinger, 1987; Jurik, Weber, and Gates, 1984; Kimball, 1983; Koch et al., 1987; Kramer and Sionit, 1987; Leverenz and Lev, 1987; Oberbauer, Strain, and Fetcher, 1985; Oberbauer et al., 1986; Oechel and Riechers, 1986; Wigley, Jones, and Kelly, 1980). In such experiments green plants do grow faster when the CO_2 concentration is increased.

We have added a carbon dioxide environmental response surface to the model. As is usual, the major problem is the lack of appropriate data to determine the shape of this response surface and the parameters. Although there are many studies of the direct effects of carbon dioxide on plants, most are for crops or other annuals. All studies grow plants with ample water and fertilizers (Hardh, 1986; Krizek, 1969). As mentioned earlier, those concerned with woody plants focus on individual stems grown in greenhouses or other artificial facilities (Bryan and Wright, 1976; Funsch, Mattson, and Mowry, 1970; Jurik, Weber, and Gates, 1984; Kramer and Sionit, 1987; Oberbauer et al., 1985; Rogers, Thomas, and Bingham, 1983; Sionit, Strain, and Helmers, 1985; Tolley and Strain, 1985). However, there is a lack of standardized methods, so that comparison between studies is often not possible. One set of plants is grown at one high light intensity and one level of humidity, while another experiment has other light intensities and other humidity levels.

We found a few studies where published data were sufficient to allow calculation of the CO_2-fertilization effect for seedlings and saplings of four tree species, each representing a set of species with similar ecological successional characteristics:

$Y = -0.050 + 0.0042X - 0.000002X^2$	(*Populus deltoides*) (Carlson and Bazzaz, 1980)
$Y = 0.840 + 0.0005X - 0.0000002X^2$	(*Platanus occidentalis*) (Carlson and Bazzaz, 1980)
$Y = 0.457 + 0.0018X - 0.0000008X^2$	(*Pinus banksiana*) (Yeatman, 1970)
$Y = 0.625 + 0.0012X - 0.0000005X^2$	(*Picea glauca*) (Yeatman, 1970)

where X is the CO_2 concentration (ppm) and Y is the ratio between the growth of the control at 340 ppm CO_2 and the growth of the treatment at the enhanced CO_2 concentration. Early successional broadleafs are represented by *Populus deltoides*; late successional broadleafs by *Platanus occidentalis*; early successional conifers by *Pinus banksiana*; and late successional conifers by *Picea glauca*. Estimation of increases in CO_2 concentrations follows that of the GISS climate model projections:

$$X = 340 + 1.2373t + 0.0205t^2 - 0.00004678t^3$$

where X is the CO_2 concentration at year t (years after 1980) (Botkin and Nisbet, 1990; Hansen et al., 1988).[1]

We carried out simulations for the sites in Virginia, Minnesota, used above. We found no effect on the projections. In fact, when we increased the fertilization effect to 50 and 100-percent increase in growth per year, over what the growth would have been otherwise, we found no effect except for sugar maple, which showed a small increase under global warming at this site. What these projections tell us is that if water and temperature are the problems, fertilizers are not the solution. It is as if a farmer, faced with a drought, kept adding nitrogen fertilizer to the soil and wondered why the crops did not grow.

These results are similar to those we reported long ago for the effects of CO_2 fertilization in a steady-state climate (Botkin et al., 1973b), and to results from another study of Michigan forests, using a model derived from JABOWA-I that considers climate interaction and immediate doubling of CO_2 (Solomon and West, 1987).

Sensitivity of the Projected Climate Change Effects to Choice of Weather Records

One other concern with the projected effects of global warming on forests has to do with how trees respond to the variability of weather. To study this, we have investigated how sensitive the model's projections are to weather patterns (Botkin and Nisbet, 1992).

All tests were made for the same standard plot used in Chapter 6 near Virginia, Minnesota, with soil depth 1 m, depth to the water table 0.8 m, and soil moisture-holding capacity 250 mm/m and on which is a mature stand of balsam fir with 0.081 ± 0.002 stems/m^2 occupying 27.9 ± 1.5 cm^2/m^2 basal area, and a moderate soil fertility. As I have said before, such a stand is typical of an undisturbed forest on a moderately wet site in the boreal forest.

Projections of mean monthly rainfall and temperature were obtained from the GISS global circulation model of NASA Goddard Space Flight Center, New York, for a transition from current CO_2 levels to a gradual doubling of CO_2 concentration during the next 90 years (Hansen et al., 1983). The experimental control climate is the 1951–1980 weather record from Virginia, Minnesota, the nearest weather station to the Superior National Forest and the BWCA.

The sensitivity of the model is tested against four alternatives to the control weather patterns. These alternatives are assembled from different combinations of Virginia weather records, and are: (1) the 1951–1980 30-year record with iterations starting with various years; (2) the warmest decade (1952–1961); (3) the coldest decade (1942–1951); and (4) an 80-year (1900–1980) record.

For conciseness, results are restricted to the response of balsam fir and sugar maple, the first representing the boreal forest and the second representing the northern hardwood forest. Their responses in terms of magnitude and timing are typical of other species, although shorter-lived species tend to respond faster and longer-lived species respond slower.

Two kinds of tests are reported here. In the first, effects of the choice of weather records are shown for "normal" conditions. In the second, the projected global warming transient scenario is compared with its equivalent normal scenario. For example, the 80-year "normal" weather record is compared with the 80-year "transient" weather projection. From the first kind of test we can learn about the projected quantitative response of species to choice of "normal" weather patterns. From the second kind of test we can learn about the projected timing of the changes.

Of the four experimental weather patterns, only one, the coldest decade, leads to a statistically significant difference in the projected basal area of balsam fir, and in that case only for the years 2040 and later, where the difference is minor. Use of the coldest weather record slightly delays the decline in balsam fir, and use of the warmest or the random weather record slightly accelerates the decline of balsam fir. Use of the 30-year control weather record yields a relatively conservative estimate of change in the basal area of balsam fir. Other factors, including competition for light among the tree species and soil-water conditions, also act to moderate the responses of balsam fir.

Projections for sugar maple tell a similar story. When the 80-year weather record is used, the forest model projects a statistically significant decline in this species compared with the control. This appears to be the result of an increase in evapotranspiration and a decrease in the availability of soil water. The slight drying under still cool conditions appears to favor balsam fir at the expense of sugar maple.

The degree-day minimum for sugar maple is 2200 and the maximum is 6300, so that maximum growth occurs at a degree-day value of 4450. The use of the warmest decade shifts growing degree-days toward the optimum for this species, and leads to a large and rapid increase in sugar maple abundance. The use of the coldest decade leads to a significant decline in sugar maple, similar to that found with use of the 80-year weather record.

Are the projections of the effects of climatic change on forests likely to be wrong because the selection of a "normal" weather record has biased our results? The simple answer is no. The choice of weather records has no effect unless we select the coldest or warmest decade of the twentieth century as standard, and even in this case the effect is to change the relative abundances of species—there is only a slight effect on the projected timing of major changes in the forests.

The robustness of the results against the choice of normal weather records is the result of the great magnitude and rate of the projected global warming. To put the projected response of forests to this warming in perspective, when "little ice age" climates were applied to the forest model, no significant change in any species occurred, consistent with the pollen records (Davis and Botkin, 1985). Thus the projected global warming would have a much greater effect on forests than that classic cooling episode.

Effects of Acid Rain on Forests

Another important large-scale environmental problem is acid rain. Whether acid rain has an important effect on forests remains controversial. This is not to say that acid rain might never affect forests. Certainly, there are locally acute situations where rain of high acidity will kill trees. Typically, however, such acid rain is produced from a nearby metal-ore smelter or coal-burning power plant. Such a facility also releases heavy metals that are also toxic to trees, and it is difficult, if not impossible, in the field to distinguish the effects of acid rain from the effects of heavy metals. The more characteristic ambiguity has to do with observed widespread declines in forests that are subjected to chronic, low levels of acid rain. These declines have been observed in locations as distant as Europe (Miller, 1980) and New England.

An additional difficulty in determining the effects of acid rain on forests is due to the longevity of trees and their ability to withstand many forms of disturbance. Acid rain could affect trees adversely in three ways: (1) direct damage to leaf tissue, decreasing photosynthetic rates of individual plants (Tamm and Cowling, 1977); (2) indirect effects of the loss of leaf tissue on the forest community; and (3) leaching of soil nutrients by the acidic water (Aber et al., 1982; Norton, Hanson, and Campana, 1980). Since acid rain is formed primarily from sulfur and nitrogen oxides, it is also possible that these elements may enrich the soil and fertilize tree growth. In fact, some experiments have shown this positive response (Woods and Bormann, 1977).

The possible effect of direct damage to leaf tissue was tested for forests in the northeast using the forest model and estimates of the percent leaf mortality for major species, based on field measurements from the mid-1970s on Long Island east of New York City (Dr. Lance Evans, personal communication, 1978). At that time, the percent leaf mortality due to acid rain damage was small: spruce suffered 0.5-percent leaf mortality; balsam fir 1.0 percent; white birch 2.5 percent; yellow birch 2.5 percent; pin cherry 4.0 percent; and sugar maple 5.5 percent.

To investigate possible effects of this level of leaf damage to the forest, the forest model was modified. Each year the amount of leaf tissue on a tree was decreased by reported percentages of leaf mortality. Two kinds of experiments were conducted: (1) Leaf tissue killed by acid rain was removed from the tree, decreasing the photosynthetic rate of the affected tree and opening the canopy to more sunlight. (2) Dead leaf tissue remained on the tree, decreasing the photosynthetic rate of the tree, but not changing the light conditions.* In both cases, the development of a forest from a clearing was projected for 600 years using Hubbard Brook Experimental Forest climatic conditions.

Reductions in leaf tissue of the percentages representing the best estimates of current acid rain damage lead to no statistically significant change in the forest nor in the abundance of any species (Fig. 7.9). Apparently, the percentage leaf mortality is too small. This result is consistent with natural history observations of an experienced wildlife biologist. For example, in the research in which I participated on Isle Royale National Park, it was difficult to measure changes in production of woody plants resulting from removal of 10 or 20 percent of leaf tissue from browsing by moose. It was even difficult to devise experiments to test the effect adequately. The model's response seems therefore to be consistent with natural history observations.

In another way, however, the model gives counterintuitive results (at least counterintuitive for me). Contrary to expectations, experiments of type 2 and

---

FIGURE 7.9. Simulated Response of Forest Biomass to Acid Rain. Control is the forest with default parameters. "Best" means the best available estimate of the effects of prevailing acid rain on each species, as given in the text. YB refers to yellow birch, P to pin cherry, G means the acid rain affects growth; L means acid rain affects light, by assuming lead tissue killed falls away from the rest of the leaf, permitting light to penetrate where there is no leaf tissue. "Best × 10" means that the best estimates of actual acid rain effects on leaf mortality are multiplied by 10. "YB (0.75 G)" means that the effect on yellow birch is reduced to 75% of the estimated actual leaf mortality due to acid rain, and so forth. (Redrawn from D. B. Botkin, and J. D. Aber, 1979, some potential effects of acid rain on forest ecosystems: implications of a computer simulation, Brookhaven National Laboratory Publication BNL 50889, National Technical Information Service, Springfield, Va. (A) is from Figure 8 of that reference; (B) from Fig. 3, and (C) from Fig. 1).

---

*This material was published in Botkin, D. B. and J. D. Aber, 1979. Some potential effects of acid rain on forest ecosystems: Implications of a computer simulation. Report to Brookhaven National Laboratory BNL 50889 UC-11, National Technical Information Center, Springfield, Va., 12 pp.

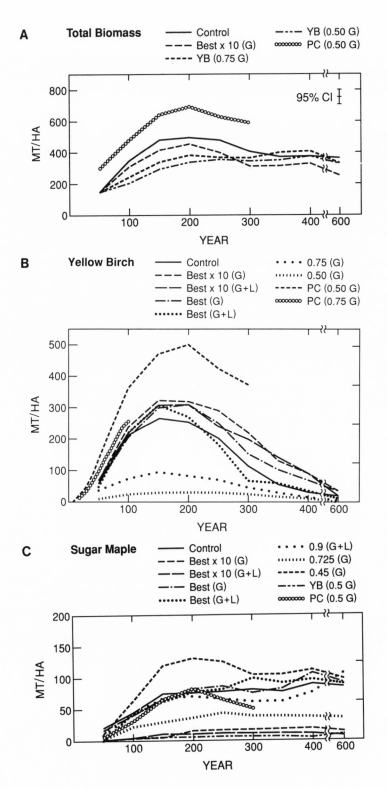

199

type 1 lead to identical results: the increase in light available at the forest floor does not increase the growth of small trees sufficiently to affect biomass, either of a single species or of the forest. Perhaps this is because the percent leaf mortality is so small that the increase in light is trivial.

Additional experiments were conducted to determine what level of leaf damage would alter the abundances of species. A growth reduction of 10 times the best estimates of actual acid rain damage decreases total forest biomass by 10 percent, which is not statistically significant at the 95-percent confidence level from the control forest (Fig. 7.9). Interestingly, there were significant effects on total biomass when the effects were limited to a single species. For example, a 50-percent reduction in pin cherry growth led to a large increase in total biomass for 300 years following clearing (Fig. 7.9(A)). While the control forest reached a peak biomass of 400 metric tons (mt) per hectare, the forest with only pin cherry subjected to growth reduction reached a peak biomass of more than 600 mt/ha. A reduction of 25 percent and 50-percent increase in yellow birch growth led to a statistically significant reduction in total forest biomass, with peak biomass remaining below 400 mt/ha. The timing of the peak biomass did not change, but occurred at year 200. By year 350, the total biomass for all treatments approached that of the control.

Sugar maple biomass increases greatly through year 400, when yellow birch is subjected to growth reduction of 50 percent, but the same reduction in pin cherry has little effect on sugar maple, and none after year 100 (Fig. 7.9(C)). However, a decrease of 50 percent in yellow birch growth leads to a reduction of more than 90 percent in the biomass of that species. The explanation for these results lies with the assumptions of the model. As explained earlier in this book, pin cherry dominates the first stages of succession under normal conditions in a New England forest. When the abundance of pin cherry is reduced, yellow birch has a competitive advantage and becomes more abundant than in the control forest (Fig. 7.9(B)). When acid rain damages pin cherry, yellow birch benefits. Because yellow birch is highly productive and comparatively long-lived, its increased abundance changes the entire forest for two centuries (Fig. 7.9(A)). The decline in yellow birch also leads to a significant decline in total biomass.

The acid rain experiments have a use beyond the question of pollution effects. These experiments give us insight into the complexities of competition among more than three species of trees, each species composed of many size classes. One interesting insight is that the measured decline in photosynthetic output may not be a good indicator of the percentage decline in biomass of that species in the forest. There are a few other applications of models, including ones derived from JABOWA, to investigate the possible effects of acid rain on forests (Dale and Gardner, 1987; Kercher and Axelrod, 1984a; Krupa and Kickert, 1987).

There are potentially important effects of acid rain on forest ecosystems not considered in these applications of the forest model. For example, sulfur and nitrogen oxides, the primary pollutants that make rain acid, can also function as fertilizers, and it is said that in some areas there is no longer any need to add sulfur as an agricultural fertilizer because so much is deposited from acid rain. Acid rain can also make soil less fertile, by increasing the solubility of cations and increasing their loss from soil in water runoff. A complete test of acid rain effects on forests would take these factors into account.

It has long been my belief that some of the most important insights in ecology have come from analysis of what are generally referred to as applied problems. Perhaps this is because we lack so much information about ecosystems that applied problems focus our attention on important phenomena and guide our studies. Readers familiar with the history of science will know many precursors to this connection between applied and fundamental research. For example, advances in the nineteenth century in thermodynamics were motivated by a desire to develop more efficient engines. Understanding of the science of aerodynamics certainly gained much from the Wright brothers' development of the first airplane. The typical academic prejudice against applied problems is not well-supported by the history of science. The motivations for the development of JABOWA were both practical and fundamental, and many of the insights that I have personally obtained from it derived from attempts to examine environmental problems.

Summary

In this chapter I have applied the JABOWA model to the question of the effects of global environmental changes on forests. The model suggests that projections of global warming would have rapid and severe effects, and that changes in air temperature and evapotranspiration would overwhelm the direct fertilization effects of carbon dioxide on tree growth and forest conditions. I also examined the sensitivity of these projections to choice of weather records. In general, the projections are not sensitive to available weather records. In contrast to the effects of global warming, the model suggests that chronic acid rain that acts only directly on leaf tissue would have small effects on forests. Application of the model to these applied problems yields insights into the dynamics of forest ecosystems. These applications of the model illustrate its utility for large-scale environmental tissues. However, the present versions of the JABOWA model discussed in this book lack feedback to the environmental conditions—changes in the forest do not lead to changes in soil water or soil fertility such as has been added by Pastor and Post (1985), evapotranspiration, reflectance, or surface roughness. The model provides the basic information from which such feedback processes could be developed, and this is one direction that future extensions of the model could usefully take.

Note

1. Hansen et al. (1988) gave DT_0, the rate of change of annual mean global surface air temperature, between 1960 and 2050 as a function of x, the CO_2 concentration:

$$DT_0(x) = f(x) - f(x_0)$$

where

$$f(x) = \ln(1 + 1.2x + 0.005x^2 + 1.4*10^{-6}x^3)$$

x is the projected CO_2 concentration and $x_0 = 340$ ppm (estimated CO_2 concentration in 1980).

From their Figure 2, we determined CO_2 concentration as a function of temperature:

$$Y = 340 + 1.2373x + 0.0205x^2 - 0.00004678x^3$$

where Y is the CO_2 concentration at year x expressed as years after 1980.

8

Use of the Model in
Natural Resource Management

Out of fact comes the law.

Maxim of Roman Jurisprudence

Destructive torrents are generally formed when hills are stripped of the trees
that formerly confined and absorbed the rains.

Pliny, Natural History, Book 31, chapter 30, VIII, 4171
(quoted by G.P. Marsh in Man and Nature, 1864, p. 188)

With the disappearance of the forest, all is changed... the climate becomes
excessive and the soil is alternately parched by the fervors of summer, and
seared by the rigors of winter....

George Perkins Marsh, 1864

Forests and Civilization

The history of civilization repeats one story: civilizations depend on wood for
construction materials, for fuel, and for material for art and paper, and civilizations developed where wood was available. As civilizations progressed,
forests were cleared and deforestation occurred widely. In the beginning of a new
civilization, the clearing of the forest's dark wilderness was taken as a part of
civilizing and taming the world, making it safe and comfortable for people. The
preface to the U.S. Census of 1810 referred to "bothersome trees" that "encumber
a rich soil··· and prevent its cultivation" (Perlin, 1989). The story of civilization
is simultaneously the story of continued deforestation—of dependence on wood
yet destruction of the sources of timber—and exploration and conquest to
provide new sources of timber. This history has been documented by John
Perlin in *A Forest Journey* (1989). The destructive effects of civilization on
forests and forest soils, and the negative consequences of this destruction for
the persistence of civilization, were lamented long ago by George Perkins Marsh
in his landmark nineteenth-century book *Man And Nature* (Marsh, 1864).

In the first chapter I discussed briefly the value people place on forests and
the uses of forest ecosystems and forest products. In preparation for discussion
of the application of JABOWA to natural resource management, it is useful
to consider these in greater detail. Forests and woodlands are important not
only as commercial crops for lumber, paper, and pulp, but also for their role

in soil conservation and erosion control; water supply; habitat for wildlife; maintenance of streams in a form and shape that can support populations of fish; recreation for camping, fishing, hunting, and nature viewing; aesthetic values; and biological conservation. Woodlands containing endangered species, and forests designated as wilderness preserves, are of special legal concern. In cities, trees ameliorate the climate, shade houses and thereby reduce air-conditioning demands, trap dust, provide habitat for birds and small mammals, convert a harsh environment into a pleasing one, and, in larger parks, create an environment for recreation.

As stresses from air and stream pollution impose greater effects on existing forests, and as the total area of forests declines in many parts of the world, it is important that we build a firm foundation for forest management. As I have tried to suggest throughout this book, sound theory and realistic models are a part of good forestry management. In recent years there has been an increase in the application of JABOWA, models derived from it, and other modeling approaches to natural resource management.

Sustainability as a Goal of Forest Management

Sustainability is the commonly stated goal of twentieth-century forestry. A sustainable timber harvest is one that does not endanger future harvests of the same quantity. As Talbot (1990) has pointed out,* there are two concepts embedded in the term *sustainability*: sustainability of timber yield and sustainability of the ecosystem. Talbot goes on to explain that *sustainability of the ecosystem* refers to sustaining the integrity of a natural forest in terms of its structure, composition (i.e., species composition and biological diversity), and ecological processes, along with the environmental services it provides. *Sustainability of timber yield* refers to sustaining a yield of timber from the forest area (Botkin and Talbot, 1992). This implies maintenance of *a* forest but not necessarily *the original* forest. We can also distinguish between a *sustainable original harvest* and a *sustainable disturbance harvest*. A sustainable original harvest is when the level of harvest obtained from an old-growth, previously uncut forest can be sustained. A sustainable disturbance harvest is when a forest, previously cut, is cut again at a lower amount, and that second harvest level can be sustained.

Foresters speak of a "rotation period" or "harvest rotation," which is the time betweeen cuts. Rotations have become shorter during the twentieth century, but the rotation period for harvests of mature trees is long in terms of human lifetimes and economic planning. Twenty years would be a short rotation for mature trees, and would be suitable only for fast-growing species. In some cases,

* The paragraphs that follow are from Botkin, D. B., and L. M. Talbot, 1991, *Biological Diversity and Forests*, Report to the World Bank; and Talbot, L. M., 1990, *A Proposal for the World Bank's Policy and Strategy for Tropical Moist Forests in Africa*, Report to the World Bank.

rotations are shorter—10 or 15 years—but in these cases the harvest is not of boards from mature trees, but of pulp and fiber.

Beginning with an original forest, at least two rotations (three harvests) are required to calculate whether harvests are sustainable. Obviously, this would be the minimal criterion of sustainability. In this case a forest would be called minimally sustainable if harvest 3 equals harvest 1 (in which case one could say there was a *sustainable original harvest*), or if harvest 3 equals harvest 2 (in which case one could say that there was a *sustainable disturbance harvest*).

The concept of sustainable forestry originated in Germany, and it has been attempted elsewhere in Europe and in North America (Talbot, 1990). However, as the history of deforestation described by Perlin (1989) suggests, it seems rare if ever that a *sustainable original harvest* has ever been achieved. Equally unlikely is the sustainability of the original ecosystem. What is most likely is a sustainable disturbance harvest, but there is little evidence that even this kind of harvest has ever been achieved.

In the state of Michigan more than 7 million hectares of original white pine were clear-cut between 1840 and 1920.* It is said that the foresters believed they would never run out of white pine, because the resource appeared huge, and by the time the last hectare was cut the first would have regrown. In fact, many of the hectares never regenerated, but have become depauperate "stump barrens," open fields of grasses, lichens, and shrubs, where no white pine or any large trees grow. Where regeneration of pines has taken place, the original mature size of the forests is never found. These forests developed over long periods, subject only to natural rates of disturbance by fire and windstorms, and not to clearings over large areas in which intense fires occurred as the result of the large amounts of fuel, in the form of the parts of the trees considered waste, left on the ground by the loggers. In some of these areas, red pine plantations were established in the 1930s, with great expectations for future sustainable harvest. But as these trees have approached maturity, insect and disease outbreaks in homogeneous, single-species stands have caused serious problems. Thus even in north temperate zone forests where sustainability is believed to be achievable there are many counter examples of a failure in sustainability.

Even if a forest appeared sustainable after three harvests, in the sense that the yield of harvest three was equal or greater than that of harvest two, the forest might not be sustainable thereafter. Secondary effects of the harvest that might not be expressed after three harvests, could decrease fertility of the soil, decrease organic content of the soil, and compact the soil. Erosion and soil loss could increase. Lack of sufficient seed-bearing trees could lead to a decline in regeneration.

Regeneration of original forests is made more complex because of climate change (as discussed in Chapter 7). Over 400 years the climate can change considerably, so that a forest that developed to maturity in one location in a

---

* The four paragraphs following are taken verbatim from Botkin, D. B. and L. M. Talbot, 1991, *Biological Diversity and Forests*, Report to the World Bank.

previous climate may not regenerate in the present climate. Forests modify the climate near the ground. Not only is it cooler and more moist under the shade of a forest canopy during the day time; it is also warmer during the night and during cold seasons. An established forest can persist under climate change, while the same forest may no longer be able to regenerate. Some existing forests, perceived as the kind of forest that will regenerate and assumed to be sustainable, may be remnants of past climatic conditions. For example, some teak forests in Zimbabwe, which regenerate poorly when affected by frost, may persist in the present climate because mature trees protect young stems from frost. The failure of white pine forests of Northern Michigan to regenerate may be the result of a similar response of trees to climate change.

Such possibilities emphasize the need to establish *prior to logging* whether the forest type to be managed can regenerate under present conditions of soils, climate, and vegetation.*

Thus it is important to investigate the effects of various timber harvest practices, including rotation period, clear-cutting, or selective cutting, on the sustainability of forests. *Selective cutting* means that either certain sizes or certain species are cut. Trees of all species might be cut to some minimum size, with the thought that remaining saplings might respond more rapidly to the opening of the forest than seeds, and that the subsequent forest growth may be greater. Or all trees below a minimum might be cut, so that only very large, seed-bearing trees remain, with the thought that the largest trees might provide the best seeds for future forest growth. In other cases, species of special interest might be cut. In New England, yellow birch, a species valuable for dowels, is sometimes logged selectively from a stand. Sugar maple is also selectively cut and used for furniture. As another example, on one farm in southern New Hampshire with which I am familiar, about 50 acres were managed for firewood over a 30-year period. Sugar maple, oaks, and ash were cut selectively, leaving aspen, a tree with comparatively poor firewood qualities, to regenerate. The best burning woods were harvested, so that the stand was used advantageously in the short-run. The long-term result was to convert the woodland to aspen, so that the stand became less useful for production of firewood. Since this conversion took more than 30 years, and since there were other stands under the same ownership that were not harvested during that time and that could provide good firewood for a future user, this strategy did not appear disadvantageous to the landowner. This example shows how complex the choices are and how intricate the responses of the forest may be to selective logging.

Just as a flight simulator is useful in training pilots, and computer-assisted design software is useful in designing and testing aspects of new aircraft, so a forest model can be useful in training foresters and investigating effects of specific policies. The utility of a model depends on its realism and accuracy, and therefore on the faith we place in it.

* End of Botkin and Talbot material.

Use of the Model to Consider Harvesting a New England Forest
to a Minimum Diameter at a Specified Rotation

A hypothetical forest in the White Mountains of New Hampshire at 909 m (3,000 ft) elevation above sea level and near the Bethlehem, New Hampshire weather station is allowed to grow from a clearcut and then harvested on a 50-year rotation. The site is good—a clay loam of moderate fertility and good soil depth. At this elevation, boreal forest species dominate. The model is used to investigate two questions: Can the harvest be sustained at 50-year intervals? Is a better harvest obtained from clear-cutting or from leaving all stems larger than 12.7 cm (5 in.)? The latter is not arbitrary—it has sometimes been used in practice as a form of selective cutting. Results are shown for screen output for a single plot (Table 8.1) and for averages for 50 replicates (Table 8.2).

Table 8.1 shows the screen output when the model is used in an exploratory mode to compare clear-cutting with cuts in which trees smaller than 12.7 cm are left standing. The forest begins from a clearing and small cherries and birches occur on the plot at year 10. By year 10 both fir and white birch dominate the plot. Several interesting results are projected: (1) On average, the forest continues to aggrade biomass after the first cut (Table 8.2). Remaining trees continue to grow; they provide a second harvest larger than the first. At year 50, 12 balsam fir provide 2.1 kg/m² biomass and 3 white birch provide 7.0 kg/m² biomass. Only two fir and the three birches have stems larger than 12.7 cm. Only these are cut, leaving most of the trees and more than one-half the biomass. At year 100 (50 years after the first selective logging), five of ten balsam fir have stems larger than 12.7 cm in diameter and are removed.

In this example, the model projects that a forest regrown from clear-cutting and then logged to a minimum diameter of 12.7 cm (5 in.) is not opened enough to promote the regeneration of early successional species. The display at year 60 shows that there are no birch or cherry trees present. This result is consistent with the natural history experience of silviculturalists (David Smith, *personal communication*, 1972).

Using the model in an exploratory mode—following a single plot with screen output—allows us to see the general trends resulting from different harvesting practices and to find cases that are interesting to us. Simulation tests involving multiple plots with the same initial conditions can be analyzed to provide statistically useful comparisons. In my experience, the model is used most efficiently in this two-stage procedure, first examining single plots on the screen and then conducting experiments with multiple plots for cases of special interest.

When we run the model for 50 iterations to study logging to 12.7 cm stem diameter we find the results similar to those in the single-plot test (Table 8.2). After the first harvest in year 50, the forest continues to aggrade, so that a much larger biomass occurs at year 100 than was available at year 50 and a much larger harvest results. The forest continues to maintain sufficient canopy cover to prevent a general resetting of the successional sequence. White birch and late successional species continue to dominate.

TABLE 8.1. Forest harvests at 909 m elevation in New Hampshire: Iteration 1; Plot 1

(A) Screen output[a]

Species	Pop./ 100 m^2	Basal Area (cm^2/m^2)	Biomass (kg/m^2)	Volume (ft^3/acre)
Year 10; Degree Days = 2470.5°F				
Pin cherry	6	0.0	0.0	0.0
Choke cherry	2	0.0	0.0	0.0
Balsam fir	3	0.1	0.0	0.0
Red spruce	5	0.0	0.0	0.0
White birch	11	1.1	0.4	0.0
Total	27	1.2	0.4	0.0
Year 50; Degree Days = 2678.2°F				
Sugar maple	3	0.0	0.0	0.0
Beech	4	0.0	0.0	0.0
Mountain maple	2	0.0	0.0	0.0
Striped maple	1	0.0	0.0	0.0
Balsam fir	12	5.3	2.1	49.0
Red spruce	13	0.5	0.1	0.0
White birch	3	10.2	7.0	388.8
Total	38	16.0	9.3	437.8

(B) Trees logged from the plot

Species	Pop./ 100 m^2	Basal Area (cm^2/m^2)	Biomass (kg/m^2)	Volume (ft^3/acre)
Year 50; Degree Days = 2678.4°F				
Balsam fir	1	1.5	0.6	25.7
White birch	2	6.5	4.5	241.5
Total	3	8.0	5.1	267.2
Year 60; Degree Days = 2516.9°F				
Sugar maple	4	0.1	0.0	0.0
Beech	4	0.1	0.0	0.0
Mountain maple	1	0.0	0.0	0.0
Balsam fir	10	3.7	1.5	36.0
Red spruce	13	0.7	0.2	0.0
Total	32	4.5	1.7	36.0
Year 70; Degree Days = 2470.5°F				
Sugar maple	2	0.1	0.0	0.0
Beech	3	0.1	0.0	0.0
Balsam fir	16	4.6	1.8	38.4
Red spruce	14	1.1	0.3	0.0
Total	35	5.8	2.2	38.4
Year 80; Degree Days = 2678.2°F				
Sugar maple	2	0.2	0.1	0.0
Beech	3	0.1	0.0	0.0
Striped maple	2	0.0	0.0	0.0
Balsam fir	18	8.3	3.4	104.0

TABLE 8.1. (*Continued*)

Species	Pop./ 100 m²	Basal Area (cm²/m²)	Biomass (kg/m²)	Volume (ft³/acre)
Red spruce	18	1.5	0.5	0.0
Total	43	10.1	4.0	104.0
	Year 90; Degree Days = 2516.9°F			
Sugar maple	2	0.3	0.1	0.0
Beech	3	0.1	0.0	0.0
Balsam fir	16	11.8	5.1	216.0
Red spruce	18	2.1	0.7	0.4
Total	39	14.3	5.9	216.4
	Year 100; Degree Days = 2470.5°F			
Sugar maple	2	0.0	0.0	0.0
Beech	3	0.0	0.0	0.0
Balsam fir	10	13.9	6.7	441.7
Red spruce	22	1.8	0.6	0.0
Total	37	15.8	7.3	441.7
	Year 100; Degree Days = 2470.4°F			
Balsam fir	5	12.1	5.8	364.6
Total	5	12.1	5.8	364.6
	Year 110; Degree Days = 2678.2°F			
Sugar maple	4	0.4	0.2	0.0
Beech	3	0.1	0.0	0.0
Balsam fir	11	5.0	2.0	56.9
Red spruce	23	3.5	1.2	4.5
Total	41	9.0	3.4	61.4

[a]Site conditions include soil depth of 1.0 m; water table depth 0.6 m; soil texture 250 mm/m; and soil nitrogen 50 kg/ha.

On average, total biomass and timber volume at year 100 are *not* negatively affected by a harvest to 12.7 cm stem diameter at year 50. The biomass of the control forest is 21 kg/m², the biomass in the selectively logged forest is 21.3 kg/m². The previously harvested and unharvested forests are indistinguishable at year 100 in terms of total biomass, although, of course, a visitor to the harvested forest would see stumps and other remnants of the logging process.

Timber volume in the control forest (without logging) is 3,186.7 ft³/acre, while the timber volume in the selectively logged forest is 3,114 ft³/acre. Biomass of white birch at year 100 is 12.4 kg/m² in the unharvested forest and 11.3 kg/m² in the selectively logged forest, statistically indistinguishable abundances. The biomass of fir at year 100 is 8.1 kg/m² in the unharvested forest and 8.8 kg/m² in the selectively logged forest, again statistically indistinguishable abundances.

TABLE 8.2. Forest harvests at 909 m elevation in New Hampshire

(A) Control: No harvest[a]

Tree Species (Common name)	Pop./ 100 m^2	95CI	Basal Area (cm^2/m^2)	95CI	Biomass (kg/m^2)	95CI	Vol. (ft^3/acre)	95CI
Year 10								
Balsam fir	4.0	0.6	0.1	0.0	0.0	0.0	2.3	0.5
White birch	0.2	0.4	0.0	0.0	0.0	0.0	0.0	0.0
Total	4.2	0.2	0.1	0.0	0.0	0.0	2.4	0.1
Year 50								
Sugar maple	3.5	0.6	0.1	0.0	0.0	0.0	1.3	0.3
Yellow birch	3.1	1.0	0.2	0.1	0.0	0.0	3.2	1.1
Balsam fir	13.6	0.8	5.7	0.6	2.3	0.2	254.8	28.9
Red spruce	11.0	1.0	0.4	0.0	0.1	0.0	8.4	1.2
White birch	19.0	2.2	9.8	1.1	5.0	0.6	581.6	72.2
Mountain ash	0.0	0.1	0.0	0.0	0.0	0.0	0.1	0.1
Red maple	0.4	0.2	0.0	0.0	0.0	0.0	1.0	0.5
Total	50.6	0.9	16.1	0.4	7.5	0.2	850.4	21.7
Year 100								
Sugar maple	2.8	0.5	0.2	0.0	0.1	0.0	3.3	0.7
Beech	0.9	0.3	0.0	0.0	0.0	0.0	0.1	0.1
Yellow birch	0.6	0.3	0.1	0.1	0.0	0.0	3.1	1.6
Balsam fir	16.8	1.0	16.8	1.6	8.1	0.8	1135.4	130.6
Red spruce	16.0	1.4	1.3	0.1	0.4	0.0	42.5	5.0
White birch	4.6	0.8	17.2	2.6	12.4	1.9	2001.4	304.0
Red maple	0.1	0.1	0.0	0.0	0.0	0.0	0.7	0.8
Total	41.9	0.8	35.6	0.8	21.0	0.5	3186.7	78.5

(B) 50-Year rotation; harvest to minimum of 12.7 cm (5 in.)

Tree Species (Common name)	Pop./ 100 m^2	95CI	Basal Area (cm^2/m^2)	95CI	Biomass (kg/m^2)	95CI	Vol. (ft^3/acre)	95CI
Year 50 (before logging)								
Sugar maple	5.4	0.8	0.1	0.0	0.0	0.0	2.1	0.1
Yellow birch	8.0	0.2	0.4	0.1	0.1	0.0	8.8	1.3
Pin cherry	0.1	0.1	0.0	0.0	0.0	0.0	0.1	0.1
Balsam fir	24.4	4.7	10.0	1.8	4.0	0.7	437.7	75.0
Red spruce	23.7	4.5	0.7	0.1	0.2	0.0	16.8	3.0
White birch	33.0	5.8	15.7	2.7	7.9	1.3	900.2	149.2
Mountain ash	0.3	0.2	0.0	0.0	0.0	0.0	0.7	0.4
Red maple	0.8	0.2	0.1	0.0	0.0	0.0	2.5	0.9
Total	95.7	1.6	27.1	0.6	12.2	0.3	1368.9	33.6
Year 50 (amount removed in harvest)								
Balsam fir	—	—	—	—	0.8	0.2	103.9	24.4
White birch	—	—	—	—	0.7	0.3	89.8	42.8
Total	—	—	—	—	1.5	0.5	193.7	67.2
Year 60								
Sugar maple	3.5	0.6	0.1	0.0	0.0	0.0	2.0	0.5
Beech	2.8	0.6	0.0	0.0	0.0	0.0	0.3	0.1

TABLE 8.2. (*Continued*)

Tree Species (Common name)	Pop./ 100 m²	95CI	Basal Area (cm²/m²)	95CI	Biomass (kg/m²)	95CI	Vol. (ft³/acre)	95CI
Yellow birch	3.4	1.0	0.3	0.1	0.1	0.0	7.4	2.5
Pin cherry	0.0	0.0	0.0	0.0	0.0	0.0	0.0	0.1
Balsam fir	13.2	0.8	6.4	0.6	2.6	0.2	298.9	29.9
Red spruce	13.2	0.9	0.5	0.1	0.2	0.0	14.0	1.7
White birch	12.1	1.8	10.4	1.3	5.8	0.7	722.9	96.7
Red maple	0.3	0.2	0.1	0.0	0.0	0.0	2.0	1.2
Total	48.6	0.8	17.9	0.4	8.7	0.2	1047.6	26.3
Year 100 (before logging)								
Sugar maple	6.3	0.8	0.3	0.0	0.1	0.0	5.1	0.4
Beech	1.9	0.2	0.0	0.0	0.0	0.0	0.3	0.1
Yellow birch	1.7	0.3	0.4	0.1	0.2	0.0	11.3	2.5
Balsam fir	28.6	5.4	19.4	2.6	8.8	1.0	1175.6	100.7
Red spruce	33.7	6.5	2.9	0.5	0.9	0.2	96.1	17.2
White birch	4.3	0.8	15.7	2.7	11.3	1.9	1824.7	312.7
Red maple	0.1	0.1	0.0	0.0	0.0	0.0	1.3	1.3
Total	76.6	1.5	38.7	0.9	21.3	0.5	3114.4	74.0
Year 100 (amount removed in harvest)								
Balsam fir	—	—	—	—	6.2	0.7	897.3	111.0
White birch	—	—	—	—	11.2	1.9	1817.1	310.8
Total	—	—	—	—	17.4	0.5	2714.4	75.8
Year 110								
Sugar maple	3.5	0.7	0.2	0.0	0.1	0.0	3.3	1.0
Beech	0.8	0.3	0.0	0.0	0.0	0.0	0.1	0.1
Yellow birch	0.7	0.3	0.2	0.1	0.1	0.0	6.3	3.0
Balsam fir	13.8	1.1	5.9	0.6	2.4	0.2	266.2	27.6
Red spruce	18.7	1.0	2.0	0.2	0.6	0.1	68.5	7.1
White birch	1.4	1.0	0.2	0.2	0.1	0.1	12.5	15.6
Red maple	0.0	0.0	0.0	0.0	0.0	0.0	0.4	0.8
Total	39.0	0.8	8.4	0.2	3.2	0.1	357.3	9.4

[a]Normal climate using the 1951–1980 weather record for Hubbard Brook, New Hampshire. Replicates: 50; soil moisture capacity: 250.0 mm/m; water table depth: 0.6 m;Weather station elevation: 576.7 m; soil depth: 1.0 m; soil nitrogen: 50.0 kg/ha.

The primary conclusion we can draw from these tests is that, in terms of the timber volume of the dominant species and total timber volume, selective logging to 12.7 cm at a 50-year rotation appears sustainable.

Earlier in this book I showed that a forest in the northeastern United States requires on the order of 400 years or more to settle into the range of abundances that characterize the long-term pattern of variability. Therefore, it seems advisable to analyze sustainability for more than 400 years, an analysis which, of course, is open to us only through simulation. In theory such an analysis might be possible through historical records, but adequate records do not exist.

Regeneration of a forest at the same site as used in Tables 8.1 and 8.2, beginning with a clearcut and not subject to logging, is shown in Figure 8.1. Total available merchantable timber peaks about year 150 after the clear-cut and then declines to a range between 2,000 and 2,500 ft^3/acre. When the same forest is harvested to a minimum of 12.7 cm (5 in.) every 50 years, peak biomass occurs earlier and is slightly smaller. The forest achieves a disturbance sustainability between year 150 and 550, with the merchantable timber ranging between 1,500 and 2,000 ft^3/acre. Clear-cutting every 50 years is also sustainable, but at a much lower amount, below 1,000 ft^3/acre.

Timber removed during each cut is shown in Figure 8.2. Total harvest is sustainable for both selective logging to 12.7 cm (5 in.) and clear-cutting. Except for harvests prior to year 150, the two practices lead to equivalent results.

In this case, which practice is preferable—clear-cutting or selective logging to a minimum size—depends on the use of the timber and the species of interest. Balsam fir quantity is greatest under no harvest and least under clear-cutting (Fig. 8.1(B)). Red spruce abundance is greatest under selective logging to 12.7 cm and is zero when the plot is clear-cut (Fig. 8.1(D)). The abundance of white birch is the mirror image of the abundance of red spruce (Fig. 8.1(C)). White birch achieves a sustainable abundance under clear-cutting, essentially disappearing as an economically viable product under no harvesting and under a regime of cutting to a minimum of 12.7 cm. The harvest of merchantable balsam fir is greater under selective cutting to a minimum of 12.7 cm than under clear-cutting (Fig. 8.2(B)), whereas the harvest of white birch is greatest under clear-cutting, except at year 100 (Fig. 8.2(C)). The reasons for the patterns of abundance of birch and red spruce should be clear from discussions of succession in previous chapters. Clear-cutting every 50 years maintains the forest in a relatively early (but not the earliest) successional stages and favors white birch. No cutting leads to a mature forest dominated by spruce.

One of the most important results of this analysis is that *reliance on three harvests can be deceptive as a way to determine whether a harvesting practice is sustainable*. In practice, foresters will tend to rely on information from three harvests, simply because the time required for longer sets of harvests is prohibitive. After the second rotation (third cut including the first clear-cut at year zero), a forester would conclude that white birch production under selective logging to 12.7 cm minimum was sustainable and produced a large quantity of timber, and could provide greater return than the yield from either balsam fir or red spruce. This conclusion is erroneous in two ways. First, the greatest long-term yield is from balsam fir, whose harvest averages just under 1,000 ft^3/acre, while the long-term average for white birch is about 500 ft^3/acre. Second, selective cutting does not reset the successional process, but maintains many of the white birch present before year 50, which originated in the original clearcut. Subsequently, these grow rapidly. Figure 8.2(C) also suggests that a greater long-term yield of white birch could be obtained if the selective cutting to 12.7 cm minimum every 50 years was alternated with clear-cutting every 100 years. Harvesting of red spruce under either regime is low and probably not economically productive. It is also instructive that the results for a single plot (Table 8.1) are not necessarily representative of all average conditions.

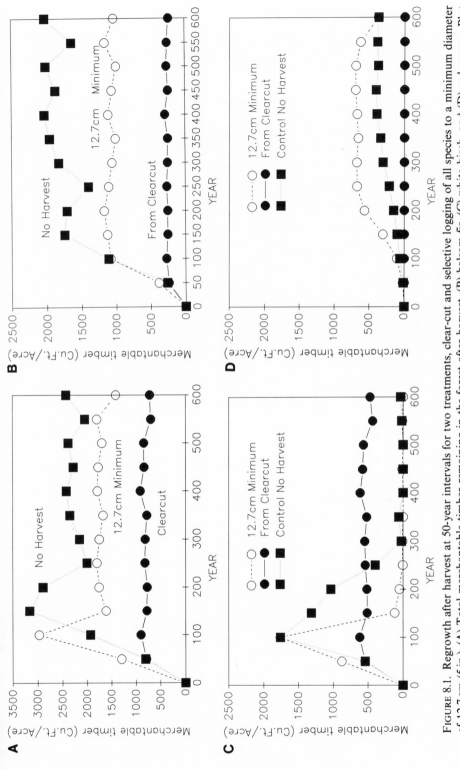

FIGURE 8.1. Regrowth after harvest at 50-year intervals for two treatments, clear-cut and selective logging of all species to a minimum diameter of 12.7 cm (5 in.). (A) Total merchantable timber remaining in the forest after harvest, (B) balsam fir, (C) white birch, and (D) red spruce. Plot conditions are given in Table 8.1.

213

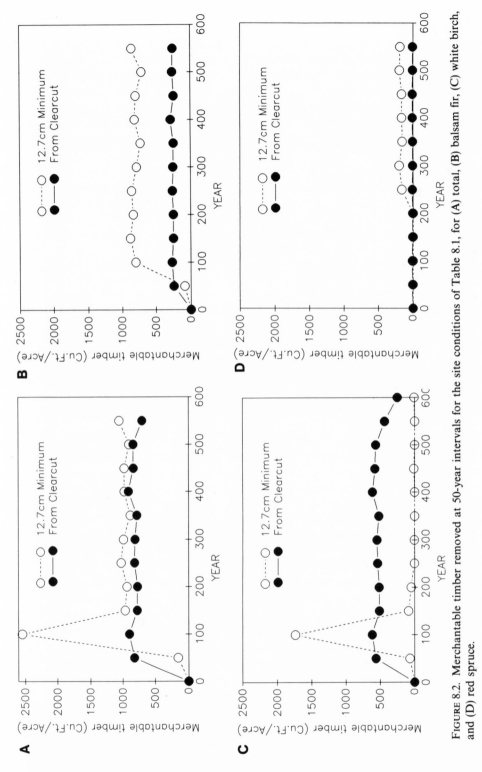

FIGURE 8.2. Merchantable timber removed at 50-year intervals for the site conditions of Table 8.1, for (A) total, (B) balsam fir, (C) white birch, and (D) red spruce.

214

These results show that useful insights into harvesting policies can be obtained with the model, results that could not be obtained in practice by direct measurements. The reader can see that many additional cases could be considered, each of which might yield different and interesting results. Readers can try these with the program available for use on microcomputers. Fortunately, new high-speed minicomputers and microcomputers allow easy application of the model to many cases. For example, we have found that simulated forest growth of 50 replicates over a 90-year period can be obtained in approximately 5 minutes on a Digital Equipment Corporation Model 5100 computer. If the model were to be used by a commercial company, the model could be embedded in an optimizing program that would automatically search for specific results. The major limitation for the user is that the model generates a tremendous amount of output. An automated optimizing program would reduce the amount of output.

It is important to note that these projections assume that harvesting has no impact on soil fertility or seed availability, that harvesting techniques do not damage the soil; destroy seeds, seedlings, or saplings; or decrease future production. Any of these assumptions could be violated in practice. Of course, there are many other limitations to these projections, which should be obvious to a reader of the earlier chapters in this book. They will not be repeated here.

Use of the Model to Consider Effects of Harvesting Regimes on Forest Floor Dynamics

A model of forest floor dynamics (not part of the software included as a companion to this book) was connected to an earlier version of the JABOWA model (Aber et al., 1978) (Fig. 8.3). That version of the JABOWA model was the first to include the nitrogen response function. The decomposition module projected decomposition of dead roots, stem wood, and leaves, and the quantity of available nitrogen produced by that decomposition. The model of forest floor dynamics is a different kind of model from JABOWA. It is a compartment model of a kind discussed in Chapter 1, with flux rates that are independent of environmental conditions. Transfer rates among compartments in the forest floor model are based on data from the Hubbard Brook Ecosystem Study (Aber, 1976; Bormann, Likens, and Melillo, 1977; Dominski, 1971) and other literature (McFee and Stone, 1966; Norton and Young, 1976; White, 1974). Nitrogen dynamics are calculated by assigning a nitrogen concentration to each input, output, and stored amount of organic matter. These calculations assume that nitrogen is added during the process of decomposition by nitrogen-fixing bacteria that feed on wood. Although this compartment model is of quite a different kind than JABOWA, it has some utility here as an ancillary to the forest model. The utility lies in cases in which the ecological interest is not in the species interactions in the forest floor, but with the forest trees. If an ecologist had the same kind of interest in the species in the forest floor as I have with trees, then the simple compartment model would not be particularly helpful.

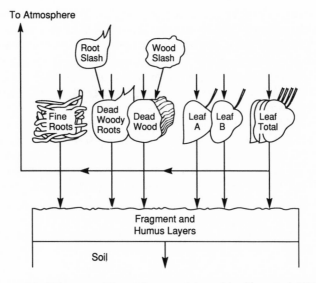

FIGURE 8.3. Diagram of forest floor model used in Aber et al. (1978).

In that case, a model in the spirit of JABOWA might be more appropriate and more useful. Referring back to the discussions of models in the first chapter, we can begin to distinguish between the kinds of utilities of different kinds of models applied to forest ecosystems. Compartment models, generally limited in the insights they provide, can be useful in generating boundary conditions for an ecologically based model.

Projections of the model of forest floor dynamics compare favorably with field studies in the Hubbard Brook Forest by Covington (1977), except that the model predicts a higher concentration of nitrogen in the forest floor. The difference is apparently due to a greater rate of decay in the real forest of plant material exposed to the sun after logging. In other words, real decomposition is a function of environmental conditions. Since the simple compartment model of forest floor dynamics does not incorporate biological responses to environmental conditions, this response is not intrinsic in the model. In the real forest, rates of decay are time variant and dependent on environmental conditions, contradicting assumptions of this simple forest floor model. For purposes of improving the realism of the calculations, but not the realism of the forest floor model, this was corrected with a linearly decreasing function representing a 30-percent increase in decomposition after logging that declines to the original value in 38 years. Obviously, that is a simple fix. It could be improved on by making decomposition a function of light available at the forest floor, so that the forest floor dynamics were coupled directly to the dynamics of JABOWA.

The resulting combined model, the JABOWA-forest floor compartment model, can be used to compare harvesting regimes. We did this for three rotations of 30, 45, and 90 years, and for three kinds of harvest: (1) traditional clear-cutting, in which only the main boles of the trees are removed from the forest and the rest of the material is allowed to decay in place, which is the traditional forestry practice; (2) whole-tree harvest, in which all biomass above the ground is removed; and (3) complete forest utilization, in which all above-ground biomass and live woody roots and dead wood on the forest floor are removed. These practices were proposed in the 1960s by Harold Young of the University of Maine as a way to increase the efficiency of utilization of forests, apparently with the belief that such practices could be done without detriment to the forest ecosystem and could be sustainable (Norton and Young, 1976). We were motivated by his suggestions to carry out these theoretical tests.

The model projects that forest floor biomass declines immediately after clear-cutting, but recovers to prelogging amounts by year 90 under the 90-year rotation (Fig. 8.4(A)). Forest floor biomass does not recover in either of the shorter rotations. Under the 30-year rotation, forest floor biomass varies between one-third and two-thirds of the prelogging amount. Under the 45-year rotation, forest floor biomass recovers to about 80 percent of the prelogging amount by the end of the rotation, only to decline again after logging (Fig. 8.4(B)).

Nitrogen never recovers to prelogging amounts under any of the treatments. After an initial rise in availability following each cut, nitrogen declines. Under the 90-year rotation, nitrogen returns to approximately one-half of the

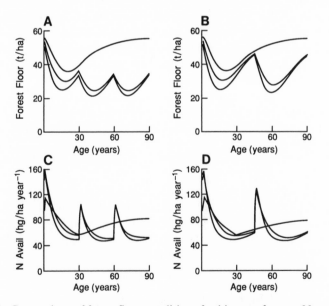

FIGURE 8.4. Comparison of forest floor conditions for biomass for one 90-year clearcut with (A) three 30-year whole-tree, two 45-year whole tree, and three 30-year complete-tree harvests, and (B) two 45-year complete-tree harvests. *Whole tree harvest* means that all biomass above the ground is removed. Complete-tree harvest means that in addition to above-ground biomass, live woody roots and dead wood on the forest floor are removed. Available nitrogen is shown in (C) and (D) for the same groupings. (From Figs. 7 and 8, Aber et al., 1978.)

original level (Fig. 8.4(C) and (D)). Under the 30-year rotation, nitrogen increases to approximately three-quarters of the original value after each cut, then declines rapidly to less than one-third of the original value (Fig. 8.4(C)). Of more concern is the gradual decline in the *peak* amount of nitrogen following each clear-cut under the 30-year rotation. In contrast, the peak value of nitrogen *increases* from cut 1 to cut 2 in the 45-year rotation. This suggests that a 30-year rotation is not sustainable but a 45-year rotation might be.

The model of forest floor dynamics was combined by Aber and colleagues with JABOWA to predict the effects of different harvesting regimes on productivity and yield (Aber et al., 1979). Available nitrogen was assumed to be that available in the forest floor. While shorter rotations and more complete utilization may seem efficient in the short-run, the resulting decline in soil fertility has a long-term detrimental effect on production and yield. In the end, the maximum total yield over a 90-year period is obtained from the 90-year rotation. For this period, maximum total net production occurs under whole-tree harvesting for a 90-year rotation; maximum timber yield is obtained under complete forest removal for a 90-year rotation (Fig. 8.5). However, traditional clear-cutting and bole removal results in production only slightly less than the maximum.

From these results we can conclude that reduction in soil nitrogen content can have an important effect on timber yield, but that this effect can be delayed and not apparent until after two or three rotations. Even longer observations

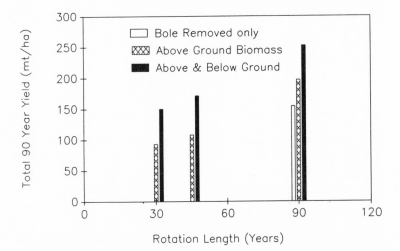

FIGURE 8.5. Simulated total harvest obtained after 90 years from a clear-cut New Hampshire forest. Three rotations and three kinds of logging are compared. One 90-year rotation in which the boles only were removed (open box) functions as a control. Removal of essentially all above-ground biomass (called whole-tree harvest) is shown as cross-hatched boxes, and removal of above-ground biomass and available root material (called complete-tree harvest) is shown as solid boxes. Each box shows *the total harvest after 90 years*. Thus the total 90-year return from three 30-year complete-tree harvests is 150 mt/ha, while the total from one 90-year complete-tree harvest is 250 mt/ha.

could show greater declines. For obvious reasons—both that sustainable-yield forestry has not been practiced long enough to permit such observations and that such observations are not readily maintained anyway—a practicing forester relying only on field observations could easily miss these implications. Some later related work can be found in Aber et al. (1982), work that explores fiber yield as affected by rotation length and other factors. Pastor and Post (1985) have extended a version of the forest model to include soil and tree feedbacks through the nitrogen cycle; more work along those lines is desirable.

Use of the Model for the Analysis of Wildlife Habitats

Recent ecological reasearch suggests that forested ecosystems depend on change and that succession and disturbance are continual and natural processes, as I have discussed earlier in this book and in greater depth elsewhere (Botkin, 1990; Botkin and Sobel, 1975). An ecological (i.e., a living) landscape is dynamic. The processes of succession offer evolutionary opportunities—niches to be filled for long enough times, and for large enough percentages of a landscape at any one time, to be important in the evolution of trees. Because tree species have evolved and adapted to each stage in succession, it is not always desirable to manage an ecosystem so that it progresses totally and always to the mature state. Even redwood forests of California appear to require fires and flooding so that the trees undergo natural regeneration from seeds. Mature redwoods generally are not damaged by moderate fires, although they may be damaged by the most severe ones. Another good example of the importance of disturbance is the history of the Kirtland's warbler, an endangered species of the north woods, which I will discuss next.

The Case of the Kirtland's Warbler

The forest model can be useful in the analysis of habitat and ecosystems for wildlife that live in forests. Elsewhere, we have shown that JABOWA can be part of a larger ecosystem model, a model of vegetation, moose, and wolves of Isle Royale (Botkin and Levitan, 1977). An interesting recent application of the model concerns management of the Kirtland's warbler habitats.

The Kirtland's warbler nests only in jack pine forests of Michigan (Figs. 8.6 and 8.7).* Jack pine, as will be clear to a reader of earlier chapters of this book, is an early successional species that regenerates only after fire. Jack pine trees are intolerant of shade, and seedlings and saplings are unable to grow under a forest canopy. Cones open and release seeds only after they have been heated in a forest fire (Fowells, 1965). Thus without fire, jack pine does not germinate and, if it could, it would not grow within a mature forest. Like many other species

---

*This section is taken from Botkin, D. B., D. A. Woodby, and R. A. Nisbet, 1991, Kirtland's warbler habitats: A possible early indicator of climatic warming. *Biological Conservation* 56 (1): 63–78.

☐ **Townships with Kirtland's Warbler nests**

▨ **Grayling sand**

FIGURE 8.6. Map of Michigan's lower peninsula showing the townships where nests of warblers were observed from 1975 to 1985, and the distribution of the Grayling sand soil type. Most nesting sites are restricted to this one soil type. Adapted from Byelich et al. (1985). (From D. B. Botkin, D. A. Woodby, and R. A. Nisbet, Kirtland's warbler habitats: A possible early indicator of climate warming. *Biological Conservation 56(1):* 63–78, Fig. 1.)

of pine, it has persisted in areas subject to frequent, recurring fires. For example, in northern Minnesota the average fire recurrence in jack pine habitats prior to the twentieth century was on the order of 15 to 30 years (Heinselman, 1981). With fire suppression in the twentieth century, the estimated fire recurrence intervals in the same area have increased to as long as 800 years (Hall et al., 1991).

When fires were suppressed in Michigan during the first part of the twentieth century, jack pine began to disappear and so did the Kirtland's warbler. Ornithologists wanting to preserve this species recognized that fire and fire-generated forest succession were needed to save the warbler. As a result, management practices have changed to promote light fires rather than to suppress them.

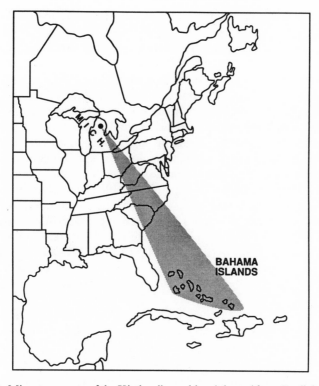

FIGURE 8.7. Migratory route of the Kirtland's warbler. Adapted from Byelich et al. (1985). (From D. B. Botkin, D. A. Woodby, and R. A. Nisbet, Kirtland's warbler habitats: A possible early indicator of climatic warming. *Biological Conservation 56(1)*: 63–78, Fig. 1.)

Peters and Darling (1985) suggested that climatic shifts brought on by global warming, as discussed in Chapter 7, may disrupt habitats of endangered species and cause the extinction of species with highly restricted ranges. In part this would be a result of rapid movement of climatic zones. For example, it has been predicted that climate zones in the mid-latitudes of eastern North America will move northward several hundred kilometers over the next century at approximately 10 times the rate characteristic of the Holocene and Pleistocene (Davis and Zabinski, in press; Schneider, 1989b) so that present vegetation types may be replaced by those characteristic of more southern regions, as discussed in Chapter 6. But as Peters and Darling (1985) suggested, these negative effects could be made worse by the patchiness of the contemporary landscape, within which habitats of endangered species have become ecological islands locally isolated from one another. Prior to the work reported here, which began in the late 1980s, the extent to which global warming might affect extinction rates of endangered species had been a subject only for qualitative speculation.

The Kirtland's warbler has been the object of considerable management effort and scientific study, and is therefore an important example of the possible effects of global warming on endangered species. It was the first songbird subject to a complete census, and the population history of this species is comparatively well known (Mayfield, 1960). The numbers dropped from an estimated 432 males in 1951 (Mayfield, 1960) to 201 males in 1971, recovering only to 212 males in 1989 (J. Weinrich, *personal communication*, 1989). This rapid decline had two important results: it stimulated scientific study of the species and it led to a management plan for its conservation that is now under way (Byelich et al. 1985). These plans include extensive habitat management through controlled burning to increase the acreage of young jack pine. Possible rapid climate change raises questions as to the outcome of this habitat management.

Jack pine occurs throughout a large part of the North American boreal forest (Botkin and Simpson, 1990b; Fowells, 1965). However, the Kirtland's warbler nests only in young (6–21-year-old) jack pine stands and almost exclusively where these grow on a single soil type, Grayling sands, a coarse sand found only at the southern edge of the jack pine's range in the northern lower peninsula of Michigan (Byelich et al., 1985) (Fig. 8.6). Hence the concern is that, if jack pine fails to regenerate in this area, the warbler will be unable to find suitable habitat elsewhere in North America, and will become extinct.

Climate Projections. To explore the consequences of the projected rapid climatic warming for jack pine in Michigan (Botkin et al., 1991), we used the climate projections for a transition from current conditions to that under twice-CO_2 atmospheric concentration provided by the general circulation model of NASA Goddard Space Flight Center, New York, as discussed in Chapter 7. The methods for creating a "normal" and "global warming" climate are also described in Chapter 7. The 1951–1980 weather records from Grayling, Michigan, served as

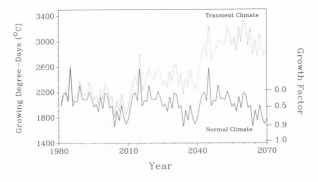

FIGURE 8.8. Projected pattern of growing degree-days (°C) from 1980 to 2070 for "normal" and global warming "transient" climates for Grayling, Michigan. The normal pattern is the 1950–1980 temperature pattern repeated. The global warming pattern is the 1950–1980 pattern modified by the projections of the NASA GISS climate model (Hansen et al., 1983, 1988). See text for explanation. (From D. B. Botkin, D. A. Woodby, and R. A. Nisbet, 1991, Kirtland's warbler habitats: A possible early indicator of climate warming. *Biological Conservation 56(1)*: 63–78, Fig. 2.)

the "normal" climate, applied in repeating sequence to project climate for 1980–2070 (Fig. 8.8)

Management by controlled burning was simulated by clearing all the plots at 30-year intervals. It is important to note. that version II of the model used here simulates no other effect of fire except the clearing of trees. Effects of fire such as the burning of organic matter, resulting in a release of some chemical elements that become immediately available for plant growth, but are also readily lost to fluvial erosion, and possible subsequent reduction in cation exchange capacity and water-holding capacity of the upper layers of the soil, and changes in buried seeds, are not considered.

Under the normal climate, the forest model projects that jack pine increases steadily between each fire, just as has occurred in the past (Fig. 8.9). The response of jack pine to the global warming climate is surprisingly rapid. By the projected year 1985, basal area of jack pine is significantly lower than under normal climate ($p < 0.95$; t-test). At first, other tree species become dominant. By year

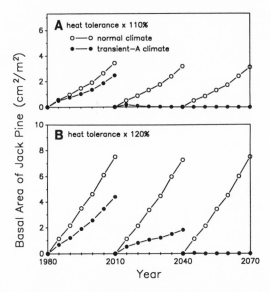

FIGURE 8.9. Projected basal area of jack pine (*Pinus banksiana*) near Grayling, Michigan, under normal climate (open circles) and transient-A climate (filled circles) for the 90-year period 1980–2070, representing three periods of growth between controlled burns. Basal area projections with thermal tolerance (maximum growing degree-days) of jack pine increased by (A) 10 percent and (B) 20 percent, Basal area is the cross-sectional area of the tree stems, summed by species for all trees in a sample. The soil particle size is a coarse sand to represent the Grayling sand where the Kirtland's warbler nests. Soil moisture-holding capacity is defined as 50 mm water/m depth of soil. Soil depth is 0.4 m; depth to the water table is 1 m; nitrogen content of the soil is 42 kg/ha. Initial conditions are a clearing with no trees but adequate seeds available for regrowth of any of the species for which the climate is suitable. (From D. B. Botkin, D. A. Woodby, and R. A. Nisbet, 1991, Kirtland's warbler habitats: A possible early indicator of climate warming. *Biological Conservation 56(1):* 63–78, Fig. 4.)

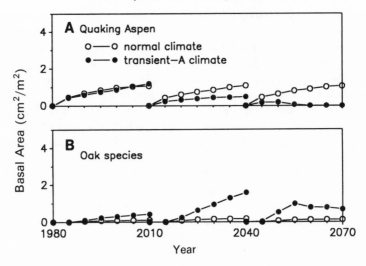

FIGURE 8.10. Projected basal area of (a) quaking aspen and (b) oaks for the conditions given in Figure 8.9.

60, quaking aspen (*Populus tremuloides*) and oaks (*Quercus* spp.) are dominant, but only small stems of oaks are common by year 90 (Fig. 8.10). Although aspen and oak replace jack pine, a mature forest never develops. Basal area is small, indicating an open woodland of small stems. A forest composed solely of the native species now present in the area would not persist to the end of the next century.

These projected rapid declines in existing forests are counterintuitive, at least for me. Previous work with M. B. Davis suggests that the last major climatic event, the "little ice age," was too small a change for too short a time to produce a statistically significant change in the forests of eastern North America. This projection is consistent with the lack of observed changes in the pollen records of the little ice age (Davis and Botkin, 1985). On this basis, I did not expect that a smooth transient from present conditions over 90 years would result in dramatic differences.

How robust are these projections? The most vulnerable part of the model is the estimation of parameters for specific equations that relate tree growth and regeneration to environmental conditions, as discussed in Chapter 6. Of particular concern in this case are the parameters for the temperature response of jack pine. Under the normal climate regime at Grayling, Michigan, near the Kirtland's warbler habitat, simulated growth of jack pine is highly sensitive to the estimated heat tolerance of the species. The temperature regime of Grayling, Michigan, from 1951 to 1980 is only slightly cooler than the estimated maximum tolerated by jack pine. Our initial simulations with jack pine default parameters projected no jack pine growth. Realistic growth of jack pine was achieved by increasing the maximum heat tolerance of jack pine by 10 percent to 2,222 growing degree-days. This estimate served as our control. This result suggests that the actual geographic range of jack pine extends farther south than has been generally observed in the twentieth century, or else that ecotypes of jack pine occurring near the southern edge of the range have a greater temperature tolerance than the average for the species.

We tested the sensitivity to error in estimation of jack pine's heat tolerance by increasing the maximum tolerance an additional 10 percent to 2,444 growing degree-days. Even with this more extreme heat tolerance, statistically significant reductions in basal area are projected to occur within the first five years (by 1985) after an initial fire ($p > 0.95$, t-test). Stark reductions in jack pine growth are still predicted to occur in the second 30-year period (after 2010) (Fig. 8.9(B)), so that the decline is delayed. The rapidity with which jack pine declines, and therefore the speed with which the breeding grounds of the warbler might disappear, suggests that the warbler may be subject to a greatly increased risk of extinction. This in turn implies that managers of the warbler's habitat may need to adjust their policies quickly, before the species undergoes a serious decline. To test the realism of these projections, jack pine stands, managed as habitat for the Kirtland's warbler, could be monitored to provide one of the earliest biological indicators of global climate warming. As long as the growth of jack pine in the Kirtland's warbler breeding habitat remains in the normal range, there would be no significant indication of a biological effect whatever the change in climate that might be occurring. However, if jack pine undergoes a rapid decline, then the effects of global warming may be appearing.

Results are restricted to effects of climate change. No other independent human-induced effects on tree growth were taken into account, such as acid rain effects, discussed earlier, or gaseous oxidants.

Use of the Model to Consider the Fate of Forest Preserves

Recently there has been considerable concern with the fate of old-growth forests around the world. The focus of this concern has been logging, especially clear-cutting, of these forests. One goal of conservation is to increase the number and size of preserves for old-growth forests. However, there is another, subtler problem with the conservation of old-growth forests, which should be clear to a reader of this book. Most forest preserves are "set aside" in the sense that stands are placed in a preserve and left alone or made a center for tourism and recreation. Little management is devoted to the need to maintain these preserves as dynamic ecological systems. It is tacitly assumed that nature knows best and will take care of itself (Botkin, 1990). But this is not to be.

One example of the problem in recent history in the conservation of old-growth forest is the last remaining stand of original white pine in the lower peninsula of Michigan, which exists in Hartwick Pines State Park, Michigan. This was set up to preserve a small stand of mature white pine that originally covered about 36 ha (90 acres) but now covers only 20 ha (50 acres). In the park, visitors trample the grounds near the pines, compacting the soils and decreasing the chances for seedling germination and sapling survival. To my knowledge

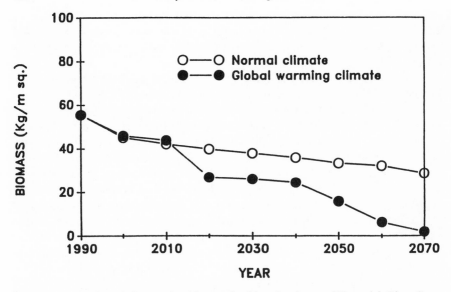

FIGURE 8.11. Projected changes in old-growth white pine forest of Hartwick Pines State Park, Michigan. Biomass is shown for the average of 3 plots in the center and three-plots in the periphery of a 65-acre white pine preserve at the park under normal conditions (open circles) and global warming conditions (filled circles). The size of the circle is equal to or greater than the 95 percent confidence limits for the samples, so a lack of overlap of open and closed circles at the same year indicate a significant difference in biomass at the 95 percent level of confidence.

there is no program to promote regeneration of white pine either within the existing mature stands or in adjacent areas. Because forests are dynamic, undergoing change as trees reach old age and die, such a program must become part of the policy for any forested natural area. In addition, global warming may damage this preserve.

To study the possible changes in the white pine stands of Hartwick Pines, we made field measurements to obtain initial conditions in 1990. Standard 10- by 10-m plots were established, both in the center of the old growth and along the periphery. Projections were made for normal and global warming climates, as described in Chapter 7 and described earlier for the Kirtland's warbler habitat. The model projects that the old-growth stands will decline with or without global warming (Fig. 8.11). Under the normal climate, dominant trees approach their maximum longevity in the twenty-first century and mortality increases. The canopy begins to break apart by year 2010, and suffers a further decline in year 2050. Under global warming, the forest decline is hastened: the forest parallels that under the normal climate until year 2060, after which white pine decreases rapidly. Under normal climate it would persist.

The model provides insight into possible changes that are not taken into consideration in the management of many forest preserves.

Summary

Since the time of the classical Greek civilization, undesirable effects of deforestation have been known, but civilization has continued to depend on forest resources and on cleared forest land for agriculture and settlement and has continued to destroy forests. In this century, one answer to the problem has been the search for sustainable forestry. Although this idea is discussed frequently, and forest plans are put into place with sustainability as a goal, there are few cases in which sustainability of timber yield, of forest production, or of the forest ecosystem, has been demonstrated.

In this chapter I have shown that the forest model can be a useful tool in exploring sustainability of forests, as well as other uses of forests, including conservation of biological diversity. The model reveals many complexities of the responses of forests to specific management actions. The standard operational definition of sustainability—that it can be measured over three harvests—is shown to be suspect. Sustainability might seem plausible from only three harvests for situations in which long-term sustainability will not be realized.

Forests respond in complex ways to what seem to be simple harvesting policies, in part because of the interactions among species with adaptations to different stages in ecological succession, and in part because of the effects on soil fertility. A few comparatively simple cases of changes in forest floor fertility were considered in this chapter. One important implication is that short rotations tend to lead to decreases in soil fertility, which may make harvests unsustainable.

The conservation of old-growth forests as habitats for endangered species will require more and more active management, especially if global warming occurs as projected by global climate models. The jack pine habitat of the Kirtland's warbler exemplifies problems that might occur with global warming for an endangered species. The decline in old-growth white pine in Hartwick Pines State Park, Michigan, illustrates the need for active management of forest preserves, management that takes into account an understanding of the dynamic qualities of forest ecosystems. Given that the model is realistic and reasonably accurate, as shown in earlier chapters, the forest model becomes a powerful tool for management of forests.

9

General Implications of the Model

Now that we have considered the assumptions and the equations that make up the model and examined a few aspects of its sensitivity and comparison with observations, where do we find ourselves? One answer is that we have a tool that allows us to do something no one has been able to do before: to sit at a computer and "grow" a forest for a century or a thousand years and see what happens to it; to change conditions of the soil and climate and even the characteristics of the trees and see how that affects which trees survive, which prevail, and which win out. People have used the model to examine questions of minimum viable populations and the likelihood of extinction; to look far back in time and consider how forests may have migrated across parts of North America since the end of the ice age. Others have used it to examine practical questions about harvesting rotations and the effects of pollutants on forests.

But is this all "true" or "real"?, one is motivated to ask, given this new kind of computer tool. Although this seems to be the question to ask, perhaps it is not the right question. The right question is: What does the model demonstrate to us?, and another is whether the model is useful.

In answer to the first question, the model in the computer is simply a demonstration of the implications of the assumptions we have made about the dynamics of a forest, about how an individual tree grows in shape and form and how that growth is affected by limiting conditions in the environment; about how trees compete and what determines success, failure, and dominance in competition. Better then to say, in reviewing the results of the use of the model, "so this is what my assumptions mean." The static properties of a forest are easy to see, but it is very difficult to see the dynamic properties of a forest and to understand interactions between processes, or between a process and an effect. It is even harder to know if what we think we understand accounts for what we can observe, without a tool that demonstrates the implications of that knowledge. When we ask this question, we can begin to improve our understanding of dynamics in a complex system such as a forest.

What has been the basis for the assumptions of the forest model described in this book? The basis is the wealth of empirical studies and of natural history information accumulated over many decades, a kind of knowledge that has been difficult to put into context so that one could understand its implications. In this sense, the model becomes a way to bring the complexities and varieties of observations about physiology and ecology of forest trees into a theoretical construct. In the twentieth century, many people have spent many years studying the shape, form, and physiology of plants and how these are affected by environmental conditions. The attempt made with the model described in this book was to abstract from this wealth of empirical material some general statements about the relationships and to form some generalized shapes for the response functions. This intentionally and invariably involves simplification. I have made this point at specific locations throughout the book, but here we can consider it in a more general context. For example, hundreds of studies of biochemical reactions, of individual plant cells, and of whole plants show that biochemical, physiological, and net growth processes increase with temperature to a maximum and then decrease beyond that optimum temperature. There are explanations for this general response for each level of analysis—biochemical, physiological, and whole organism. By taking this wealth of material and simplifying it into a general response function for an entire species, we have a part of a formulation of a theoretical construct of the dynamics of populations of trees in a forest. With a similar approach to responses to light, soil moisture, and soil nitrogen, we build more of this theoretical construct.

Is a model necessary? Yes. Essentially all scientific work involves a model, but often the model is implicit and one may not be aware of it, nor of its assumptions and consequences. If we are unaware of our assumptions, we may be led to assume things contrary to our knowledge, or to accept two mutually contradictory assumptions. That this has happened in the history of ecology I have reviewed elsewhere (Botkin, 1990). In this book we have examined an explicit model. Operated on a computer, it demonstrates the consequences of what we believe we know; this is what it does, and this is part of its utility.

Empirical knowledge without some kind of framework in which to relate it to other knowledge and understand it is not very useful. The importance of models is not restricted to complex systems such as we find in ecology. An analogy can be made from research in chemistry. For example, a listing of every chemical reaction known would not help much if one wants to predict the results of a new reaction. In this case a theoretical construct is necessary.

Such constructs can be useful and helpful even if we know that they are not true in an absolute sense. For example, we know now that Newton's equations of motion are not true in an absolute sense, but are an approximation of what we understand to be a more exact description. However, these equations have been sufficient for us to determine the trajectories required to send people to the moon and back and send space probes to more distant objects in our solar system. In chemistry, orbitals of electrons and the "spin" of an electron are now recognized as not true in a physical sense, but the idea of them has allowed chemists to make huge advances in their science.

Is this modeling then science? If science is a practice that is open to disproof, and in principle the model is directly open to such tests, then developing and using the model is a part of the scientific method. It is setting down in specific form the assumptions (and therefore the hypotheses) that we have held inside our heads about forest ecosystems. This is a surprising thing about the study of ecology—not the lack of theory, but the lack of systematic collections of data that serve as a proper test for a theory. To me, activities to provide those data—long-term monitoring of forest conditions—are one of the essential needs for real progress in ecology.

Another weak link in the development of models such as the one described in this book is the lack of adequate experiments from which one can achieve a reliable quantitative estimate of parameters. As I mentioned earlier, there has been a lack of standardization of experimental conditions in the appropriate physiological studies, for the simple reason that researchers have worked separately; this has meant that sometimes their results are less useful in a larger context than they might be, so that a project that integrated the activities of the physiologist and the modeler would have benefits to both.

One of the cases I have emphasized in earlier chapters, but which is representative, is the lack of adequate data to determine the degree-day minimum and maximum for a species. The required experiments are straightforward to describe but difficult to execute: physiological (laboratory and greenhouse) and physiological-ecological (field and controlled environment) measures of photosynthesis and net growth of trees as a function of temperature, conducted for a sufficient sample of individuals to characterize a population. Studies of this kind have been done by Bazzaz (see, for example, Parrish and Bazzaz [1982]), by Ledig (see, for example, Ledig and Korbobo [1983]), and by many other physiological ecologists and forest geneticists. It would be helpful to have a coordinated program in which such scientists worked with those involved with a model so that the experiments could be directly useful in a larger context.

Although I have reviewed here the example of obtaining parameters for the thermal limits of a species, the point clearly extends to other aspects, including parameters for soil nitrogen and soil-water response functions.

As another example, in recent years there have been a number of laboratory and controlled environment studies of the effects of elevated carbon dioxide levels on photosynthesis and growth of plants. However, each scientist has chosen his own set of conditions within which to do these experiments. One experiment involves certain levels of light, humidity, and fertilization, while another has another set of these conditions. In some cases, some factors are carefully controlled (such as light intensity) while others (such as soil properties or humidity) may not be; a second experiment may have different properties carefully controlled. While the results of each experiment are useful, it is difficult to compare the results done in different laboratories. This means that it is not easy to extrapolate from the experiments in laboratories to the possible effects of increases in carbon dioxide concentration that is occurring due to human activities on natural ecosystems. One might say this is a limitation of the model or a limitation of the experiments. But a more useful perspective is to recognize that an attempt to add this CO_2 fertilization to the forest model makes one aware of the lack of standardization. The attempt to develop a new response function for the model becomes a messenger that tells us a limitation to our knowledge and understanding.

A cooperative research program in which experimentalists and theoreticians worked together to create a set of standard environments for such studies would be of mutual benefit. (Of course, sometimes the experimentalist and theoretician can be the same person, as with Morris [1982]).

Having said this much about the lack of precise quantitative data on which to develop parameters, one finds that, in spite of such limitations, the model described in this book reproduces with considerable realism the dynamics of forests as we know them. Beyond this, it has given insights not recognized previously, some of which have been later reinforced by observation. This suggests that, in spite of the great complexity of ecosystems, it does seem possible to develop reasonable and realistic simulations. The approach implied by this book includes the following:

1. Do not impose a stringent equilibrium condition on the system (that is, do not force it to be what Soros [1989] has called a closed system), but instead develop assumptions about how components of the system operate and let the computer program show, as part of the outcome, what kind of stability is a consequence of the assumptions.
2. Include stochastic processes.
3. Stay with Occam's razor (always use the simplest assumption consistent with observation).
4. Rely on observations and natural history experience as a source of assumptions.

All this may seem well and good for education and for academic research, but what then of the practical applications of this model? If we cannot be assured of the exactness of all parameters, can we rely on the model? Perhaps again this is not phrasing the question quite in the right way. The management and conservation of our living resources are processes, as are all other human activities, not fixed endpoints. In medicine, a doctor does not stop trying to treat a patient because he or she knows that knowledge is incomplete, nor is medical research abandoned because some diseases can be cured. The two activities proceed together. At any one time, a doctor uses the best available method of treatment, with the hope that in the future there will be better methods.

So it is with ecological systems and their conservation and management. At any one time the humanly best thing we can do is to use the best available tools at our disposal. One part of these tools is an ability to confront un-flinchingly the implications of our assumptions. It is important for us to know that we have failed to measure certain things with an accuracy that seems relevant to a specific problem. For example, the limits of the thermal ranges of a species seems, in hindsight, a rather basic and obvious characteristic. Why do we not know it better than we do? In part because we lacked a general framework within which the importance of such a piece of basic information was revealed— that is, a model. One can also say, of course, that this is also a case where the need to gather important basic information was not taken seriously until a practical issue that required that knowledge rose to the surface.

Thus an important use of the model is in demonstrating our ignorance. As another example, consider the application of the model to effects of acid rain

on forests, discussed in Chapter 8. The results suggest that there were no significant effects in the location tested. We may desire to believe that acid rain damage
to tree leaves will kill trees; but we cannot sidestep the result that what was at
the time the best measure of the percent mortality of leaf tissue led to an
immeasurable change in the model forest. If we do not like that result, then we
have a series of assumptions that we can examine, revise, or reject. Otherwise
we have to accept the results, like them or not.

The quantitative limits of the model, as I have described them, are the limits
set by our observations. The model only confronts us with the implications of
these limits. It is the messenger that tells us that we do not know as much as
we would like. It tells us what assumptions we need to be most concerned about.

There are some other deeper and more difficult issues about the extent to
which we should rely on the projections of such a model as the one described
in this book. First, there is the problem of multiple causality. It is possible that
two separate lines of argument lead to a similar conclusion from different
assumptions. In the context of forest ecosystems, it is possible that trees really
compete through their roots for water more than they compete through their
leaves for light. While this is a true problem, it is a problem with our knowledge
and understanding, of which the model is simply an expression. Of course, this
is a problem for science in general, not just for ecology.

Some may believe that computer simulation can only be useful if it is
extremely precise and accurate. But forests are intrinsically quite vairable, as I
have explained before, and a model of forest dynamics need not (and perhaps
should not) be more precise in its trajectories than a real forest. Moreover, often
the points that are keys in the application of this model to practical questions
have to do with directions and general rates of change, rather than details. For
example, that global warming would lead to a rapid decline in the boreal forests
in Minnesota and Michigan was a surprise to me; whether this occurs exactly
in one decade or two or three is much less important than that the implication
of all our assumptions is that global warming will have more rapid effects on
forests than did the little ice age or any period for which we can reconstruct
the dynamics of forests.

Over the years I have found that the nonlinear responses of the model have
led to results that at first seemed counterintuitive, but that on reflection were
the kind of thing I wished I could have guessed simply from knowledge of
natural history. An example was our early application of the model to analyze
the effects of increased CO_2 concentration on a forest. The result was that carbon
dioxide fertilization of trees in a mixed-species, mixed-age-class forest would
not behave like fertilization of a single plant in a well-watered pot, simply causing
an increase in net production. Instead, the increase in CO_2 concentration led
to an increase in growth that resulted in more leaves on more trees, such that
less light penetrated into the canopy. Succession was speeded up, and the
fast-growing, early-successional species were suppressed earlier than they were
in the normal forest. As a result, the forest experienced a smaller increase in
production than a simple extrapolation from a single plant in a pot.

Another example is the model's projection of how to maximize short-term timber harvest under global warming. The result is that one interested solely in this maximization would be motivated to cut down as many of the trees as possible in as short a time as possible. Before doing this experiment, the result was not obvious, but afterward it seemed something I wish I could have perceived directly. These examples show again that a use of the model lies in demonstrating implications of assumptions, some of which might be understood directly by natural history experts.

Thus the demand to place on a model is not that it must be completely precise or contain all of the details of the real world. On the contrary, as I discussed in the first chapter, a model is by intention less than the real world—it is a purposeful abstraction of nature so that some parts of it can be examined.

To say that the model described in this book has uses in education, research, and applied problems is not to say that it is an endpoint or should be treated as a static entity. On the contrary, the continual development of models is a fundamental part of science and should be a basic part of our approaches to resource management. The model is not so much true as useful; it is not so much good or bad as up to the level of our current understanding or not. It is part of the process whereby we seek to improve our understanding of nature and our relationship with our natural resources.

How General Is the Model?

Some who are not familiar with the model understand that it was developed originally for forests of New Hampshire and wonder how it can be applied beyond that geographic area. From the beginning the model was designed to be a general expression of the dynamics of a community of trees of different species. The fundamental growth equation and the environmental response functions are based on an understanding of the general physiological responses of trees as constrained by structure and form. As much as possible, methods for determining the parameters of the environmental response functions depend on the entire range of a species. It was the original hope of Janak, Wallis, and myself that the model would project local growth rates from a generalized set of relationships. How well does the model succeed?

The maximum size that a tree of a certain species can reach is a single parameter that is obtained from the literature. It represents the maximum known for a tree of a species growing anywhere in its range. Version I of the model provided output of the size and age of each tree at death, an output that has been little used and that rapidly consumes much space in computer storage. However, as I discussed before but is worth reemphasizing here, in some early tests we compared the maximum size at death of trees as projected by the model for the environmental conditions of New Hampshire with measured maximum tree sizes in that state. There was close comparison. The maximum sizes observed in New Hampshire for northern hardwoods, such as sugar maple, are considerably smaller than the maxima observed at the center of the ranges of these species, to the south. The combination of the decrease in growth resulting from suboptimal environmental conditions and the effects of small growth rates on mortality led to a decrease in the maximum projected sizes of trees for a site not at the center of the range of a species. This is an example of how the model is general and local at the same time.

A Generalized Approach Implied by the Model

Now that experience with the model suggests that it is realistic and reasonably accurate, the question naturally arises: Is there a technique implied by this model that could be applied more generally in ecology, either to widely dispersed forests or to nonforest ecosystems? The answer seems to be a qualified yes. Over the years the model and variants of it have been applied to a wide variety of forests, including: forests of Europe (Kienast and Kuhn, 1989; Oja, 1983); oak forests of the American Appalachians (Harrison and Shugart, 1990); Louisiana swamp forests (Conner and Brody, 1989); forests of Northwest America (Dale and Franklin, 1989; and Keane, Arno, and Brown, 1990a, b); riparian woodlands (Hanson, Malanson, and Armstrong, 1990); Australian rain forests (Shugart, Hopkins, Burgess, and Mortlock, 1980); forests of New Zealand (Develice, 1988).

The approach seems somewhat general beyond forest ecosystems for a set of competing autotrophic species, with some possibility that it could be generalized to any set of competing species. My colleagues and I have applied the approach to the analysis of forests as ecological systems described in Chapters 1 and 2 to other ecological communities and ecosystems, including a mixed assemblage of algae and their relationship to their environment (Lehman et al., 1975a, b), a materially closed ecosystem of algae and bacteria (Wilson and Botkin, 1990), and a salt marsh ecosystem (Morris et al., 1984). The extension of the forest model to a model of the Isle Royale ecosystem (Botkin and Levitan, 1977), including an age-structure model of moose and wolf populations, as well as the development by Pastor and Post (1985) of a coupled forest–soil model using a variant of the JABOWA model as the basis, suggests that the modeling approach may be incorporated into even more general theoretical constructs. In this section the general basis of this approach is described. Others have used the general approach of the JABOWA model to other kinds of ecological communities, as for example in a model of a semiarid grassland (Coffin and Lauenroth, 1990); European heathlands (Prentice, de Smidt, and van Tongeren, 1987); and coastal landscapes (Pearlstine, McKellar, and Kitchens, 1985). Some authors have discussed the possibilities and difficulties of extending this kind of approach to, for example, aquatic food webs (DeAngelis, 1988). Aspects of the approach can be found in a model for forage production (Roise, Betters, and Kent, 1981).

If the model can be generalized, then it is useful to set down a statement of the general method implied by the model. Some recent papers by others have aspects of such a discussion (see, for example, Armstrong, 1990; Smith and Huston, 1989), including some general implications of this kind of model (see, for example, Armstrong, 1988; Huston, DeAngelis, and Post, 1988).

DeAngelis and Waterhouse (1987) discuss equilibrium and nonequilibrium concepts in relation to this kind of model. First I will state some aspects of the general method in summary form. Then I will discuss these in relation to specific equations of the model.

Some aspects of the general method include the following.

Assumptions of the model define fundamental niches for species; *results* of running the model generate realized niches as well as emergent properties of ecological communities and ecosystems such as productivity, biomass, species diversity, and storage and export of materials.

Each species is characterized by a fundamental growth equation, which is the maximum growth that can be achieved by an individual of a specific size or age. For algae and bacteria, this is simply the maximum division rate, independent of size and age. Growth is then decreased by suboptimal environmental conditions, through environmental response functions. Competition is represented in the model when the availability of an environmental resource is restricted. In the versions of the forest model discussed here, trees compete for light and individual trees restrict the amount of light received by smaller trees. Mortality and regeneration are expressed explicitly. The simplest explanation consistent with observations is used in each case. The model is structured so that new algorithms can be easily substituted for existing ones.

It is especially important to recognize that methods for the calculations of environmental conditions that are made as part of the computer code are not intrinsic to the ecological concepts of the model. As an example, the model, as described in this book, employs the Thornthwaite method for the calculation of evapotranspiration. There is nothing essential or necessary in this choice, and if another method becomes available and appears preferable, substitution is simple without violation of any biological or ecological assumption of the model. In this way, the model was designed to have an open structure.

Niches and the Model

As described at the end of Chapter 4, the approach of the forest model has a basis in niche theory, and through this theory the model relates responses and interactions of populations to physical, biochemical, physiological, and genetic phenomena and observations. This approach can be used to explore theoretical aspects of species interactions including the relationship between species diversity and environmental conditions and stability; questions about niche width, overlap, and packing; and effects of competition on persistence of individual species and on the characteristics of the community and the ecosystem. With the addition of predation using the same methodology, the effects of predator–prey interactions on individual species as well as on the community and ecosystem can be investigated, as in Botkin and Levitan (1977).

Niche axes are defined in the model as simple, direct response surfaces or response functions. This is consistent with the way that niche axes have been measured elsewhere, as the concentration of a resource in the environment (such as calories of energy, grams of chemical elements per unit area or volume). (For herbivores, predators, and decomposers, the niche axes can be measured as the concentration [density] of prey, in terms of numbers of individuals or biomass, or related indices.)

Our experience suggests that few resource axes are required to account for the major dynamic behavior of complex assemblages of populations. While no restrictions are placed by the methodology on the form or number of growth responses to resource axes, empirical observations suggest that real growth responses are represented by a small family of curves, either concave (such as a parabola), or an asymptotic or pseudoasymptotic function.

The response of a species or population is separated into the three primary processes of population dynamics: growth, reproduction, and death. Stages in the life cycle of an organism are defined explicitly, denoted by age or size. These stages are used in addition to the niche axes to define fundamental growth and reproduction equations.

The model includes a fundamental growth equation for an individual with abundant (nonlimiting) resources, whose rate of growth is limited only by intrinsic (genetic) charcteristics, and may be a function of age or size. An optimal or maximum growth response is defined for each species, and this is then modified by the relative functional responses to reduce growth in relation to the availability of resources, as defined in equation (1.1) for trees, and which can be generalized as

$$\frac{\delta(B_{ij})}{\delta t} = H\{B_{ij}\} * f(\text{environment})$$

$$\frac{\delta(B_{ij})}{\delta t} = H\{B_{ij}\}, f_{i,1}(a_1), f_{i,2}(a_2), f_{i,3}(a_3) \cdots f_{i,k}(a_k) \tag{9.1}$$

where B_{ij} = biomass of jth individual of ith species
$\quad\quad j$ = stage in life history of individual of species i
$\quad H, f$ = functions
$\quad f_i(a_k)$ = response of species i to environmental resources k

Graphically, $H\{\}$ is the fundamental growth axis (dimension) and f_k is the kth resource axis (dimension).

An analogous general regeneration relationship can be defined:

$$\frac{\delta(R_{ij})}{\delta t} = H\{B_{ij}\} * f(\text{environment})$$

$$\frac{\delta(R_{ij})}{\delta t} = H\{B_{ij}\}, f_{i,1}(a_1), f_{i,2}(a_2), f_{i,3}(a_3) \cdots f_{i,k}(a_k) \tag{9.2}$$

where R is the number of propagules (number of new individuals, biomass of new individuals, or recruitment of individuals to a specific age or size class), and the other terms are as defined in the previous equation.

The model incorporates some stochastic processes, applied to reproduction and mortality.

Mortality is defined simply by

$$P_{ij} = j(i, \mu_j)$$

where P_{ij} is the probability of mortality of individual j of the ith species and μ_j is a representation of the physiological state of that individual, which may be simply the growth rate (δB_{ij}) or the current biomass (B_{ij}).

Each resource may be affected by the population, so that

$$A_k(t) = \tau[S_1(t-1), S_2(t-1), \cdots S_m(t-1)]$$

where $A_k(t)$ is the state of the kth resource at time t and $S_m(t)$ is the state of the mth species at time $t-1$, and τ is same function of the state of the species.

In the operation of the model, the response of the populations to the resource axes and the effects of the populations on these axes are handled sequentially, beginning with initial conditions for all resources and species, so that the sequence of operations is: (1) Initialize values of the resources and the populations. (2) Calculate the change in the state of the populations. (3) Calculate the change in the state of the resources. (4) Return to step 2. This methodology places no restrictions on the form of the responses or the interactions among resource axes or among populations and resource axes. The functions may be deterministic or stochastic.

The environmental response surfaces (those characterizing the niche of a species) can be used directly to investigate the potential geographic range of a species, to graph relative competitive success, and to consider other aspects of ecological interactions. In this way, the assumptions of the model have an analogy with the approaches of Tilman (1985, 1990a,b).

Possible Extensions and Improvements in the Model

Ecological Guilds

In the JABOWA model, species are defined as distinct units, but in some ecological systems, such as tropical rain forests, there are too many species to define each uniquely. In one study, 10 hectares in Borneo were found to contain 700 tree species, about the same number as found in all of North America (Wilson, 1988). In another, 570 plant species were found in 1 hectare near Kuala Lumpur in Malaysia. More typically, 40 to 100 species of trees can be found in 1 hectare in South American tropical rain forests (Reid and Miller, 1989), and that figure rises to 300 per hectare in Amazonian Peru (Peter H. Raven, personal communication, 1990). A modern high-speed computer could handle parameters for 300 species, but an ecologist would be hard pressed to obtain unique parameters for each. For such a system it may be sufficient to define groups of species as a unit, which would be made up of similar species with similar ecological roles and with environmental response functions similar to each other. For example, shade-intolerant and nitrogen-tolerant species able to grow under dry conditions, with high reproductive rates and short lifetimes, could be combined into a single group of early successional species.

In such a model, the unit defined would be an ecological guild in the sense originally meant by Root (1967), who defined the guild as "a group of species that exploit the same class of environmental resources in a similar way".

Although there have been some applications of the JABOWA model to forests of low latitudes (e.g., Doyle, 1981), the possibility of developing a JABOWA-like model based on ecological guilds has not been explored sufficiently. This seems an appropriate approach for tropical rain forests (Vanclay, 1989).

If an ecological community has a "profession" carried out by only one species, then the guild has only one member, and the species is said to be a *key species* in the ecological community, and in the model the species would be represented as a distinct unit. This links the present formulation of the model to one in which the unit is an ecological guild.

The Expression of Competition in the Model

Individuals can compete in many ways, some direct and some indirect. Ecologists generally divide competition into two categories, *interference* or *contest*, and *exploitation* or *scramble*. In interference competition, an individual competes by controlling the availability of a resource. When a resource becomes limiting, part of the population obtains an adequate amount no matter what the population size; the rest have less than adequate (or perhaps no) supply. In exploitation competition, all individuals have access to the resources; all have an adequate supply until the resource becomes limiting, at which time all suffer. In an idealized case where there was a complete scramble and all shared equally, if there were 100 meals and 101 individuals, every individual would receive 100/101 of a meal and all would be undernourished.

The forest model has interference competition for light, with the taller trees controlling the availability of that resource to smaller trees. The other environmental response surfaces are consistent with the idea of exploitation competition, but the abundance of the resource is determined by the external environment and is not decreased by the abundance of the trees. When the model was originally developed, we assumed that the expression of competition for light was sufficient as an expression of all competition. This was one of the early simplifying assumptions. We recognized that a similar kind of competition could exist for water, with the roots of larger trees extending farther than those of smaller trees and interfering with the water supply to smaller trees. But in that case we chose to assume that competition among roots for water would vary in a 1:1 manner with competition among leaves for light, and that it was unnecessary to represent both in the model. An interesting kind of extension of the model would be to make other environmental response functions into interference competition, and to have the abundance of trees affect the supply of a resource. For example, as I mentioned earlier in this book, it is important for some questions that the calculated evapotranspiration by trees decreases the water stored in the soil. This is an aspect of the model that several of us are trying to develop at this time, but this extension is not presently available. Extensions of the model to include a soil decomposition module, as done by Aber et al., (1978, 1979), or the inclusion of more complex kinds of feedback between the soil and the trees, through nitrogen cycling, as done by Pastor and Post (1985a, b), are also useful extensions.

Issues of Larger Scale and Spatial Versions of the Model

In the past two decades, considerable progress has been made in understanding the dynamics of forest succession and in modeling that process. We have made advances in understanding how forests change over time. We are much farther behind in our understanding of how forests change spatially—in the causality of spatial patterns. Perhaps it was necessary to improve our understanding of temporal processes before we could begin to focus on larger scale spatial processes. But also in the past we have been limited by methods to study and to model large-scale spatial phenomena.

There is growing recognition of the scientific importance of spatial phenomena, which have become referred to under the rubric of *landscape ecology* (Forman and Godron, 1986). Accompanying this recognition is another: that some important resource management and environmental problems must be viewed in a larger, spatial context. In particular, management in terms of fire, insect and disease outbreaks, and effects of storms on forests would seem to benefit from a larger perspective. During the past two decades, some scientists have used the JABOWA model, variations of it, or other modeling approaches to consider some of the phenomena. For example, some authors have used the JABOWA model or derivatives of it to consider effects of fire on forests (Shugart and West, 1981; Kercher and Axelrod, 1984a, b; Green, 1989; Keane, Arno, and Brown, 1990a, b). However, most of these approaches have not considered in a formal, quantitative manner the spatial interactions among forest plots.

The forest model, as described previously in this book, does not consider interactions among plots. The change in the state of any one plot is independent of the present state or the change in state of any other plot. In reality, there are several possible ways that neighboring plots could affect one another, including (1) spread of seeds from seed-bearing trees; (2) spread of fire; (3) spread of insect pests and diseases; (4) upslope plots affecting runoff, erosion, and transport of chemical elements necessary to vegetation, or toxic pollutants detrimental to trees, onto downslope plots; (5) state of the adjacent plot affecting local abundance of vertebrate wildlife, in turn affecting vegetation on a plot; (6) neighboring plots with many trees which can protect interior plots from the harsh effects of winter winds and storms, and increase the chance of survival of trees on a plot, as observed by Sprugel (1985a, b); (7) neighboring plots also changing the roughness of the surface; increased surface roughness caused by the presence of trees could decrease transport of particulates and ions, including toxic pollutants, from the wind onto a plot of interest. Clearing neighboring plots could expose a plot to a greater concentration of these pollutants.

The spread of seeds from one plot to another could be especially important in commercial logging or in large-scale catastrophes, such as the 1938 hurricane that blew down many trees in New England. Regeneration of such areas depends on buried seeds, vegetative reproduction from roots and cut stems, and the migration of seeds from remaining mature trees.

Neighboring plots could increase or decrease the chance of fire reaching a plot. If neighboring plots are toward the prevailing winds, and if they are on upland sites populated by readily burned trees such as pines, then these plots could increase the chance of fire reaching a plot of interest. In contrast, if the neighboring plots are lowlands, populated by cedar and balsam fir and wetland shrubs, and the soil is saturated with water throughout the year, these plots could decrease the chance of fire reaching the plot of interest.

There is evidence that certain local areas in forests serve as centers for the spread of epidemic diseases or outbreaks of insects. As an example, it has been speculated that periodic outbreaks of the gypsy moth in Connecticut began around suburban houses where there were woodpiles for firewood, which provided a locus for the deposition of gypsy moth eggs. Once a high population of the larvae emerged, these spread out to other areas. Holling has suggested that outbreaks of spruce budworm in Canada occur through the same kind of mechanism (Holling, 1965).

All of these phenomena suggest that a model that considers interactions among neighboring plots would be useful. Ten years ago, such a model would have been difficult to use because only very large computing systems could handle the number of calculations and the amount of data conveniently. That situation is changing rapidly with the availability of physically small but extremely powerful desktop minicomputers that have 16 megabytes of RAM memory and whose speed is an order of magnitude faster than most desktop computers.

There have been several attempts to develop spatial versions of JABOWA and its derivatives (e.g., Smith and Urban, 1988). Computer-based geographic information systems offer a powerful tool, readily accessible on small computers (Star and Estes, 1990) that could be productively combined with the forest model. In recent years a number of scientists have suggested that a promising path of development for landscape models of ecological processes is to combine models such as JABOWA with geographic information systems (Baker, 1985; Turner, 1989). During the past five years, several attempts have been made to achieve this integration but in most cases the geographic information system is used only as a back-end display device, not as an integral part of a dynamic simulation. That is, the forest model projects forest growth for a number of independent plots with a variety of site conditions; the geographic information system is used to display each plot in its proper geographic location. However, there are no mechanisms that link the change in state in one plot to the state of neighboring plots or other, more distant plots on the landscape. In one test, a JABOWA model projected the growth of a bottomland forest in response to several flooding regimes (Pearlstine, McKeller, and Kitchens, 1985).

Several of us demonstrated the feasibility of connecting JABOWA-II to a geographic information system and to a graphics display (Van Voris et al., 1990). The geographic information system provided data on real topography for a region of approximately 50 square miles. Normal and global warming climates were projected on the topography, using standard meteorological lapse rates. Fifty replicates of the forest model were run for each of the elevational bands, and each replicate was attached to randomly chosen coordinates within its elevational band. The result showed a landscape with variation among plots, giving a more realistic look to the display. There was no interaction among plots, however; the next state of a plot was not a function of the present state, or change in state, of any neighboring plot. This experiment was a technical demonstration of capabilities. Adding the mechanisms to invoke interactions among neighboring plots is by comparison straightforward.

Woodby (1991) developed a spatial version of JABOWA-II to investigate certain issues in biological conservation, primarily focusing on the question of how competition among trees of different species might change the probability of extinction of an endangered species. In his model, a set of neighboring plots was generated, and the presence of a tree of a species older than some minimum on any of the plots was recorded. If such a tree were present, or had been present within the past 15 years (the assumed length of viability of seeds for that species plus the length of time for a seed to sprout and grow to the minimum size accounted for in the JABOWA-II model), then regeneration could occur. If no such tree existed within the last 15 years on any of the neighboring plots, including the central plot, then regeneration of that species ceased.

Now that computers are readily available to handle the amount of data and number of calculations, further development of spatial versions of JABOWA-II should proceed quite rapidly.

Stability and Natural Ecological Systems

In Chapter 4 I discussed ecological stability and long-term patterns of ecological succession as predicted by the model. The major conclusion is that the model's predictions are not consistent with the classical idea of constancy in nature, nor of succession to a static climax forest, that was the predominant idea in plant ecology during much of the twentieth century. In the past few years, interest in this topic has grown, as testified to by a symposium held at the annual meeting of the Ecological Society in 1990. For a review of concepts of stability through the mid-1970s, see Van Voris (1976). The discussions of stability by May (1973a, b; 1981) have also been influential.

Use of the forest model has been influential in my own thinking about the stability of ecosystems, a topic I have explored in greater length recently (Botkin, 1990) and to which readers interested in this central issue in ecology might want to refer. Suffice it to say here that in most of the twentieth-century ecological literature, the idea of ecological stability has been vague and implicit. Where it is stated explicitly it is almost invariably equivalent to, and usually consciously borrowed from, the physical concept of the stability of a mechanical system, which has been referred to as "classical static stability" (Botkin and Sobel, 1975). A system has this kind of stability if it meets two requirements: (1) It has an equilibrium condition, that is, a single condition in which it will remain indefinitely once that condition is attained. (2) Following a disturbance, it will return to that same condition. This is the stability of a pendulum that remains vertical and at rest until pushed. Once pushed it oscillates for a while but the oscillations gradually die out and the pendulum returns to its original position. The pendulum has "negative feedback," familiar to engineers and an idea that was applied quite commonly in ecology after 1960 (Patten, 1971). For example, classic static stability and the recovery of a simple mechanical system like a pendulum underlay the helpful ideas of resistance and resilience as defined by Holling (1974).

In the first half of the twentieth century, there was a general agreement that a landscape of vegetation, if undisturbed, achieved a constant composition, form, and structure and that, if disturbed, the vegetation would return to its original constant state, that is, vegetation achieved an equilibrium and had classic static stability.

Simplicity, Complexity, Sensitivity, and Accuracy

Do we need simpler or more complex models? Over the years, ecologists interested in models have discussed this question. Some suggest that, given the limitations of our observations, we should seek to develop simpler models. Others suggest that, given the complexities of forest ecosystems, it is necessary to make more complex models. There is no single answer to the question. The useful and relevant complexity of a model depends in part on the use for which it is intended. For example, models that are to be restricted solely to projecting the growth of planted and managed even-aged stands of a single genotype of one species might make do with a simple means of projecting future harvests. The greater the management intervention, such as in thinning, fertilizing, and irrigation—that is, the more uniform the environment of each tree is made through human action—the simpler the methods of projecting future conditions can be.

On the other hand, as our empirical information increases, we can produce models of greater complexity. As concern with global environmental issues grow, there may be a need for models of greater complexity. But in all cases, as I have emphasized throughout this book, we need to be careful not to violate Occam's razor, not to seek explanations unnecessarily more complex than can be accounted for by observations.

Once we raise the question of useful complexity, we are reminded again that models are part of the process of doing science, and that they are not fixed, static structures but are tools that will change with our uses, our needs, and our knowledge.

A discussion of complexity also raises again the question about the desirable sensitivity of a model. There are two kinds of sensitivity discussed in this book: sensitivity of the model to errors in estimation of its parameters, and sensitivity to accuracy of input data. As I discussed in Chapter 6, one might expect that the minimum sensitivity to errors in estimation of parameters would be the most desirable. But a model that is completely insensitive to changes in parameters could not distinguish among species (all species would respond the same way no matter what the parameters were), while a model that is quite sensitive to changes in the estimate of parameters might be useful in analyzing competition among genotypes within a species. The conclusion reached in Chapter 6 is that moderate sensitivity is most useful—a model needs to be sensitive enough to respond to changes in parameters so that species can be differentiated and so that the model mimics the known sensitivity of trees in a real forest. Another way to state this point is that the model should not be completely stable against all changes in parameters. The conclusion is analogous to that in the design of an airplane: a completely insensitive (i.e., completely stable) airplane would fly in one direction and could not be controlled. A highly sensitive airplane would need such constant attention by the pilot that it could not be flown safely. The practical solution is to design airplanes with intermediate sensitivity. This seems to be the case also with models of forest dynamics.

Another expectation that seems plausible at the outset is a desire for great accuracy, especially in the estimation of parameters. But the model described in this book gives surprisingly reasonable results from some quite approximate methods of estimating parameters. Furthermore, there is considerable intrinsic variability in a natural forest, and there is no need for a model to be more precise in its projection of the trajectory of a variable than occurs in the real forest. Thus the necessary accuracy may be less than we once feared.

This is not to say that we do not need to improve methods for estimating parameters. On the contrary, as I have pointed out repeatedly in this book, there are specific parameters for which an improvement in the methods for estimation is highly desirable. In making this last point, it is worthwhile to emphasize that an improvement in a method to estimate a parameter could lead us to have greater confidence in that estimate, without necessarily leading to a greater accuracy in the estimate. Another way to make this last point is that we might choose a method to estimate a parameter that would have a better foundation in basic principles of physiology and population dynamics, but not reduce the variance of the estimate. The net result would be a more realistic model, but not necessarily a more accurate one.

Final Words

The forest model has now been in use for more than 20 years, and has been applied widely in a number of variations, as I discussed in the Preface, and for which Appendix 7 provides a list of references. Readers interested in applying the model to new situations should remember that the model is not a static structure, but is designed to be open to modifications, and from the beginning of its development it has been my hope that people who used it would move rapidly beyond some of our initial methods, especially for the determination of parameters. For example, as I explained in Chapters 2 and 3, the parameters that determine the warm and cold limits of the growth of a species were estimated originally from range maps of tree species and climatic maps. This was done as a matter of convenience when we were trying to get the model to work. I never expected that those who used the model in subsequent years would remain with that simple method, but that has been the case.

It has also been a surprise that most variants of the model have remained with the original list of environmental response functions. In version II, used in this book, the dimensions of environmental conditions are light, temperature, soil nitrogen content, and soil water—four dimensions that seem sufficient to account for much of what we know about forest vegetation. This is sparse, considering the number of plant species and the number of possible limiting factors. Green plants require approximately 20 chemical elements, and in theory one might expect that a complete model of vegetation dynamics would require that each of these be explicit. However, work that my colleagues and I have done on the nature of ecological theory suggests that a surprisingly small set of environmental conditions can separate large numbers of species (Botkin, Maguire et al., 1979; Maguire et al., 1980; Slobodkin et al., 1980). One might postulate a general theorem, that four environmental dimensions may be sufficient to distinguish all the tree species in a forest—to allow for unique tree niches. If correct, this theorem would suggest that the dynamics of ecological systems may be more open to analysis than seems apparent from the bewildering variety of life forms and the complexity and diversity of forest ecosystems. The real test of this theorem is obviously in a model for tropical rain forests.

Another way that the model could be improved is in the feedback between trees and the environment. For example, in reality the evapotranspiration of water by trees decreases the amount of water stored in the soil. The uptake of nitrogen from the soil depletes the soil nitrogen pool. These changes in soil affect future states of the forest. As I have mentioned several times in this book, and which is worth repeating here, the versions of the model I have described do not have such feedback. It is surprising that so much of the dynamics of forests have been successfully mimicked without such feedback. Perhaps this is because the effects of feedback processes are slow under most conditions. But clearly this is an aspect of the model that deserves development.

In this book I have focused on the JABOWA model as I have participated in its development and use. For the sake of clarity and simplicity, I have not dealt in detail with modifications and applications of the model by others. This should not be taken to imply a criticism or lack of interest on my part in these modifications. On the contrary, it is gratifying to know that the JABOWA model is one of the few computer simulations that has been used to any degree by scientists other than the authors, as attested to by the many publications and variations of models that derive from the original (Appendix 7). I hope in the future there will be more application and use of the model and an increasing understanding of its assumptions and algorithms, leading to extensions of the concepts beyond those I have conceived of myself. If this were to take place, and the explicit discussion of the model in this book were a factor in these activities, then this book will have been a success.

Some Additional Mathematical Notes

A. Derivation of Fundamental Growth Equation and the Constants b_2 and b_3

The fundamental growth equation, formulated first in terms of the change in volume (D^2H) of a tree (2.1):

$$\delta(D^2H) = R*LA\left(1 - \frac{DH}{D_{max(i)}H_{max(i)}}\right)*f(\text{environment}) \tag{2.1}$$

is converted to equation (2.6) for the change in diameter, and the parameters b_2 and b_3 are determined as follows:

The constants, b_2 and b_3 in the fundamental growth equation

$$\delta D = \frac{(G_iD[1 - D(137 + b_2D - b_3D^2)/D_{max(i)}H_{max(i)}])}{274 + 3b_2D - 4b_3D^2}*f(\text{environment}) \tag{2.6a}$$

are derived as follows (ignoring for the moment the functions represented by $f(\text{environment})$):

Beginning with the equation relating tree height to diameter:

$$H(D) = 137 + b_2D - b_3D^2 \tag{2.3}$$

and assuming that H reaches H_{max} at the same time that D reaches D_{max}, and taking the derivative of H with respect to D, then

$$\frac{dH}{dD} = 0 = b_2 - 2b_3D_{max} \tag{2.3a}$$

$$b_2 = 2b_3D_{max} \tag{2.3b}$$

and substituting in equation (2.2) above when $H = H_{max}$ and $D = D_{max}$

$$H_{max} = 137 - 2b_3(D_{max})^2 + b_3(D_{max})^2 \tag{2.3c}$$

$$H_{max} - 137 = b_3(2D_{max}^2 - D_{max}^2) = b_3D_{max}^2 \tag{2.3d}$$

$$b_3 = \frac{(H_{max} - 137)}{D_{max}^2} \tag{2.5}$$

substituting equation (2.3d) in (2.3b) gives

$$b_2 = -2D_{max}\left(\frac{H_{max} - 137}{D_{max}^{\;2}}\right)$$

$$b_2 = \frac{-2(H_{max} - 137)}{D_{max}^{\;2}} \tag{2.4}$$

This completes the derivation of parameters b_2 and b_3.

Equation (2.6) is then derived by making use of partial differentials. The chain rule of calculus tells us that we can consider that an infinitesimal change in volume is equal to a change in volume with respect to diameter, times an infinitesimal change in diameter, which means that we can set

$$\delta(D^2H) = \delta\left(\frac{D^2H}{\delta D}\right)\delta D = R*LA\left(1 - \frac{DH}{D_{max(i)}H_{max(i)}}\right) \tag{2.1}$$

Then

$$\delta D = \frac{\{R*LA(1 - DH/D_{max(i)}H_{max(i)})\}}{\delta(D^2H/\delta D)} \tag{2.1a}$$

Remember also that $LA = CD^2$ and $G = RC$, so that

$$\delta D = \frac{\{GD^2(1 - DH/D_{max(i)}H_{max(i)})\}}{\delta(D^2H/\delta D)} \tag{2.1b}$$

Now the question is how to get the denominator. This is the differential of D^2H with respect to D. We can substitute for H from equation (2.3), so that

$$\delta\left(\frac{D^2H}{\delta D}\right) = \frac{\delta\{D^2[137 + b_2D - b_3D^2]\}}{\delta D} \tag{2.1c}$$

multiplying through

$$\delta\left(\frac{D^2H}{\delta D}\right) = \frac{\delta\{137D^2 + b_2D^3 - b_3D^4\}}{\delta D} \tag{2.1d}$$

taking the derivative

$$\delta\left(\frac{D^2H}{\delta D}\right) = 274D + 3b_2D^2 - 4b_3D^3 \tag{2.1e}$$

Then substituting (2.1e) in (2.1b) gives

$$\delta D = \frac{\{GD^2(1 - DH/D_{max(i)}H_{max(i)}\}}{274D + 3b_2D^2 - 4b_3D^3} \tag{2.1f}$$

Factoring out the D in the denominator

$$\delta D = \frac{\{GD^2(1 - DH/D_{max(i)}H_{max(i)}\}}{D[274D + 3b_2D - 4b_3D^2]} \tag{2.1g}$$

so that we have equation 2.6, which is

$$\delta D = \frac{\{G_i D[1-(DH/D_{max}H_{max})]}{274+3b_2 D - 4b_3 D^2} \tag{2.6}$$

B. Integration of Growth Equation

The following, from Botkin et al. (1972b), was derived by James F. Janak. It is repeated here for completeness.

If, by definition, $x = D/D_m$ and $a = 1 - 137/H_m$, the growth equation, equation (2.6), is

$$\frac{dx}{dt} = \frac{G}{2H_m} \frac{x(1-x)(1+ax(1-x))}{[1-a(1-x)(1-2x)]} \tag{A1}$$

The value of G for each species has been arbitrarily chosen in the text so that x is approximately $\frac{2}{3}$ when $t = \text{AGEMX}/2$. To simplify determination of numerical values of G, and to study other methods of fixing the value of G, it is convenient to have the integral $x(t)$ of this equation. If x_0 is the value of x when $t = 0$, one has from equation (A1)

$$\frac{Gt}{2H_m} = \int_{x_0}^{x(t)} \frac{dx}{x(1-x)} \left[1 - \frac{a(1-x)^2}{1+ax(1-x)}\right] \tag{A2}$$

or, from tables of integrals,

$$\ln\left(\frac{x}{1-x}\right) + \frac{a}{2}\ln\left(\frac{1+ax-ax^2}{x^2}\right) - \frac{(a+a^2/2)}{\sqrt{a^2+4a}}\ln\left(\frac{2-(\sqrt{a^2+4a}-a)x}{2+(\sqrt{a^2+4a}+a)x}\right)$$

$$= \frac{Gt}{2H_m} + C \tag{A3}$$

where C is the value of the left-hand side for $x = x_0$. If $x_0 = 1/2D_m$ (i.e., $D(0) = 0.5 \text{ cm}$), the value of G giving $x = \frac{2}{3}$ when $t = \text{AGEMAX}/2$ is

$$G = \frac{4H_m}{\text{AGEMAX}} \left\{ \ln(2(2D_m - 1)) + \frac{a}{2}\ln\left(\frac{(9/4+a/2)}{4D_m^2 + 2aD_m - a}\right) \right.$$

$$\left. - \frac{a+a^2/2}{\sqrt{a^2+4a}}\ln\left[\frac{(3+a-\sqrt{a^2+4a})(4D_m+a+\sqrt{a^2+4a})}{(3+a+\sqrt{a^2+4a})(4D_m+a-\sqrt{a^2+4a})}\right] \right\} \tag{A4}$$

in which D_m and H_m are in centimeters and AGEMAX is in years.

In some cases the values of G obtained from equation (A4) give unreasonable growth rates. This is particularly true of the short-lived species, for which (A4) gives growth rates that are too large, and of beech, for which (A4) gives too small a growth rate. The values of G for these species have been adjusted in the simulator to give more reasonable growth rates.

Probably a much better way of determining the value of G for each species lies in demanding that the maximum possible annual diameter increment given by equation (2.6) be equal to some value δD_{max}, which could be determined from field observations. One finds that the required value of G is such that

$$\delta D_{max} \approx 0.2 \frac{G D_m}{H_m} \qquad (A5)$$

The value of G for beech used in the simulator corresponds to $\delta D_{max} = 1.0$ cm, whereas the value of G implied by equation (A4) for beech would lead to $\delta D_{max} = 0.7$ cm. The latter value is almost certainly too samll; this merely means that are arbitrary assumption that $D/D_{max} = \frac{2}{3}$ when $t = \text{AGEMAX}/2$ is not correct for all species.

Equation (2.7) is obtained simply by solving [A5] for G, so that

$$G \cong 5 H_{max} \left(\frac{\delta D_{max}}{D_{max}} \right) \qquad (2.7)$$

where δD_{max} is the maximum observed diameter increment.

APPENDIX II

Equations, Variables, and Constants

Equations

$$\delta(D^2 H) = R * LA\left(1 - \frac{DH}{D_{max(i)} H_{max(i)}}\right) * f(\text{environment}) \tag{2.1}$$

$$W = C_i D^2 \tag{2.2}$$

$$H(D) = 137 + b_2 D - b_3 D^2 \tag{2.3}$$

$$b_{2(i)} = \frac{2(H_{max(i)} - 137)}{D_{max(i)}} \tag{2.4}$$

$$b_{3(i)} = \frac{(H_{max(i)} - 137)}{D^2_{max(i)}} \tag{2.5}$$

$$\delta D = \frac{\{G_i D[1 - (DH/D_{max(i)} H_{max(i)})]\}}{274 + 3b_2 D - 4b_3 D^2} * f(\text{environment}) \tag{2.6}$$

$$\delta D = \left(G_i D\left\{\frac{1 - [D(137 + b_2 D - b_3 D^2)/D_{max(i)} H_{max(i)}]}{274 + 3b_2 D - 4b_3 D^2}\right\}\right) * f(\text{environment}) \tag{2.6a}$$

$$G \sim = 5H_{max}\left(\frac{\delta D_{max}}{D_{max}}\right) \tag{2.7}$$

$$AL(h) = AL_0 e^{-k\int_h^x LA(h')dh'} \tag{2.8}$$

$$AL = PHI e^{-k*SLA} \tag{2.9}$$

$$DEGD = \int Tt\, dt \tag{2.10}$$

$$DEGD = \frac{365}{2\pi}(T_{July} - T_{Jan}) - \frac{365}{\pi}\left[40 - \frac{(T_{July} + T_{Jan})}{2}\right]$$
$$+ \frac{365}{\pi}\frac{\left(40 - \frac{T_{July} + T_{Jan}}{2}\right)^2}{T_{July} - T_{Jan}} \tag{2.11}$$

$$dw/dt = P + U - E - S \tag{2.12}$$

where

$$PP_j = BASEP_j + RLAPSE*(ELEV_{site} - ELEV_{base}) \tag{2.13}$$

$$\mu_j = K_s(\tau - t) \tag{2.14}$$

$$E_0 = 16\left(\frac{10T_j}{I}\right)^a \tag{2.15}$$

$$I = \sum_{m=1}^{12}\left(\frac{T_j}{5}\right)^{1.514} \tag{2.16}$$

$$a = (0.675I^3 - 77.1I^2 + 17,920I + 492,390) \times 10^{-6} \tag{2.17}$$

$$E_{0j} = -41.947 + 3.246(T_j) - 0.0436(T_j)^2 \tag{2.18}$$

$$E = E_0 \quad \text{if} \quad w \geqslant w_k \tag{2.19a}$$

$$E = E_0(w/w_k) \quad \text{if} \quad w < w_k \tag{2.19b}$$

$$S = \frac{[0.8P^2][w]}{[E_0 + P][w_{max}]} \tag{2.20}$$

$$U_j = \frac{K*(w_{max} - w)}{w_{max}*DT} \tag{2.21}$$

$$DWT = \left(\frac{dw}{dt}\right) - \rho \tag{2.22}$$

$$v = -170 + 4.0*AVAILN \tag{2.23}$$

$$\frac{\delta D}{\delta t} = \left(\frac{\delta D}{\delta t}\right)_{opt}*MIN\{f_1, f_2, \ldots, f_j\} \tag{3.1a}$$

$$\frac{\delta D}{\delta t} = \left(\frac{\delta D}{\delta t}\right)_{opt}*f_1*f_2*\cdots f_j \tag{3.1b}$$

$$f(\text{environment}) = f_i(\text{light})*Q_i*s(BAR) \tag{3.2}$$

$$Q_i = TF_i*WiF_i*WeF_i*NF_i \tag{3.3}$$

$$f(AL)_L = A_1\{1 - e^{(A_2AL - A_3)}\} \tag{3.4a}$$

where L is the light tolerance type

$$f(AL)_1 = 2.24\{1 - e^{-1.136(AL - 0.08)}\} \quad \text{shade intolerant} \tag{3.4b}$$

$$f(AL)_3 = 1 - e^{-4.64(AL - 0.05)} \quad \text{shade tolerant} \tag{3.4c}$$

$$TF_i = max(0, TDEGD_i) \tag{3.5}$$

$$TDEGD_i = \frac{4(DEGD - DEGD_{min(i)})(DEGD_{max(i)} - DEGD)}{(DEGD_{max(i)} - DEGD_{min(i)})^2} \tag{3.6}$$

$$\text{TF}_i = e^{\{-(\text{DEGD} - \gamma)^2 / 2\sigma^2\}} \tag{3.7}$$

$$\text{WILT} = \frac{(E_0 - E)}{E_0} \tag{3.8a}$$

$$\text{WILT} = \frac{(w_k - w)}{w_k} \tag{3.8b}$$

$$\text{WiF}_i = \max\left\{0, 1 - \left(\frac{\text{WILT}}{\text{WLMAX}_i}\right)^2\right\} \tag{3.9}$$

$$\text{WeF}_i = \max\left(0, 1 - \left(\frac{\text{DTMIN}_i}{\text{DT}}\right)\right) \tag{3.10}$$

$$\lambda_N = \alpha_1[1 - 10^{-\alpha_2(\text{AVAILN} + \alpha_3)}] \tag{3.11}$$

$$\text{NF}_i = \frac{(\alpha_4 + \alpha_5 \times \lambda_N)}{\alpha_6} \tag{3.12}$$

$$s(\text{BAR}) = \frac{(1 - \text{BAR})}{\text{SOILQ}} \tag{3.13}$$

$$M_i = 1(1 - \varepsilon_i)^{\text{AGEMX}_i} \tag{3.14}$$

$$\varepsilon_i = \frac{4.0}{\text{AGEMX}_i} \tag{3.15}$$

$$E_i = \zeta * S_i * f(\text{AL})_1 * Q_i \tag{3.16}$$

$$\zeta < f(\text{AL})_1 * Q_i \tag{3.17a}$$

$$\zeta < f(\text{AL})_3 * Q_i \tag{3.17b}$$

$$E_i = \zeta * S_i \tag{3.18}$$

$$H' = -c \sum_i p_i \log p_i \tag{4.1}$$

$$\frac{\delta(_jB_i)}{\delta t} = H\{_jB_i\} * f(\text{environment}$$

$$\delta(_jB_i)\delta t = H\{_jB_i\}, f_{i,1}(a_1), f_{i,2}(a_2), f_{i,3}(a_3) \cdots f_{i,k}(a_k) \tag{9.1}$$

$$\frac{\delta(_jR_i)}{\delta t} = H\{_jB_i\} * f(\text{environment}) $$

$$\delta(_jR_i)\delta t = H\{_jB_i\}, f_{i,1}(a_1), f_{i,2}(a_2), f_{i,3}(a_3) \cdots f_{i,k}(a_k) \tag{9.2}$$

Note: These are generalizations of equations (3.1) and (3.2). Where $_jB_i$ is the biomass of the jth individual of the ith species, j is the stage in the life history of an individual of species i; H and f are functions, $f_k(a_k)$ the response of species i to environmental resources k. Graphically, $H\{ \}$ is the fundamental growth axis (dimension) and f_k is the kth resource axis (dimension) and R is the number of propagules.

Variables and Constants

A_1: Empirically derived constant in the light response function, set to 1 for shade-tolerant species and to 2.24 for intolerant species.

A_2: Empirically derived constant in the light response function, set to -4.64 for shade-tolerant species and to -1.136 for shade-intolerant species.

A_3: Empirically derived constant in the light response function, set to -0.05 for shade-tolerant species and to -0.08 for shade-intolerant species.

a: A constant in the Thornthwaite evapotranspiration equation (2.11). See equations 2.15 and 2.17.

a: A constant in the equation for nitrogen content of leaves.

α_1 through α_6: Constants in equations for nitrogen tolerance classes and nitrogen response functions.

AGEMX: Maximum age for a species.

$AINC_i$: Minimum annual diameter increment required for an individual tree to avoid being subjected to a second, high probability of mortality.

AL: The light available to a tree.

$AL(h)$: The light available at a specific height above the ground in the forest.

AL_0: The light available above the forest (hence the sunlight intensity at the base of the atmosphere).

AVAILN: The available nitrogen in the soil.

B: Biomass.

$b_{2(i)}$: A constant in the fundamental growth equation.

$b_{3(i)}$: A constant in the fundamental growth equation.

β: A constant used in the calculation of surface runoff, taken to be $0.8(P^2/(E_0 + P))$.

BAR: Actual total basal area of trees on a plot.

$BASEP_j$: The rainfall recorded at that station in month j.

BIRCH: A parameter in version I that determined the number of new saplings that could be added for intermediate light-tolerant species, set at 400 (revised in version II).

C_i: A constant of proportionality, relating leaf weight to tree diameter.

D: Tree diameter.

$D_{max(i)}$: The maximum diameter for an individual of species i.

D_{opt}: Tree diameter calculated from the fundamental growth equation with no decrease due to suboptimal environment conditions.

ΔD_{opt}: Change in tree diameter calculated from the fundamental growth equation with no decrease due to suboptimal environmental conditions.

DEGD: Growing degree-days.

$DEGD_{min(i)}$: Value of DEGD at the northern end of the range of species i.

$DEGD_{max(i)}$: Value of DEGD at the southern end of the range of species i.

DS: Depth of the soil.

DT: Depth of the water table.

$DTMIN_i$: Minimum distance to the water table tolerable for species i.

$DW\pm$: Corrected water storage (corrected for field capacity).

E: Actual evapotranspiration (evaporation + transpiration).

E_i: Expected (and calculated) number of saplings to be added to a plot in one year.

E_0: Potential evapotranspiration in millimeters per month.

E_{0m}: Potential evapotranspiration calculated between 26.5°C and 38°C and asymptotic at 185 mm above 38°C.

ε_i: The annual probability of death.

$f(AL)$: The light response function for the categories of types used in the model.

$f_i(light)$: The light response of species i.

$f(x)$: Any function of a vairable or set of variables listed between the parentheses and denoted here by "x".

G_d: A default value of G for a specific species.

G_i: A parameter in the fundamental growth equation, equal to R_iC_i, and determining the timing of the rapid rise in diameter of a tree of species i (the larger the value of this parameter, the earlier in life a tree reaches one-half its maximum diameter).

H: Tree height.

H': The Shannon–Weaver index of species diversity.

$H_{max(i)}$: The maximum tree height for species i.

$HEIGHT_i$: The calculated height of a new saplings when it is added to a plot.

η: Growth efficiency, or the net assimilation or net photosynthetic ratio (net assimilation per unit of leaf area).

η_{max}: The maximum growth efficiency.

Γ_i: Minimum light intensity for saplings of species i to be added to a plot.

I: The heat index for the month.

i: Species.

j: Month.

K_s: The snowmelt coefficient (2.7 mm/day/°C).

k: The light extinction coefficient (set at 1/6,000).

K: The coefficient of upward capillary water flux, set equal to 15.

κ: Used simply to represent a constant in an equation not otherwise given specific attributes in the text.

L: Light tolerance class ($L = 1$ is intolerant of low light; L = 3 is tolerant of low light; and L = 2 is intermediate in light tolerance).

LA: Leaf area.

LAI: Leaf area index (the dimensionless quantity that is the area of leaves about a unit area of ground surface).

λ_N: The concentration of nitrogen in leaves of trees of nitrogen tolerance class *N*.

M_i: The probability that a tree of species *i* at age 1 will reach the maximum age (default value is 0.02).

m: Standard month (30 days with 12 hours of daylight).

μ: The amount of snow melted in a month.

N: Nitrogen tolerance class ($N = 1$ is intolerant of low nitrogen; $N = 3$ is tolerant of low nitrogen; $N = 2$ is intermediate in tolerance).

n_i: The number of individuals per forest plot of species *i*.

NF_i: An index of tree response to nitrogen content of the soil.

N: The total number of compartments in a compartment model of an ecosystem.

v: Relative available nitrogen.

P: Rainfall plus snowmelt at the site.

P_i: The relative photosynthetic efficiency (scaled between 0 and 1) of the light conditions at the ground.

p_i: The probability that a random sample will yield an individual of species *i*.

PHI: Light intensity above the forest (normalized to 1).

PP_j: The rainfall recorded at that station in month *j*.

Q_i: The product of response functions for thermal properties of the environment, drought (wilt factor), soil-water saturation (soil wetness factor), and soil fertility.

R: A constant in the generalized form of the fundamental growth equation. *R* is not actually used or calculated because it is replaced by G_i, where that parameter is equal to RC_i.

RANDOM(DSEED): A random number between 0 and 1 (identical with ζ).

RLAPSE: Average lapse rate for the difference in elevation between the base weather station and the plot.

ρ: Combined term for loss of water that is above field capacity via runoff and downward percolation.

S: Moisture surplus, combining runoff and downward percolation.

S_i: The maximum number of saplings that can be added in one year to a plot; this represents recruitment (see Glossary of Technical Terms) of new individuals to the population of species *i*.

s(BAR): The proportion unoccupied of the maximum potential stem area on the site, equivalent to the forestry concept of a site quality index.

SLA_n: "Shading leaf area," actually the leaf weight above tree *n*, the sum of the weight of the leaves on all trees taller than tree *n*.

Snow-D-D: The number of snow degree-days that is summed from a base of 0°C for the snowmelt calculation (see Appendix III).

Snow-MELT: Snow melted in millimeters per day.

Snow-S: The amount of water stored in snow.

SNOW-TEMP: The threshold temperature (°C) at which snowmelt begins to occur (-3.4°C).

SOILQ: Maximum basal area that the plot can support and set at a high value of 20,000.

σ: Standard deviation.

T_j: The average temperature for the month, $\geqslant 0$°C, so that $T_j = \max(0, \text{mean temperature (°C) for month } j)$.

T_{Jan}: Average January minimum temperature.

T_{July}: Average July maximum temperature.

T_s: Temperature at a site.

T_t: Average temperature for day t.

TDEGD$_i$: The degree-days factor for species i at a site.

TF$_i$: The general temperature response function.

τ: Snowmelt temperature—the minimum average monthly temperature at which some snow can be assumed to melt. In version II of the model, this is -3.4°C, an average that assumes that some parts of some daylight periods in the month will have temperatures above 0°C.

v: The average of the maximum and minimum degree-day limits for a species.

U: Upward capillary transport from water table to the root zone ($=$ depth of the soil).

W: The weight of leaves on a tree.

w: The amount of water stored in the soil between the soil surface and the base of the soil.

Water-S: See w.

WeF$_i$: A "soil wetness" factor, an index of the amount of water saturation of the soil a tree can withstand.

WiF$_i$: The "wilt" factor and index of the drought conditions that a tree can withstand.

w_k: The value of w at which water limits evaporation, taken to be $0.7w_{max}$.

w_{max}: The maximum capillary water storage or the field capacity of the soil.

W_{max}: The maximum leaf weight above a site for germination of seedlings of species i.

W_{min}: The minimum leaf weight above the site for germination and survival of seedlings of species i.

WILT: The drought water stress of the soil at a site, expressed as a dimensionless quantity that ranges from zero in swampy sites to about 30 percent in thin interior soils with a low till depth.

WLMAX$_i$: The maximum wilt tolerable by species i.

Ω: w/w_{max}.

χ_i: A compartment in a compartment model; the amount of energy or some material stored in the ith compartment.

ζ: A random number between zero and number, chosen uniformly (with equal probability that any number within this range will be chosen). Also referred to as RANDOM(DSEED).

APPENDIX III

Additional Information on the Calculation of Environmental Conditions

The calculation of environmental conditions is in some ways the most intricate and complex of all the calculations used with the model. Because these calculations are to some extent external to the biology of a forest, it is appropriate to remove the details from the main body of the text and make them available as an appendix. The purpose of this appendix is to provide information sufficient for readers to understand all of the calculations regarding the environmental aspects of the site in the model. The most complex calculations have to do with the water budget. A few notes will be useful.

The generalized water balance equation given in Chapter 2 is

$$\frac{dw}{dt} = P + U - E - S \tag{2.12}$$

where w = amount of water held in capillary storage in root zone
P = liquid precipitation plus snowmelt at site
U = upward capillary transport from water table to root zone
E = evapotranspiration (evaporation + transpiration)
S = combined term for water loss via runoff and downward percolation

For the first year of climate for a site, the initial January soil moisture level is unknown. It is calculated by an iterative process with arbitrary initial conditions. The goal is to achieve a water balance such that the amount of water stored in the soil at the end of December in one year of the iteration is identical (within some small fraction) to the amount of water sorted on the first day of January of the next year. This iterative approach was first applied to the model by Levitan (Botkin and Levitan, 1977). The time-dependent solution to the water balance depends on whether $w \geqslant w_k$ or $< w_k$, where $w_k = 0.7 w_{max}$, as explained in the text. Evapotranspiration (E) occurs at its maximum rate(E_0) whenever the soil-water storage equals or exceeds w_k. When w is less than w_k, E is a function of the water storage as given below. These two evapotranspiration regimes result in two forms for the water balance differential equations, and lead to different solutions to the iterative procedure that results in the water storage at the end of December of one year equaling the water storage at the first day of January of the next year. For these two cases, E is

$$\text{Case 1: } E = E_0 \quad \text{if} \quad w \geqslant w_k \tag{2.19a}$$

$$\text{Case 2: } E = E_0\left(\frac{w}{w_k}\right) \quad \text{if} \quad w < w_k \tag{2.19b}$$

As discussed in Chapter 2, the other water balance fluxes can be represented by the following equations, repeated here for convenience.

$$S = \frac{[0.8P^2][w]}{[E_0 + P][w_{max}]} = \beta w \tag{2.20}$$

so that β is used as a simplification and equal to

$$\beta = \frac{[0.8P^2]}{[E_0 + P][w_{max}]}$$

$$U = \frac{K(w_{max} - w)}{(w_{max}\tau)} = \frac{K}{\tau} - \frac{K}{(w_{max}\tau)w} \tag{2.21}$$

where τ is the depth of the water table (referred to elsewhere in the body of the text as variable DT, but for readability referred to as τ here), and K is an empirically derived proportionality constant for upward capillary water transport.

K was estimated by Levitan to be approximately 15 mm of water transported upward per month per meter depth of soil (Botkin and Levitan, 1977). This was based on information on page 177 in Sellers (1965) for conditions in a North Carolina forest where measurements had been made. With these equations, it is now possible to formulate the instantaneous change in water storage for both cases:

$$\frac{dw}{dt} = P - E_0 + \frac{K}{\tau} - \left(\beta + \frac{K}{(w_{max}\tau)}\right)w \quad \text{if} \quad w \geqslant w_k \tag{2.12b}$$

$$\frac{dw}{dt} = P + \left(\frac{K}{\tau}\right) - \left[\frac{E_0}{w_k} + \beta + \frac{K}{(w_{max}\tau)}\right]w \quad \text{if} \quad w < w_k \tag{2.12c}$$

Both (2.12b) and (2.12c) have the form

$$\frac{dw}{dt} = b - aw \tag{2.12d}$$

whose exact solution is

$$w(t) = \frac{b}{a} - \left(\frac{b}{a} - w_0\right)e^{-a(t - t_0)} \tag{2.12e}$$

where $w_0 = w(t)$, is that is, w_0 is w at time zero and $t - t_0$ is the time elapsed since time zero. From these equations (2.12b–2.12e, 2.19a, 2.19b, 2.20, 2.21), the expanded solutions to (2.12) for the two cases are:

Case 1, when $w \geqslant w_k$:

$$a = \beta + \frac{K}{(w_{max}\tau)}$$

$$b = P - E_0 + \frac{K}{\tau}$$

and substituting in (2.12e)

$$w(t) = \frac{P - E_0 + (K/\tau)}{\beta + [K/(w_{max}\tau)]} - \left\{ \frac{P - E_0 + (K/\tau)}{\beta + [K/(w_{max}\tau)]} - w_0 \right\} e^{[\beta + (K/(w_{max}\tau))](t - t_0)}$$

Case 2, when $w < w_k$:

$$a = \frac{E_0}{w_k} + \beta + \frac{K}{(w_{max}\tau)}$$

$$b = P + \frac{K}{\tau}$$

and, substituting in (2.12e)

$$w(t) = \frac{P + (K/\tau)}{(E_0/w_k) + \beta + [K/(w_{max}\tau)]}$$
$$- \left\{ \frac{P + (K/\tau)}{(E_0/w_k) + \beta + [K/(w_{max}\tau)]} - w_0 \right\} e^{-[(E_0/w_k) + \beta + (K/(w_{max}\tau))](t - t_0)}$$

The monthly evapotranspiration, which is the integral of Edt, is different for the cases:

Case 1, for the time period when $w \geqslant w_k$:

$$\int Edt = E_0(t_2 - t_1)$$

Case 2, for the time period when $w < w_k$:

$$\int Edt = \left\{ \left(\frac{b}{a} \right)(t_2 - t_1) - \left(\frac{b}{a} - w_0 \right) \frac{(1 - e^{-a(t_2 - t_1)})}{a} \right\} \left(\frac{E_0}{w_k} \right)$$

where a and b are defined as in Case 2 above, that is:

$$a = \frac{E_0}{w_k} + \beta + \frac{K}{(w_{max}\tau)}$$

$$b = P + \frac{K}{\tau}$$

Latitude Correction for E_0

The amount of daylight available is corrected for latitude following C. W. Thornthwaite (1948): An approach toward a rational classification of climate, *Geographical Review* 38, Table V, p. 93.

TABLE A.III.1. Correction of available daylight for latitude

| Month | \multicolumn{8}{c}{Latitude} |
	00.0	10.0	20.0	30.0	35.0	40.0	45.0	50.0
January	1.04	1.00	0.95	0.90	0.87	0.84	0.80	0.74
February	0.94	0.91	0.90	0.87	0.85	0.83	0.81	0.78
March	1.04	1.03	1.03	1.03	1.03	1.03	1.02	1.02
April	1.01	1.03	1.05	1.08	1.09	1.11	1.13	1.15
May	1.04	1.08	1.13	1.18	1.21	1.24	1.28	1.33
June	1.01	1.06	1.11	1.17	1.21	1.25	1.29	1.36
July	1.04	1.08	1.14	1.20	1.23	1.27	1.31	1.37
August	1.04	1.07	1.11	1.14	1.16	1.18	1.21	1.25
September	1.01	1.02	1.02	1.03	1.03	1.04	1.04	1.06
October	1.04	1.02	1.00	0.98	0.97	0.96	0.94	0.92
November	1.01	0.98	0.93	0.89	0.86	0.83	0.79	0.76
December	1.04	0.99	0.94	0.88	0.85	0.81	0.75	0.70

The above applies for latitudes below 50 degrees, because the Thornthwaite method has not been tested for conditions above 50 degrees latitude. Beyond that, the calculation is corrected for daylight as $LT_m = DLCF_m$, as explained below.

The value of E_0 is calculated at monthly intervals, following equation (2.15), which is given in the body of the text but is repeated here for convenience:

$$E_{0m} = 16\left(\frac{10T_m}{I}\right)a \qquad (2.15)$$

where m = standard month (30 days with 12 hours of daylight)
  T_m = average temperature for month, so that
  T_m = max (0, mean temperature (°C)
  I = heat index for month, so that I for year is

$$I = \sum_{m=1}^{12}\left(\frac{T_m}{5}\right)^{1.514} \qquad (2.16)$$

$$a = (0.675I^3 - 77.1I^2 + 17{,}920I + 492{,}390) \times 10^{-6} \qquad (2.17)$$

and E_0 is in millimeters per month.

Note also as explained in the text that this equation is considered valid between 0 and 26.5°C. Between 26.5°C and 38°C the relationship of potential evapotranspiration follows a parabolic function expressed by

$$E_{0m} = -41.947 + 3.246(T_m) - 0.0436(T_m)^2 \qquad (iii.1)$$

Above 38°C, potential evapotranspiration asymptotes at 18.5 cm.

The correction for latitude is implemented by modifying T_m. Table III.1 (Thornthwaite) gives correction factors of a small number of latitudes. The correction factor for a specific latitude is interpolated from table of latitude correction values (Table III.1), according to a simple linear interpolation. For a specific latitude between two values in the table, the latitude correction factor is

$$\text{FACTOR} = \frac{\text{LI}(I) - \text{Latitude}}{\text{LI}(I) - \text{LI}(I - 1)} \tag{iii.2}$$

where variable "Latitude" is the actual latitude, and $\text{LI}(I)$ and $\text{LI}(I - 1)$ are the values in Table A.III.1 for the latitudes that bound the actual latitude.

Then a latitudinal month, LT_m, (calculated for month m as

$$\text{LT}_m = \text{DLCF}_m - \left[\frac{\text{LI}(I) - \text{Latitude}}{\text{LI}(I) - \text{LI}(I - 1)}\right] * (\text{DLCF}_m - \text{DLCF}_{m-1}) \tag{iii.3}$$

The latitudinal month is then used in the calculation of E_0. The parabolic function was inferred from data provided by Thornthwaite (1947).

Details of the water balance calculations were devised by Levitan, and given in Botkin and Levitan (1977).

APPENDIX IV

Initial Stand Used
in Tables 6.1 and 6.2

The initial stand used in Tables 6.1 and 6.2 has the tree population given in Table A-IV.1. This represents an old-growth forest near Virginia, Minnesota. Note that the model assumes a pool of seeds of all species whose site quality is greater than zero. Thus there are traces of basswood at this site that would not be expected to persist and, given biogeographic considerations, would probably not have seeds present.

The initial population used in Table 6.5 to represent a jack pine stand in central Michigan is given in Table A-IV.2. The site is a coarse, sandy soil near Grayling, Michigan. Weather records are from Grayling, Michigan. The soil depth is 1.0 m and depth to the water table is 1.2 m. Elevation: 900 ft; water table depth 1.2 m; soil texture 50.0 mm/m; percent rock 0.0; available nitrogen 55.0 kg/ha.

TABLE A-IV.1. Tree population of initial stands used in Tables 6.1 and 6.2

Species	DBH[a]	Height	Species	DBH[a]	Height
Sugar maple	20.0	849.7	Balsam fir	6.3	439.6
Sugar maple	13.6	630.6	Balsam fir	6.2	436.0
Sugar maple	10.8	531.7	Balsam fir	5.8	416.0
Sugar maple	4.6	309.9	Balsam fir	4.8	371.5
Sugar maple	4.4	301.2	Balsam fir	4.6	360.5
Sugar maple	4.3	296.8	Balsam fir	4.2	342.8
Sugar maple	4.1	289.4	Balsam fir	4.1	339.2
Sugar maple	3.7	275.3	Balsam fir	4.1	336.6
Sugar maple	3.5	267.7	Balsam fir	3.9	327.2
Sugar maple	2.2	220.5	Balsam fir	3.6	315.4
Sugar maple	2.1	217.3	Balsam fir	3.6	313.7
Sugar maple	1.7	200.5	Balsam fir	2.8	273.3
Striped maple	0.3	152.9	Balsam fir	2.3	250.7
Striped maple	0.0	139.2	Balsam fir	2.1	241.7
Striped maple	0.4	170.2	Hemlock	0.7	170.0
Balsam fir	36.1	1566.2	Hemlock	0.7	169.1
Balsam fir	34.1	1507.1	Basswood	0.5	164.3
Balsam fir	9.0	562.6	Basswood	0.4	161.7
Balsam fir	7.2	483.3			

[a] DBH = diameter in centimeters at breast height.

TABLE A-IV.2. Initial population of jack pine stand used in Table 6.5

Species	DBH[a]	Height	Species	DBH[a]	Height
Jack pine	4.1	462.7	Tr. aspen	5.1	289.4
Jack pine	4.1	460.0	Tr. aspen	4.8	272.9
Jack pine	4.1	458.4	Tr. aspen	4.8	272.9
Jack pine	4.0	445.5	White pine	3.4	291.8
Jack pine	4.0	442.4	White pine	3.3	288.9
Jack pine	3.9	440.1	White pine	3.3	288.9
Jack pine	3.9	439.1	Red oak	1.6	90.4
Jack pine	3.9	437.7	Red oak	1.4	78.6
Jack pine	3.9	437.0	Red oak	1.3	75.1
Jack pine	3.9	435.4	Red oak	1.2	72.4
Jack pine	3.9	434.0	Red oak	1.2	71.4
Jack pine	3.9	432.4	Red oak	1.1	66.6
Jack pine	3.8	429.4	Red oak	1.6	90.4
Jack pine	3.8	429.0	Red oak	1.4	78.6
Jack pine	3.8	428.5	Red oak	1.3	75.1
Jack pine	3.8	428.1	Red oak	1.2	72.4
Jack pine	3.8	427.5	Red oak	1.2	71.4
Jack pine	3.8	427.0	Red oak	1.1	66.6
Jack pine	3.8	426.6	Red oak	1.1	66.6
Jack pine	3.8	426.0	Red oak	1.1	66.6
Jack pine	3.8	425.6	Red oak	1.2	72.4
Jack pine	3.8	425.1	Red oak	1.2	71.4
Jack pine	3.8	424.7	Red oak	1.1	66.6
Jack pine	3.8	423.9	Red oak	1.1	66.6
Jack pine	3.7	413.0	Red oak	1.1	66.6
Jack pine	3.5	395.1	Red oak	1.2	72.4
White oak	8.6	526.9	Red oak	1.2	71.4
white oak	8.4	512.3	Red oak	1.1	66.6
Tr. aspen	5.3	299.6	Red oak	1.1	66.6
Tr. aspen	5.2	295.7	Red oak	1.1	66.6
Tr. aspen	5.2	293.6			

[a] DBH = diameter in centimeters at breast height.

APPENDIX V

Field Measurements Required to Provide Initial Conditions for JABOWA-II Forest Model

DANIEL B. BOTKIN

LLOYD G. SIMPSON

This appendix provides information about the collection of initial forest conditions from which the model can be run. It does not explain how to collect biomass data separated by parts of the tree, which is necessary to obtain parameters for the biomass calculations and for parameters in the fundamental growth equation of the model. Those methods, known as dimension analysis, have been described elsewhere (Woods, K. D., A. H. Fieveson, and D. B. Botkin, 1991).

Necessary Initial Conditions

1. Select the area for which projections are desired. This can be a specific stand representing a forest condition of interest (for example, a particular old-age stand), or it can be a randomly chosen sample.
2. Within that area, lay out a set of five plots arranged with one at the center and the other four contiguous to it, each of the outer plots facing one of the cardinal compass directions (north, east, south, west). Each plot is a square, 10- by 10-m. (Obviously, the sides of the central plot will run north–south and east–west.)
3. For the entire set, obtain the following background information:
   a. The elevation of the plot above sea level and above the nearest weather station.
   b. Any available history of land use, especially whether the area has been plowed or subject to any pollution of the soil or is at present subject to any known air pollution.
   c. Any readily available history of logging or clearing.
4. At one location within the entire set, do the following:
   a. Dig a soil pit to 1-m depth.
      (1) Describe the soil profile. We recommend using Olson, G. W. 1976,

Criteria for making and interpreting a soil profile description, *State geological Survey of Kansas Bulletin* 212, as a guide.

(2) Determine the depth of the water table if less than 1 m.

(3) Determine the presence and depth of mottling.

(4) If bedrock or a rock is reached before 1-m depth, record that depth.

(5) Determine whether there is a hard pan, and if so record its depth.

(6) Note the percent coverage of rocks on the soil profile face.

(7) Determine the texture of the upper 30 cm of the soil by feel.

5. Within each 10- by 10-m plot, record the following.

   a. Write a verbal description of the plot.

   b. Record species, diameter, and height of each tree whose diameter at breast height is equal to or greater than 2 cm.

   c. Obtain a thin probe, approximately 1-m long (such as a length of welding rod). At 10 randomly chosen points in the plot, push the rod into the ground as far as it will go. If the rod meets a rock, record the depth to that rock. The average of these measurements will represent the soil depth.

   d. At one point within each plot, obtain measurements for bulk density. The method is as follows:

      (1) Clear away the litter and humus layer of the soil over a small patch, approximately 20 cm in diameter.

      (2) With a blade such as a machete, cut an approximately square hole in the soil approximately 10 cm on a side, and to a depth of approximately 30 cm.

      (3) Remove the soil, placing it in a plastic bag; mix the soil well.

      (4) When the soil has been emptied, push a plastic bag into it, so that the bag fits against all of the edges (bottom and side) of the hole.

      (5) Fill the bag with water, and measure the volume of the water with graduated cylinder.

      (6) Weight the soil removed from the hole.

      (7) Checking that the soil removed from the hole has been well mixed, take a sample of approximately 100 g.

         (a) Weight the sample.

         (b) Bring the sample back to the laboratory.

         (c) Oven-dry the sample.

         (d) Weigh the dried sample.

         (e) Submit the sample to analysis for total nitrogen.

Desirable Additional Information

1. For the two largest trees in the plot: core each at breast height and send the increment core to us. We will age the tree from core.

2. Determine the pH of the soil horizons described in the field. This could be done with a portable pH meter or chemical pH kit in the field.

3. Determine the particle size distribution (texture) of the upper 30 cm of soil by laboratory analysis.

APPENDIX VI

Publications That Cite the JABOWA Model or One of Its Descendants

The emphasis in this bibliography is on publications that have appeared since 1980. Some earlier references are provided to give the reader historical continuity. While I have tried to check the literature fully, my apologies to anyone whose paper I have inadvertantly omitted.

1970

Botkin, D. B., J. F. Janak, and J. R. Wallis, 1970, A simulator for northeastern forest growth: A contribution of the Hubbard Brook Ecosystem Study and IBM Research, *IBM Research Report* 3140, Yorktown Heights, New York 21 pp.

1971

Botkin, D. B., J. R. Janak, and J. R. Wallis, 1971, A simulation of forest growth. In Proceedings of the Summer Computer Simulation Conference, Board of Simulation Conferences, Denver, pp. 812–819.

1972

Botkin, D. B., J. R. Janak, and J. R. Wallis, 1972a, Rationale, limitations and assumptions of a northeast forest growth simulator. *IBM Journal of Research and Development* 16:101–116.

Botkin, D. B., J. F. Janak, and J. R. Wallis, 1972b, Some ecological consequences of a computer model of forest growth. *Journal of Ecology* 60:849–872.

1973

Botkin, D. B., J. F. Janak, and J. R. Wallis, 1973, Estimating the effects of carbon fertilization on forest composition by ecosystem simulation. In G. M. Woodwell and E. V. Pecan, eds., *Carbon and the Biosphere*, Brookhaven National Laboratory Symposium No. 24, Technical Information Center, U.S.A.E.C., Oak Ridge, Tenn, pp. 328–344.

1974

Botkin, D. B. and R. S. Miller, 1974, Complex ecosystems: Models and predictions. *Amercan Scientist*. 62:448–453.

1975

Botkin, D. B., 1975, A functional approach to the niche concept in forest communities. In G. S. Innes, ed., *New Directions in the Analysis of Ecological Systems*, Simulation Council Proceedings, Society of Computer Simulation, La Jolla, Calif., pp. 149–158.

Botkin, D. B. and M. J. Sobel, 1975, Stability in time-varying ecosystems. *American Naturalist* 109:625–646.

1976

Botkin, D. B., 1976, The role of species interactions in the response of a forest ecosystem to environmental perturbation. In B. C. Patten, ed., *System Analysis and Simulation in Ecology*, vol. IV, Academic Press, New York, pp. 147–171.

1977

Botkin D. B., 1977. Life and death in a forest community: the computer as an aid to understanding. In C. Hall and J. Day, eds., *Models as Ecological Tools: Theory and Case Histories*, Wiley, New York, pp. 213–234.

Botkin, D. B. and R. E. Levitan, 1977, Wolves, moose and trees: An age specific trophic-level model of Isle Royale National Park. *IBM Research Report in Life-Sciences RC* 6834, 64 pp.

Shugart, H. H. and D. C. West, 1977, Development of an Appalachian deciduous forest succession model and its application to assessment of the impact of the chestnut blight. *Journal of Environmental Management* 5:161–179.

1978

Aber, J. S., D. B. Botkin, and J. M. Melillo, 1978, Predicting the effects of different harvesting regimes on forest floor dynamics in northern hardwoods. *Canadian Journal of Forest Research* 8:306–315.

Mielke, D. L., H. H. Shugart, and D. C. West, 1978, A stand model for uplands forests of southern Arkansas. Report ORNL/TM-6225, Environmental Sciences Div. Pub. No. 1134, Oak Ridge National Laboratory, Oak Ridge, Tenn.

1979

Aber, J. D., D. B. Botkin, and J. M. Melillo, 1979, Predicting the effects of different harvesting regimes on productivity and yield in northern hardwoods. *Canadian Journal of Forest Research* 9:10–14.

Bormann, F. H. and G. E. Likens, 1979, *Pattern and Process in a Forested Ecosystem.* Springer-Verlag, New York.

Botkin, D. B. and J. D. Aber, 1979, Some potential effects of acid rain on forest ecosystems: Implications of a computer simulation. Brookhaven National Laboratory Publication BNL 50889, National Technical Information Service, Springfield, Va.

Finn, J. T., 1979, A model of succession and nutrient cycling in a northern hardwoods forest. Bulletin of the Ecology Society of America 60:123.

Reed, K. L. and S. G. Clark, 1979, Succession Simulator: A coniferous forest simulator. Model documentation, Coniferous Biome Ecosystem Analysis Bulletin Number 11, University of Washington, Seattle.

Shugart, H. H. and D. C. West, 1979, Size and pattern of simulated forest stands. *Forest Science* 25:120–122.

1980

Aber, J. S., G. R. Hendrey, D. B. Botkin A. J. Francis, and J. M. Melillo, 1980, Simulation of acid precipitation effects on soil nitrogen and productivity in forest ecosystems. Brookhaven National Laboratory Publications BNL 28658, Associated Universities, Inc, New York.

Barden, L. S., 1980, Tree replacement in small canopy gaps of a *Tsuga-canadensis* forest in the southern Appalachians, Tennessee, *Oecologia* 44(1):141–142.

Buongiorno, J. and B. R. Michie, 1980, A matrix model of Uneven-aged forest management. *Forest Science* 26(4):609–625.

Huschle, G. and M. Hironaka, 1980, Classification and ordination of serial plant-communities. *Journal of Range Management* 33(3):179–182.

Kessell, S. R. and M. W. Potter, 1980, A quantitative succession model for 9 Montana Forest Communities. *Environmental Management* 4(3):227–240.

McNab, W. H. and M. B. Edwards, 1980, Climatic factors related to the range of saw-palmetto (Serenoa-Repens (Bartr) Small). *American Midland Naturalist* 103(1):204–208.

Peet, R. K. and N. L. Christensen, 1980, Succession—a population process. *Vegatatio* 43(1–2):131–140.

Reed, K. L., 1980, An ecological approach to modeling growth of forest trees. *Forest Science* 26(1):33–50.

Sestak, Z. and J. Catsky, 1980, Bibliography of reviews and methods of photosynthesis 48. *Photosynthetica* 14(2):284–304.

Shugart, H. H., I. P. Burgess, M. S. Hopkins, and A. T. Mortlock, 1980. The Development of a succession model for sub-tropical rain-forest and its application to assess the effects of timber harvest at Wiangaree State Forest, New South Wales. *Journal of Environmental Management* 11(3):243–265.

Shugart, H. H., D. L. DeAngelis, W. R. Emanuel, and D. C. West, 1980, Environmental gradients in a simulation model of a beech-yellow-poplar stand. *Mathematical Biosciences* 50(3–4):163–170.

Shugart, H. H. and D. C. West, 1980, Forest succession models. *Bioscience* 30(5):308–313.

Solomon, A. M., T. J. Blasing, H. R. Delcourt and D. C. West, 1980, Testing a simulation model for reconstruction of prehistoric forest stand dynamics. *Quaternary Research* 14(3):275–293.

West, D. C., S. B. Mclaughlin, and H. H. Shugart, 1980, Simulated forest response to chronic air pollution stress. *Journal of Environmental Quality* 9(1):43–49.

1981

Acevedo, M. F., 1981, On Horn-Markovian model of forest dynamics with particular reference to tropical forests. *Theoretical Population Biology* 19(2):230–250.

Austin, M. P. and L. Belbin, 1981, An analysis of succession along an environmental gradient using data from a lawn. *Vegetatio* 46(7):19–30.

Barden, L. S., 1981, Forest development in canopy gaps of a diverse hardwood forest of the Southern Appalachian Mountains, *Oikos* 37(2):205–209.

Botkin, D. B., 1981, Causality and succession. Chapter 5 in West, Shugart, and Botkin, eds., *Forest Succession: Concepts and Applications.* Springer-Verlag, New York, pp. 36–55.

Covington, W. W., 1981, Changes in forest floor organic matter and nutrient content following clear cutting in northern hardwoods. *Ecology* 62(1):41–48.

Evans, L. S., A. J. Francis, G. R. Hendrey, D. W. Johnson, and G. J. Stensland, 1981, Acidic precipitation—considerations for an air-quality standard, water air and soil pollution. *Vegetatio* 16(4):469–509.

Harwell, M. A., W. P. Cropper and H. L. Ragsdale, 1981, Analyses of transient characteristics of a nutrient cycling model, *Ecological Modelling* 12(1–2):105–131.

Hasse, W. D. and A. R. Ek, 1981, A simulated comparison of yields for even-aged versus uneven-aged management of northern hardwood stands. *Journal of Environmental Management* 12(3):235—246.

Kessell, S. R., 1981, The challenge of modeling post-disturbance plant succession. *Environmental Management* 5(1):5–13.

Roise, J. P., and D. R. Betters, and B. M. Kent, 1981, An approach to functionalizing key environmental factors—forage production in Rocky Mountain aspen stands. *Ecological Modelling* 14(1–2):133–146.

Runkle, J. R., 1981, Gap regeneration in some old-growth forests of the eastern United States. *Ecology* 62(4):1041–1051.

Shugart, H. H. and I. R. Noble, 1981, A computer model of succession and fire response of the high-altitude Eucalyptus forest of the Brindabella Range Australian Capital Territory. *Australian Journal of Ecology* 6(2):149–164.

Shugart, H. H. and D. C. West, 1981, Long-term dynamics of forest ecosystems. *American Scientist* 69(6):647–652.

Usher, M. B., 1981, Modeling ecological succession, with particular reference to Markovian models. *Vegetatio* 46(7):11–18.

West, D. C., H. H. Shugart, and D. B. Botkin, eds., 1981, *Forest Succession: Concepts and Applications.* Springer-Verlag, New York, 517 pp.

1982

Aber, J. D., D. B. Botkin, A. J. Francis, G. R. Hendrey and J. M. Melillo, 1982, Potential effects of acid precipitation on soil Nitrogen and productivity of forest ecosystems. *Water, Air and Soil Pollution* 18(1–3):405–412.

Aber, J.D., C.A. Federer and J. M. Melillo, 1982, Predicting the effects of rotation length, harvest intensity, and fertilization on fiber yield from northern hardwood forests in New England. *Forest Science* 28(1):31–45.

Aber, J. D., G. R. Hendrey, A. J. Francis, D. B. Botkin, and J. M. Melillo, 1982, Potential effects of acid precipitation on soil nitrogen and productivity of forest ecosystems. In F. M. D'itri, ed., *Acid Precipitation: Effects on Ecological Systems.* Ann Arbor Science, Ann Arbor, Mich., pp. 411–433.

Evans, L. S., 1982, Biological effects of acidity in precipitation on vegetation—a review. *Environmental and Experimental Botany* 22(2):155–169.

Green, D. G., 1982, Fire and stability in the post-glacial forests of southwest Nova-Scotia. *Journal of Biogeography* 9(1):29–40.

Lacey, C. J., I. R. Noble, and J. Walker, 1982, Fire in Australian tropical savannas. *Ecological Studies* 42:246–272.

Samuels, M. L. and J. L. Betancourt, 1982, Modeling the long-term effects of fuelwood harvests on pinyon-juniper woodlands. *Environmental Managenent* 6(6):505–515.

1983

Allen, T. F. H. and H. H. Shugart, 1983, Ordination of simulated complex forest succession—a few tests of ordination methods. *Vegetatio* 51(3):141–155.

Brand, G. J. and M. R. Holdaway, 1983, Users need performance information to evaluate models. *Journal of Forestry* 81(4):235.

Buchman, R. G. and S. R. Shifley, 1983, Guide to evaluating forest growth projection systems. *Journal of Forestry* 81(4):232.

Green, D. G., 1983, The ecological interpretation of fine resolution pollen records. *New Phytologist* 94(3):459–477.

Harcombe, P. A. and P. L. Marks, 1983, 5 Years of tree death in a *Fagus-Magnolia* forest, southeast Texas. *Oecologia* 57(1–2):49–54.

Hibbs, D. E., 1983, 40 Years of forest succession in central New England. *Ecology* 64(6):1394–1401.

Lorimer, C. G., 1983, A test of the accuracy of shade-tolerance classifications based on physiognomic and reproductive traits. *Canadian Journal of Botany* 61(6):1595–1598.

McMurtrie, R. and L. Wolf, 1983, Above-ground and below-ground growth of forest stands—a carbon, budget model. *Annals of Botany* 52(4):437–448.

McNaughton, S. J., 1983, Serengeti grassland ecology—the role of composite environmental factors and contingency in community organization. *Ecological Monograhs* 53(3):291–320.

Rauscher, H. M., T. L. Sharik, and D. W. Smith, 1983, A forest management gaming model of the nitrogen cycle in Appalachian upland oak forests. *Ecological Modelling* 20(2–3):175–199.

Reed, K. L., C. S. Bledsoe, J. S. Shumway, and R. B. Walker, 1983, Evaluation of the interaction of 2 environmental factors affecting Douglas fir seedling growth—light and nitrogen. *Forest Science* 29(1):193–203.

1984

El-Bayoumi, M. A., H. H. Shugart, and R. W. Wein, 1984, Modeling succession of eastern Canadian mixedwood forst. *Ecological modelling* 21(3):175–198.

Evans, L. S., 1984, Acidic precipitation effects on terrestrial vegetation. *Annual Review of Phytopathology* 22:397–420.

Evans, L. S., 1984, Botanical aspects of acidic precipitation. *Botanical Review* 50(4):449–490.

Kercher, J. R. and M. C. Axelrod, 1984, Analysis of silva—a model for forecasting the effects of SO_2 pollution and fire on western coniferous forests. *Ecological Modelling* 23(1–2):165–184.

Kercher, J. R. and M. C. Axelord, 1984, A process model of fire ecology and succession in a mixed-conifer forest. *Ecology* 65(6):1725–1742.

Koppel, A. and T. Oja, 1984, Regime of diffuse solar radiation in an individual Norway spruce. *Photosynthetica* 18(4):529–535.

Malanson, G. P., 1984, Linked Leslie matrices for the simulation of succession. *Ecological Modelling* 21(1–2):13–20.

Reiners, W. A., D. Y. Hollinger, and G. E. Lang, 1984, Temperature and evapotranspiration gradients of the White Mountains, New Hampshire, USA. *Arctic and Alpine Research* 16(1):31–36.

1985

Costanza, R. and F. H. Sklar, 1985, Articulation, accuracy and effectiveness of mathematical models—a review of fresh-water wetland applications. *Ecological Modelling* 27(1–2):45–68.

Dale, V. H., T. W. Doyle, and H. H. Shugart, 1985, A comparison of tree growth models. *Ecological Modelling* 29(1–4):145–169.

Davis, M. B. and D. B. Botkin, 1985, Sensitivity of cool-temperate forests and their fossil pollen record to rapid temperature change. *Quaternary Research* 23(3):327–340.

DeAngelis, D. L., R. V. O'Neill, W. M. Post and J. C. Waterhouse, 1985, Ecological modeling and disturbance evaluation. *Ecological Modelling* 29(14):399–419.

Delcourt, H. R. and P. A. Delcourt, 1985, Comparison of taxon calibrations, modern analog techniques, and forest-stand simulation models for the quantitative reconstruction of past vegetation. *Earth Surfaces Processes and Landforms* 10(3):293–304.

Lemee, G., 1985, Functions of shade intolerant trees in the dynamic of a natural beech-wood (Fontainebleau Forest). *Acta Oecologica-Oecologia Plantarum* 6(1):3–20.

Lippe, E., J. T. de Smidt, D. C. and Glenn-Lewin, 1985, Markov models and succession—a test from a heathland in the Netherlands. *Journal of Ecology* 73(3): 775–791.

Pearlstine, L., W. Kitchens, and H. McKellar, 1985, Modeling the impacts of a river diversion on bottomland forest communities in the Santee River floodplain, South Carolina. *Ecological Modelling* 29(1–4):283–302.

Sharpe, P. J. H., H. I. Wu, J. Walker, and L. K. Penridge, 1985, A physiologically based continuous-time Markov approach to plant growth modeling in semi-arid woodlands. *Ecological Modelling* 29(1–4):189–213.

Shugart, H. H. and W. R. Emanuel, 1985, Carbon-dioxide increase—the implications at the ecosystem level. *Plant Cell and Environment* 8(6):381–386.

Solomon, A. M. and T. Webb, 1985, Computer-aided reconstruction of late quarternary landscape dynamics. *Annual Review of Ecology and Systematics* 16:63–84.

Steinhorst, R. K., P. Morgan, and L. F. Neuenschwander, 1985, A stochastic-deterministic simulation model of shrub succession. *Ecological Modelling* 29(1–4):35–55.

Tilman, D., 1985, The resource-ratio hypothesis of plant succession. *American Naturalist* 125(6):827–852.

1986

Bartlein, P. J., I. C. Prentice, and T. Webb, 1986, Climatic response surfaces from pollen data for some eastern North-American taxa. *Journal of Biogeography* 13(1):35–57.

Dale, V. H., J. Franklin, and M. Hemstrom, 1986, Modeling the long-term effects of disturbances on forest succession, Olympic Peninsula, Washington. *Canadian Journal of Forest Research* 16(1):56–67.

Delcourt, H. R. and P. A. Delcourt, 1986, Comparison of taxon calibrations, modern analog techniques, and forest-stand simulation models for the quantitative reconstruction of past vegetation. *Earth Surface Processes and Landforms* 11(6):687–691.

Gore, J. A. and W. A. Patterson, 1986, Mass of downed wood in northern hardwood forests in New Hampshire—potential effects of forest management. *Canadian Journal of Forest Research* 16(2):335–339.

Hamilton, D. A., 1986, A logistic model of mortality in thinned and unthinned mixed conifer stands of northern Idaho. *Forest Science* 32(4):989–1000.

Mäkela, A., 1986, Implications of the pipe model theory on dry-matter partitioning and height growth in trees. *Journal of Theoretical Biology* 123(1):103–120.

Mäkela, A. and P. Hari, 1986, Stand growth model based on carbon uptake and allocation in individual trees. *Ecological Modelling* 33(2–4):205–229.

Pastor, J. and W. M. Post, 1986, Influence of climate soil moisture, and succession on forest carbon and nitrogen cycles. *Biogeochemistry* 2(1):3–27.

Prentice, I. C., 1986, Vegetation responses to past climatic variation. *Vegetatio* 67(2):131–141.

Solomon, A. M., 1986, Transient response of forests to CO_2-induced climate change—simulation modeling experiments in eastern North America. *Oecologia* 68(4):567–579.

van Tongeren, O. and I. C. Prentice, 1986, A spatial simulation model for vegetation dynamics. *Vegetatio* 65(3):163–173.

Waldrop, T. A., E. R. Buckner, C. E. McGee, and H. H. Shugart, 1986, FORCAT—a single tree model of stand development following clearcutting on the Cumberland Plateau. *Forest Science* 32(2):297–317.

1987

Busing, R. T. and E. E. C. Clebsch, 1987, Application of a spruce fir forest canopy gap model. *Forect Ecology and Management* 20(1–2):151–169.

Dale, V. H. and T. W. Doyle, 1987, The role of stand history in assessing forest impacts. *Environmental Management* 11(3):351–357.

Dale, V. H. and R. H. Gardner, 1987, Assessing regional impacts of growth declines using a forest succession model. *Journal of Environmental Management* 24(1):83–93.

DeAngelis, D. L. and M. A. Huston, 1987, Effects of growth rates in models of size distribution formation in plants and animals. *Ecological Modelling* 36(1–2):119–137.

DeAngelis, D. L. and J. C. Waterhouse, 1987, Equilibrium and nonequilibrium concepts in ecological models. *Ecological Monographs* 57(1):1–21.

Huston, M. A. and D. L. DeAngelis, 1987, Size bimodality in monospecific populations—a critical review of potential mechanisms. *American Naturalist* 129(5):678–707.

Huston, M. and T. M. Smith, 1987, Plant succession—life history and competition. *American Naturalist* 130(2):168–198.

Krupa, S. and R. N. Kickert, 1987, An analysis of numerical models of air pollutant exposure and vegetation response. *Environmental Pollution* 44(2):127–158.

Leemans, R. and I. C. Prentice, 1987, Description and simulation of tree-layer composition and size distributions in a primeval picea-pinus forest. *Vegetatio* 69(1–3):147–156.

Pastor, J., W. M. Post, R. H. Gardner and V. H. Dale, 1987, Successional changes in nitrogen availability as a potential factor contributing to spruce declines in boreal North America. *Canadian Journal of Forest Research* 17(11):1394–1400.

Prentice, I. C., J. T. de Smidt, and O. van Tongeren, 1987, Simulation of heathland vegetation dynamics. *Journal of Ecology* 75(1):203–219.

Shugart, H. H., 1987, Dynamic ecosystem consequences of tree birth and death patterns. *BioScience* 37(8):596–602.

Walker, L. R. and F. S. Chapin, 1987, Interactions among processes controlling successional change. *Oikos* 50(1):131–135.

Wu, H., E. J. Rykiel, and P. J. H. Sharpe, 1987, Age Size population dynamics—derivation of a general matrix methodology. *Computers & Mathematics with Applications* 13(9–11):759–766.

1988

Armstrong, R. A., 1988, The effects of disturbance patch size on species coexistence. *Journal of Theoretical Biology* 133(2):169–184.

Dale, V. H., H. I. Jager, R. H. Gardner, and A. E. Rosen, 1988, Using sensitivity and uncertainty analyses to improve predictions of broad-scale forest development. *Ecological Modelling* 42(3–4):165–178.

DeAngelis, D. L., 1988, Strategies and difficulties of applying models to aquatic populations and food webs. *Ecological Modelling* 43(1–2):57–73.

Develice, R. L., 1988, Test of a forest dynamics simulator in New Zealand. *New Zealand Journal of Botany* 26(3):387–392.

Fralish, J. S., 1988, Predicting potential stand composition from site characteristics in the Shawnee Hills forest of Illinois. *American Midland Naturalist* 120(1):79–101.

Huston, M., D. L. DeAngelis and W. M. Post, 1988, New computer models unify ecological theory—computer simulations show that many ecological patterns can be explained by interactions among individual organisms. *BioScience* 38(10):682–691

Iwasa, Y., 1988, Free fitness that always increases in evolution. *Journal of Theoretical Biology* 135(3):265–281.

Kimmins, J. P., 1988, Community organization—methods of study and prediction of the productivity and yield of forest ecosystems. *Canadian Journal of Botany* 66(12):2654–2672.

McAuliffe, J. R., 1988, Markovian dynamics of simple and complex desert plant communities. *American Naturalist* 131(4):459–490.

Miyanishi, K. and M. Kellman, 1988, Ecological and simulation studies of the responses of Miconia Albicans and Clidemia Sericea populations to prescribed burning. *Forest Ecology and Management* 23(2–3):121–137.

Orloci, L. and M. Orloci, 1988, On recovery, Markov chains, and canonical analysis. *Ecology* 69(4):1260–1265.

Parks, P. J. and R. J. Alig, 1988, Land base models for forest resource supply analysis—a critical review. *Canadian Journal of Forest Research* 18(8):965–973.

Sievanen, R. T. E. Burk, and A. R. Ek, 1988, Construction of a stand growth model utilizing photosynthesis and respiration relationships in individual trees. *Canadian Journal of Forest Research* 18(8):1027–1035.

Smith, T. M. and D. L. Urban, 1988, Scale and resolution of forest structural pattern. *Vegetatio* 74(2–3):143–150.

1989

Armstrong, R. A., 1989, Fugitive coexistence in sessile species—models with continuous recruitment and determinate growth. *Ecology* 70(3):674–680.

Bonan, G. B. and M. D. Korzuhin, 1989, Simulation of moss and tree dynamics in the boreal forests of interior Alaska. *Vegetatio* 84(1):31–44.

Conner, W. H. and M. Brody, 1989, Rising water levels and the future of southeastern Louisiana swamp forests. *Estuaries* 12(4):318–323.

Dale, V. H. and J. F. Franklin, 1989, Potential effects of climate change on stand development in the Pacific Northwest. *Canadian Journal of Forest Research* 19(12):1581–1590.

Green, D. G., 1989, Simulated effects of fire, dispersal and spatial pattern on competition within forest mosaics. *Vegetatio* 82(2):139–153.

Harrison, E. A., B. M. McIntyre, and R. D. Dueser, 1989, Community dynamics and topographic controls on forest pattern in Shenandoah National Park, Virginia. *Bulletin of the Torrey Botanical Club* 116(1):1–14.

Hatton, T. J., 1989, Spatial analysis of a subalpine heath woodland. *Australian Journal of Ecology* 14(1):65–75.

Kienast, F. and N. Kuhn, 1989, Simulating forest succession along ecological gradients in southern central Europe. *Vegetatio* 79(1–2):7–20.

Moore, A. D., 1989, On the maximum growth equation used in forest gap simulation Models. *Ecological Modelling* 45(1):63–67.

Roberts, D. W., 1989, Analysis of forest succession with fuzzy graph theory. *Ecological Modelling* 45(4):261–274.

Smith, T. M. and M. Huston, 1989, A theory of the spatial and temporal dynamics of plant communities. *Vegetatio* 83(1–2):49–69.

Turner, M. G., 1989, Landscape ecology—the effect of pattern on process. *Annual Review of Ecology and Systematics* 20:171–197.

Urban, D. L. and T. M. Smith, 1989, Microhabitat pattern and the structure of forest bird communities. *American Naturalist* 133(6):811–829.

Vanclay, J. K., 1989, A growth model for North Queensland Rainforests. *Forest Ecology and Management* 27(3–4):245–271.

Webb, S. L., 1989, Contrasting windstorm consequences in 2 forests, Itasca State Park, Minnesota. *Ecology* 70(4):1167–1180.

Wein, R. W., M. A. El-Bayoumi, and J. Dasilva, 1989, Simulated predictions of forest dynamics in Fundy National Park, Canada. *Forest Ecology and Management* 28(1):47–60.

1990

Armstrong, R. A., 1990, Flexible model of incomplete dominance in neighborhood competition for space. *Journal of Theoretical Biology* 144(3):287–302.

Bonan, G. B., 1990, Carbon and Nitrogen cycling in North American boreal forests. 1. Litter quality and soil thermal effects in interior Alaska. *Biogeochemistry* 10(1):1–28.

Bonan, G. B., 1990, Carbon and Nitrogen cycling in North American Boreal Forests. 2. Biogeographic patterns. *Canadian Journal of Forest Research* 20(7):1077–1088.

Bonan, G. B., H. H. Shugart, and D. L. Urban, 1990, The sensitivity of some high-latitude boreal forests to climatic parameters. *Climate Change* 16(1):9–29.

Botkin, D. B., 1990, *Discordant Harmonies: A New Ecology for the 21st Century.* Oxford University Press, New York.

Chertov, O. G. 1990, Specom—a single tree model of pine stand raw humus soil ecosystem. *Ecological Modelling* 50(1–3):107–132.

Coffin, D. P. and W. K. Lauenroth, 1990, A gap dynamics simulation model of succession in a semiarid grassland. *Ecological Modelling* 49(3–4):229–266.

Costanza, R., F. H. Sklar, and M. L. White, 1990, Modeling coastal landscape dynamics. *BioScience* 40(2):91–107.

French, N. R., 1990, The utility of models in the study of mountain development and tranformation. *Mountain Research and Development* 10(2):141–149.

Hanson, J. S., G. P. Malanson, and M. P. Armstrong, 1990, Landscape fragmentation and dispersal in a model of riparian forest dynamics. *Ecological Modelling* 49(3–4):277–296.

Harrison, E. A. and H. H. Shugart, 1990, Evaluating performance of an Appalachian oak forest dynamics model, *Vegetatio* 86(1):1–13.

Keane, R. E., S. F. Arno, and J. K. Brown, 1990, Simulating cumulative fire effects in ponderosa pine Douglas fir forests. *Ecology* 71(1):189–203.

Keane, R. E., S. F. Arno, J. K. Brown and D. F. Tomback, 1990, Modeling stand dynamics in whitebark pine (*Pinus albicaulis*) forests. *Ecological Modelling* 51(1–2):73–95.

Maguire, D. A. and D. W. Hann, 1990, Constructing models for direct prediction of 5-year crown recession in southwestern Oregon Douglas fir. *Canadian Journal of Forest Research* 20(7):1044–1052.

Moore, A. D. and I. R. Noble, 1990, An individualistic model of vegetation stand dynamics. *Journal of Environmental Management* 31(1):61–81.

Prentice, I. C. and R. Leemans, 1990, Pattern and process and the dynamics of forest structure—a simulation approach. *Journal of Ecology* 78(2):340–355.

Glossary

Actual evapotranspiration: The amount actually evaporated, given the water available and the energy available. Note that it is a theoretical estimate and not a measured value.

Additive Relationship among Variables: The effect of a set of variables at time t is determined by the variable with the minimum value at that time.

Basal area: The cross-section area of a tree stem, or of any population of trees. It is the area of a stump measured at breast height (1.37 m above the ground). Basal area (BA) is:

$$BA = \int_1^n (\pi * r^2) \quad \text{(where } n \text{ is the number of trees)}$$

Biomass: The quantity of organic matter, usually measured in dry weight, and usually as a density (e.g. kilograms per square meter).

Biosphere: The planetary system that includes and sustains life. The biosphere is made up of the lower atmosphere, all biota, the oceans, soils, and the solid sediments in active exchange with the biota. (The biosphere has also been used to mean total biomass on the earth, and to mean the global habitat—the totality of the places where life exists on Earth. In this book, biosphere is used only with the first meaning.)

Chronic patchiness: When ecological succession does not take place; instead, a species dominates an area for a while, disappears locally, and is replaced by another, which in turn is later replaced by the first.

Conceptual basis: An understanding of phenomena at a lower level of organization is used to explain and project phenomena at a higher level of organization.

Default parameter: The value of a parameter taken on by the model unless it is changed by the user. The default values are provided in the software available as a companion to this book.

Environmental response function: An equation (or computer algorithm) expressing in quantitative fashion the response of an individual, a population, or a species to a specific environmental factor. In the forest model, these functions are expressed as relative to a maximum or standard level.

Evapotranspiration: The sum of the water evaporated from the soil surface to the atmosphere and that transferred by transpiration from leaves.

Facilitation (as a mechanism of species replacement during succession): One species prepares the way for the second; without the presence of the first, the second could never appear.

Field capacity (of a soil): The maximum amount of water that can be held against gravity by the soil particles.

Fundamental niche: The range of conditions under which a species can persist without competition from other species.

Growing degree-days: The integral under the curve of temperature above some baseline plotted against time. In this book, the baseline is 4.4°C.

Guild (ecological): A group of species that exploit the same class of environmental resources in a similar way.

Harvest rotation: See rotation period.

Interference (as a mechanism of species replacement during succession): Individuals of one species prevent the individuals of the second from entering into the forest, but when trees of the first species have completed their life cycle and died, individuals of the second are able to enter.

Liebig's law of the minimum: The assumption that growth of a plant is controlled simply and totally by the single factor available in the least supply, so that there are no interactions among environmental resources. Generally applied only to chemical elements necessary for plant nutrition.

Model I: The first version of the forest model, with growth a function of light and temperature; regeneration determined by light, temperature, and a minimum soil moisture condition.

Model II: The current version of the forest mode, with growth and regeneration functions of light, temperature, soil nitrogen, soil dryness, and soil-water saturation.

Multiplicative relationships among variables: The effect of a set of variables at time t is calculated as the product of the values at that time.

Neighborhood (in a forest): An area small enough so that a single large tree can shade all other trees in that neighborhood.

Niche: The range of environmental conditions under which a species can persist (this definition is called the "Hutchinsonian niche").

Parameter: Used in this book to mean a constant in an equation, and therefore distinguished from the term *variable*.

Pioneer species: Species of plants that are found in early successional stands. Typically, these species have seeds that are light and are spread rapidly by wind or by animals, and individual trees are fast growing, especially when exposed to direct sunlight.

Potential evapotranspiration: The amount of water that could be transferred from the soil to the atmosphere given the amount of thermal energy present.

Prediction: This word has the connotation of an actual foretelling of future events, and its use is restricted in this book.

Projection: This word is used in this book to mean a calculation by a model of a future state of a system, given the present state (see prediction).

Realized niche: The range of conditions under which a species can persist, given the existence of competition from other species.

Recruitment: The growth of a cohort of seedlings to reach some minimum size class called saplings, at which point the saplings become listed as part of the population. In JABOWA, a sapling reaches recruitment size when it is 137 cm high, the height at which diameter at breast height is measured. Recruitment, rather than germination or birth, is often used as the initial age class in population models, including models of animal populations.

Rotation period: The time between two incidents of logging. Also called the harvest rotation.

Sensitivity test: A test that determines how great a change occurs in the value of an output variable with a change in the value of either an input variable or a parameter intrinsic to the model.

Shade-intolerant: A tree that grows poorly in the shade but very well in bright light. See **Shade-tolerant.**

Shade-tolerant: A tree that is able to grow in the shade. Typically, the light response functions of such trees show a relative high growth rate at low light intensities, but a maximum growth rate that is lower than trees of shade-intolerant species.

Sustainability: A sustainable timber harvest is one that does not endanger future harvests of the same quantity. Sustainability of an ecosystem refers to sustaining the integrity of a natural forest in terms of its structure, composition, and ecological processes, along with the environmental services it provides. Sustainability of timber yield refers to sustaining a yield of timber from a forest but not necessarily the original (preexploitation) forest.

Synergistic relation (among environmental factors): Where one environmental factor influences another, and the combination of two factors acts on tree growth as a single function to produce a new, single response curve.

Tolerance (as a mechanism of species replacement during succession): Early successional species neither increase nor decrease the rate of recruitment, growth, or survival of the later species. The process is in part simply a difference among life histories: individuals of one species appear first because they "get there quicker" and grow faster than individuals of the second. Both the "early" and "late" successional species can grow in the early environmental stages; but the late successional species are more tolerant of the limited resources that occur when competitors are abundant, regardless of their species.

Tolerant: A tree is tolerant of an environmental factor if it can grow, survive, and/or reproduce when the availability of that factor is low.

Validation: In mathematics and logic, used to mean an argument so constructed that if the premises are jointly asserted, the conclusion cannot be denied without contradiction. Some ecologists involved in ecological models use validation as a test of a model against independent data, data that have not been used in the formulation or parameterization of the model. Since these are contradictory uses, I have avoided the term as much as possible.

Verification: In mathematics and logic, evidence that establishes or confirms the accuracy or truth of something. Some ecologists use this term to mean ensuring that calculations are done properly, which is closer to the use of "validation" in logic and mathematics. Since these are contradictory uses, I have avoided this term as much as possible.

References

Aber, J. D., 1976, A computer model of canopy dynamics and competition for light and nutrients in northern hardwoods. Ph.D. thesis, Yale University, New Haven, Conn.

Aber, J. D., D. B. Botkin, and J. M. Melillo, 1978, Predicting the effects of different harvesting regimes on forest floor dynamics in northern hardwoods. *Canadian Journal of Forest Research* 8:306–315.

Aber, J. D., Botkin, D. B., and Melillo, J. M. 1979, Predicting the effects of different harvesting regimes on forest floor dynamics in northern hardwoods. *Canadian Journal of Forest Research* 9:10–14.

Aber, J. D., G. R. Hendrey, A. J. Francies, D. B. Botkin, and J. M. Melillo, 1982, Potential effects of acid precipitation on soil nitrogen and productivity of forest ecosystems. In F. M. D'itri, ed., *Acid Precipitation: Effects on Ecological Systems*, Ann Arbor Science, Ann Arbor, Mich., pp. 411–433.

Aber, J. D., J. Melillo and C. A. Federer, 1982, Predicting the effects of rotation length, harvest intensity, and fertilization on fiber yield from northern hardwood forests in New England. *Forest Science* 28:31–45.

Aber, John D., J. M., Melillo, and C. A. McClaugherty, 1990. Predicting long-term patterns of mass loss: Nitrogen dynamics and soil organic matter formation from initial fine litter chemistry in temperate forest ecosystems. *Canadian Journal of Botany* 68:2201.

Acevedo, M. F., 1981, On Horn Markovian model of forest dynamics with particular reference to tropical forests. *Theoretical Population Biology* 19(2):230–250.

Aldred, A. H. and I. S. Alemdag, 1988, Guidelines for forest biomass inventory. Petawawa National Forestry Institute, Canadian Forest Service Information Report, PI-X-77.

Allen, T. F. H. and H. H. Shugart, 1983, Ordination of simulated complex forest succession: A new test of ordination methods. *Vegetatio* 51:141–155.

Allen, T. F. H. and E. P. Wyleto, 1983, A hierarchical model for complexity of plant communities. *Journal of Theoretical Biology* 101:529–540.

Antonovsky, M. Y., F. Z. Glebov, and M. D. Krozuhin, 1987, A regional model of long-term wetland-forest dynamics. *Working Paper WP-87-63*, International Institute for Applied Systems Analysis, Laxenburg, Austria, 51 pp.

Armstrong, R. A., 1988, The effects of disturbance patch size on species coexistence. *Journal of Theoretical Biology* 133(2):169–184.

Armstrong, R. A., 1990, Flexible model of incomplete dominance in neighborhood competition for space. *Journal of Theoretical Biology* 144(3):287–302.

Bacastow R. and C. D. Keeling, 1973, Atmospheric carbon dioxide and radiocarbon in the natural carbon cycle: II. Changes from A. D. 1700 to 2070 as deduced from a geochemical model. In G. M. Woodwell and E. V. Pecan, eds., *Carbon and the Biosphere*, Brookhaven National Laboratory Symposium No. 24, Technical Information Center, U.S.A.E.C., Oak Ridge, Tenn., pp. 86–135.

Baker, W. L., 1989, A review of models of landscape change. *Landscape Ecology* 2:111–133.

Baldocchi, D. D., Boyd A. Hutchison, and R. T. McMillen, 1984, Solar radiation with an oak-hickory forest: an evaluation of the extinction coefficients for several radiation components during fully-leafed and leafless periods. *Agricultural and Forest Meterology* 32:307–322.

Barden, L. S., 1981, Forest development in canopy gaps of a diverse hardwood forest of the southern Appalachian Mountains. *Oikos* 37(2):205–209.

Baskerville, G. L., 1965, Dry matter production in immature balsam fir stands. *Forest Science Monographs* 9, 42 pp.

Bingham, G. E., H. H. Rogers, and W. W. Heck, 1981, Responses of plants to elevated carbon dioxide in the field: Photosynthesis, photorespiration, and stomatal conductance. *Plant Physiology* 67:84.

Binkley, D., K. Cromack, and R. L. Fredriksen, 1982, Nitrogen accretion and availability in some snowbrush ecosystems. *Forest Science* 28:720–724.

Bolin, B. and E. Eriksson, 1959, Changes in the carbon dioxide content of the atmosphere and sea due to fossil fuel combustion. In B. Bolin, ed., *The Atmosphere and the Sea in Motion*. Rockerfeller Inst. Press, New York, pp. 130–142.

Bonan, G. B., 1988a, Environmental processes and vegetation patterns in boreal forests. Ph.D. Thesis, University of Virgina, Carlottesville, Va., 286 pp.

Bonan, G. B., 1988b, A simulation model of environmental processes and vegetation patterns in boreal forests: Test case Fairbanks, Alaska. *Working paper WP-88-63*, International Institute for Applied Systems Analysis, Laxenburg, Austria, 63 pp.

Bonan, G. B. and H. H. Shugart, 1989, Environmental factors and ecological processes in boreal forests. *Annual Review of Ecology and Systematics* 20:1–28.

Bormann, F. H., 1953, Factors determining the role of loblolly pine and sweetgum in early old-field succession in the Piedmont of North Carolina. *Ecological Monographs* 23:339–358.

Bormann, F. H. and M. F. Buell, 1964, Old-age stand of hemlock-northern hardwood forest in central Vermont. Bulletin of the Torrey Botanical Club 91:451–465.

Bormann, F. H. and J. C. Gordon, 1984, Stand density effects in young red alder plantations: Productivity, photosynthate, partitioning, and nitrogen fixation. *Ecology* 65:394–402.

Bormann, F. H. and G. E. Likens, 1979, Pattern and Process in a Forested Ecosystem. Springer-Verlag, New York.

Bormann, F. H., G. E. Likens, and J. M. Melillo, 1977, Nitrogen budget for an aggrading northern hardwood forest ecosystem. *Science* 196:981–983.

Bormann, F. H., G. E. Likens, T. G. Siccama, R. S. Pierce, and J. S. Eaton, 1974, The export of nutrients and recovery of stable conditions following deforestation at Hubbard Brook. *Ecology Monograph* 44:255–277.

Bormann, F. H., T. G. Siccama, G. E. Likens, and R. H. Whittaker, 1971, The Hubbard Brook ecosystem study: Composition and dynamics of the tree stratum. *Ecological Monographs* 40:373–88.

Botkin, D. B., 1969, Prediction of net photosynthesis of trees from light intensity and temperature, *Ecology* 50:854–858.

Botkin, D. B., 1977, Forests, lakes and the anthropogenic production of carbon dioxide. *BioScience* 27:325–331.

Botkin, D. B., ed. 1980, *Life from a Planetary Perspective: Fundamental Issues in Global Ecology.* Final report NASA Grant NASW-3392, 49 pp.

Botkin, D. B. 1982, Can there be a theory of global ecology? *Journal of Theoretical Biology* 96:95–98.

Botkin, D. B., 1984, The biosphere: The new aerospace engineering challenge. *Aerospace America*, July 1984, pp. 73–75.

Botkin, D. B., 1989, Science and the global environment. Chapter 1 in Botkin, D. B., M. Caswell, J. E. Estes, and A. Orio, eds., *Man's Role in Changing The Global Environment: Perspectives on Human Involvement.* Academic Press, Boston, pp. 3–14.

Botkin, D. B., 1990, *Discordant Harmonies: A New Ecology for the 21st Century.* Oxford University Press, New York.

Botkin, D. B., M. Caswell, J. E. Estes, and A. Orio, eds. 1989a, *Changing The Global Environment.* Perspectives on Human Involvement, Academic Press, Boston.

Botkin, D. B., J. F. Janak, and J. R. Wallis, 1970, A simulator for northeastern forest growth: A contribution of the Hubbard Brook Ecosystem Study and IBM Research. *IBM Research Report 3140*, Yorktown Heights, New York, 21 pp.

Botkin, D. B., J. F. Janak, and J. R. Wallis, 1972a, Rationale, limitations and assumptions of a northeast forest growth simulator. *IBM Journal of Research and Development* 16:101–116.

Botkin, D. B., J. F. Janak, and J. R. Wallis, 1972b, Some ecological consequences of a computer model of forest growth, *Journal of Ecology* 60:849–872.

Botkin, D. B., J. F. Janak, and J. R. Wallis, 1973, Estimating the effects of carbon fertilization on forest composition by ecosystem simulation. In G. M. Woodwell and E. V. Pecan, eds., *Carbon and the Biosphere*, U. S. Department of Commerce, Washington, D.C., pp. 328–344. (Available from National Technical Information Service, Springfield, Va., as CONF-720510)

Botkin, D. B. and R. E. Levitan, 1977, Wolves, moose and trees: An age specific trophic-level model of Isle Royale National Park. *IBM Research Report in Life-Sciences RC 6834*, 64 pp.

Botkin, D. B., B. Maguire, B. Moore III, H. J. Morowitz, and L. B. Slobodkin, 1979, A foundation for ecological theory. In R. de Bernardi, ed., *Biological and Mathematical Aspects in Population Dynamics.* Proceedings of the Palanza Symposium, Memorie dell'Istituto Italiano di Idrobiologia Supplement 37.

Botkin, D. B. and R. A. Nisbet, 1990, *Response of Forests to Global Warming and CO$_2$ Fertilization. Report to U.S. Environmental Protection Agency.*

Botkin, D. B. and R. A. Nisbet, in press, Projecting effects of climate change on biodiversity in forests. In R. L. Peters, ed., *Consequences of the Greenhouse Effect for Biological Diversity.* Yale University Press, New Haven, Conn.

Botkin, D. B., R. A. Nisbet, and T. E. Reynales, 1989b, Effects of climate change on forests of the Great Lake states. In J. B. Smith and D. A. Tirpak, eds., *The Potential Effects of Global Climate Change on the United States.* U.S. Environmental Protection Agency, Washington, D.C., EPA-203-05-89-0.

Botkin, D. B. and L. Simpson, 1990a, The first statistically valid estimate of biomass for a large region. *Biogeochemistry* 9:161–174.

Botkin, D. B. and L. G. Simpson, 1990b, Distribution of biomass in the North American boreal forest. In G. Lund, ed., Proceedings of the International Conference *Global Natural Resource Monitoring and Assessments: Preparing for the 21st Century,* American Society for Photogrammetry and Remote Sensing, Venice, Italy, 1989.

Botkin, D. B. and M. J. Sobel, 1975, Stability in time varying ecosystems. *American Naturalist* 109:625–646.

Botkin, D. B., and L. M. Talbot, 1990, Biological Diversity and Forests. Report to the World Bank.

Botkin, D. B., D. A. Woodby, and R. A. Nisbet, 1991, Kirtland's warbler habitats: A possible early indicator of climatic warming. *Biological Conservation* 56(1):63–78.

Botkin, D. B., Woodwell, and N. Tempel, 1970, Forest productivity estimated from carbon dioxide uptake. *Ecology* 51:1057–1060

Bowler, J. M., G. S. Hope, J. N. Jennings, G. Singh, and D. Walker, 1976, Late quaternary climates of Australia and New Guinea. *Quaternary Research* 6:359–394.

Box, E. O., 1981, *Macroclimate and Plant Forms: An Introduction to Predictive Modeling in Phytogeography.* D. W. Junk, Publisher, The Hague.

Bretherton, F. P., D. J. Baker, D. B. Botkin, K. C. A. Burke, M. Chahine, J. A. Dutton, L. A. Fisk, N. W. Hinners, D. A. Landgrebe, J. J. McCarthy, B. Moore, R. G. Prinn, C. B. Raleight, W. V. H. Reis, W. F. Wees, and P. J. Zinke, 1986, *Earth Systems Science: A Program for Global Change.* NASA Earth Systems Science Committe of the NASA Advisory Council, Washington, D.C., 48 pp. + supplements.

Broecker, W. S., T. Takahashi, H. J. Simpson, and T. H. Peng, 1979, Fate of fossil fuel carbon dioxide and the global carbon budget. *Science* 206:409–418.

Bryan, W. and R. Wright, 1976, The effect of enhanced CO_2 levels and variable light intensities on net photosynthesis in competing mountain trees. American Midl and Naturalist 95:446.

Bundy, D., and C. R. Tracy, 1977, Behavioral response of American toads (*Bufo americanus*) to stressful thermal and hydric environments. *Herpetologica* 33:455–458.

Buongiorno, J. and B. R. Michie, 1980, A matrix model of uneven-aged forest management. *Forest Science* 26(4):609–625.

Byelich, J., M. E. DeCapita, G. W. Irvine, R. E. Radtke, N. I. Johnson, W. R. Jones, H. Mayfield, and W. J. Mahalak, 1985, *Kirtland's Warbler Recovery Plan.* U.S. Fish & Wildlife Service, Rockville, M. 78 pp.

Canham, A. E., and W. J. McCavish, 1981, Some effects of CO_2, daylength and nutrition of the growth of young forest tree plants I. The seedling stage, *Forestry* 54:169–182.

Carlson, R. W. and F. A. Bazzaz, 1980, The effects of elevated CO_2 concentrations on growth, photosynthesis, transpiration, and water-use efficiency of plants. In J. Singh and A. Deepak, eds., *Proceedings of a Symposium on Environmental and Climate Impact of Coal Utilization.* Academic Press, New York, pp. 609–612.

Caswell, H., 1976, The validation problem. In B. C. Patten, ed., Vol. IV, *Systems Analysis and Stimulation in Ecology.* Academic Press, New York, pp. 313–325.

Chittenden, A. K., 1905, Forest conditions of northern New Hampshire. *Bulletin Bureau of Forestry, U. S. Department of Agriculture* 55, 100 pp.

Clark, F. E. and T. H. Rosswall, 1981, *Nitrogen Cycling in Terrestrial Ecosystems: Processes, Ecosystem Strategies, and Management Implications.* Ecological Bulletin 33, Swedish Natural Science Research Council, Stockholm, Sweden.

Coffin, D. P., and W. K. Laurenroth, 1990, A gap dynamics simulation-model of succession in a semiarid grassland. *Ecological Modelling* 49 (3–4):229–266.

COHMAP, 1988, Climatic changes of the last 18,000 years: Observations and model simulations. *Science* 241:1043–1052.

Connell, J. H. and R. O. Slatyer, 1977, Mechanisms of succession in natural communities and their role in community stability and organization. *American Naturalist* 111:1119–1144.

Conner, W. H. and M. Brody, 1989, Rising water levels and the future of southeastern Louisiana swamp forests. *Estuaries* 12(4):318–323.

Cooper, W. S., 1913, The climax forest of Isle Royale, Lake Superior, and its development. *Botanical Gazette.* 55:1–44, 115–140, 189–234.

Costanza, R. and F. H. Sklar, 1985, Articulation, accuracy and effectiveness of mathematical models—a review of fresh-water wetland applications. *Ecological Modelling* 27(1–2):45–68.

Costanza, R., F. H. Sklar and M. L. White 1990, Modeling coastal landscape dynamics. *Bioscience* 40(2):91–107.

Cottam, G. and J. T. Curtis, 1949, A method for making rapid surveys of woodlands by means of pairs of randomly selected trees. *Ecology* 30:101–104.

Covington, W. W., 1977, Secondary succession in northern hardwoods: Forest floor organic matter and nutrients and leaf fall. Ph.D. thesis, Yale University, New Haven, Conn.

Covington, W. W., 1981, Changes in forest floor organic matter and nutrient content following clearcutting in northern hardwoods. *Ecology* 62:41–48.

Crow, T. R., 1978, Biomass and production in three contiguous forests in northern Wisconsin. *Ecology* 59:265–273.

Dale, V. H., T. W. Doyle, and H. H. Shugart, 1985, A comparison of tree growth models. *Ecological Modelling* 29:145–169.

Dale, V. H. and J. F. Franklin, 1989, Potential effects of climate change on stand development in the Pacific Northwest. *Canadian Journal of Forest Research* 19(12):1581–1590.

Dale, V. H. and R. H. Gardner, 1987, Assessing regional impacts of growth declines using a forest succession model. *Journal of Environmental Management* 24:83–93.

Dale, V. H. and M. Hemstrom, 1984, CLIMACS: A computer model of forest stand development for western Oregon and Washington. *Research paper PNW-327, Pacific Northwest Forest and Range Experiment Station,* U.S. Department of Agriculture, Forest Service.

Dale, V. H., M. Hemstrom, and J. Franklin, 1986, Modeling the long-term effects of disturbances on forest succession, Olympic Peninsula, Washington. *Canadian Journal of Forest Research* 16:56–67.

Dale, V. H., H. I. Jager, R. H. Gardner, and A. E. Rosen, 1988, Using sensitivity and uncertainty to improve predictions of broad-scale forest development. *Ecological Modelling* 42:165–178.

Davis, D. M., 1985, *The Practice of Silviculture.* 8th ed. John Wiley, New York, 527 pp.

Davis, M. B., 1976, Pleistocene biogeography of temperate deciduous forests. *GeoScience and Man* 13:13–26.

Davis, M. B., 1981, Quaternary history and the stability of forest communities. In D. C. West, H. H. Shugart, and D. B. Botkin, eds., *Forest Succession: Concepts and Applications.* Springer-Verlag, New York, pp. 312–153.

Davis, M. B., 1983, Holocene vegetational history of the eastern United States. In H. E. Wright Jr., ed., *Late Quarternary Environments of the United States.* Vol. 2. *The Holocene.* University of Minnesota Press, Minneapolis, pp. 166–181.

Davis, M. B. and D. B. Botkin, 1985, Sensitivity of the cool-temperate forests and their fossil pollen to rapid climatic change. *Quarternary Research* 23:327–340.

Davis, M. B. and C. Zabinski, in press, Rates of dispersal for North American trees: Implications for response to climatic warming. In R. L. Peters, ed., *Consequences of the Greenhouse Effect for Biological Diversity.* Yale University Press, New Haven, Conn.

DeAngelis, D. L., 1988, Strategies and difficulties of applying models to aquatic populations and food webs. *Ecological Modelling* 43(1–2):57–73.

DeAngelis, D. L., R. V. O'Neill, W. M. Post, and J. C. Waterhouse, 1985, Ecological modelling and disturbance evaluation. *Ecological Modelling* 29(1–4):399–419.

DeAngelis, D. L. and J. C. Waterhouse, 1987, Equilibrium and nonequilibrium concepts in ecological models. *Ecological Monographs* 57(1):1–21.

Delcourt, P. A. and H. R. Delcourt, 1977, The Tunica Hills, Louisiana-Missisippi: Late glacial locality for spruce and deciduous forest species. *Quaternary Research* 7:218–237.

Develice, R. L., 1988, Test of a forest dynamics simulator in New Zealand. *New Zealand Journal of Botany* 26:387–392.

Dickson, B. A. and R. L. Crocker, 1953, A chronosequence of soils and vegetation near Mt. Shasta, California. II. The development of the forest floors and the carbon and nitrogen profiles of the soils. *Journal of Soil Science* 4:142–154.

Dominski, A. S., 1971, Nitrogen transformations in a northern hardwood podzol on forested and cutover sites. Ph. D. thesis, Yale University, New Haven, Conn.

Downton, W. J. S., O. Björkman, and C. S. Pike, 1980, Consequences of increasing atmospheric concentrations of CO_2 for growth and photosynthesis of higher plants. In Pearman, G. I., ed., *CO_2 and climate*, Australian Research Australian Academy of Science, Canberra, Australia, pp. 143–151.

Doyle, T. W., 1981, The role of disturbance in the gap dynamics of a montane rain forest: An application of a tropical forest succession model. pp. 56–73. In D. C. West, H. H. Shugart, and D. B. Botkin, eds., *Forest Succession: Concepts and Applications.* Springer-Verlag, New York, pp. 56–73.

Doyle, T. W., 1983, Competition and growth relationships in a mixed-age, mixed species forest community. Ph.D. thesis, University of Tennessee, Knoxville.

Drury, W. H. and I. C. T.. Nisbet, 1973, Succession. *Journal of Arboretum* 54:331–368.

du Cloux, H. C. and R. J. Vivoli, 1984, Effect of increased atmospheric carbon dioxide on growth, photosynthesis, photorespiration, and water use efficiency. In C. Sybesma, ed., *Advances in Photosynthetic Research, Proc. VI International Congress on Photosynthesis. Vol V.* Martinus Nijhoff, Dr. W. Junk, The Hague, Netherlands, pp. 213–216.

Dunne, T. and L. B. Leopold, 1978, *Water in Environmental Planning.* W. H. Freeman, San Francisco.

Ekdahl, C. A. and C. D. Keeling, 1973, Atmospheric carbon dioxide and radiocarbon in the natural carbon cycle. 1. Quantitative deductions from records at Mauna Loa Observatory and at the South Pole. In G. M. Woodwell and E. V. Pecan, eds., *Carbon and the Biosphere*, Brookhaven National Laboratory Symposium No. 24. Technical Information Service, Oak Ridge, Tenn.

Elton, C., 1924, Periodic fluctuations in the numbers of animals: Their causes and effects. *Journal of Experimental Biology* 2:119–163.

Elton, C. S., 1927, *Animal Ecology* Sigwick and Jackson, London, 207 pp.

Esau, K., 1977, *Anatomy of Seed Plants.* John Wiley, New York.

Evans, L. S., 1982, Biological effects of acidity in precipitation on vegetation—a review. *Environmental and Experimental Botany* 22(2):155–169.

Evert, F., 1985, Systems of equations for estimating ovendry mass of 18 Canadian tree species. Petawawa Forestry Institute, Canadian Forest Service Information Report, Rpt. PI-X-59.

Federer, C. A., 1984, Organic matter and nitrogen content of the forest floor in even-aged northern hardwoods. *Canadian Journal of Forest Research* 14:763–767.

Finn, J. T., 1979, A model of succession and nutrient cycling in a northern hardwood forest. *Bulletin of the Ecology Society of America* 60:123.

Flexner, S. B., ed., 1987, *Random House Dictionary of the English Language*. Random House, New York.

Flint, R. F., 1971, *Glacial and Quaternary Geology*. John Wiley, New York, 892 pp.

Forbes, S. A., 1925, The lake as a microcosm. *Peoria Scientific Association Bulletin*, pp. 77–87; reprinted in *Illinois Natural History Survey Bulletin* 15:537–550.

Forman, R. T. T. and M. Godron, 1986, *Landscape Ecology*. John Wiley, New York.

Fowells, H. A., 1965, *Silvics of Forest Trees of the United States*. Agriculture Handbook 271, U.S. Government Printing Office, Washington, D.C.

Fownes, J. H. and R. A. Harrington, 1990, Modelling growth and optimal rotations of tropical multipurpose trees using unit leaf rate and leaf area index. *Journal of Applied Ecology* 27:886–896.

French, N. R., 1990, The utility of models in the study of mountain development and transformation. *Mountain Research and Development* 10(2):141–149

Fritts, H. C., 1976, *Tree Rings and Climate* Academic Press, New York.

Fryer, J. H. and F. T. Ledig, 1972, Microevolution of the photosynthetic temperature optimum in relation to the elevational complex gradient. *Canadian Journal of Botany* 50:1231–1235.

Fuller, L. G., D. D. Reed, and M. J. Homes, 1987, Modeling northern hardwood diameter growth using weekly climatic factors in northern Michigan. In *Forest Growth Modeling and Prediction*, Vol. I. U.S. Department of Agriculture Forest Service, General Technical Report NC-120, pp. 467–474.

Funsch, R. W., R. H. Mattson, and G. R. Mowry, 1970, Carbon dioxide-supplemented atmosphere increases growth of *Pinus strobus* seedlings. *Forest Science* 16:459–460.

Gause, G. F., 1934, *The Struggle for Existence*, Williams and Wilkins, Baltimore, Md, 163 pp.

Gates, D. M., 1962, *Energy Exchange in the Biosphere* Harper's, New York.

Gates, D. M., 1980, *Biophysical Ecology*, Springer-Verlag, New York, 611 pp.

Ghan, S. J., J. W. Lingaas, M. E. Schlesinger, R. L. Mobley, and W. L. Gates, 1982, *A Documentation of the OSU Two-Level Atmospheric General Circulation Model*. Rep. No. 35, Climate Research Institute, Oregon State University, Corvallis, 395 pp.

Goel, N. S., S. C. Maitra, and E. W. Montroll, 1971, On the Volterra and other non-linear models of interacting populations. *Review of Modern Physics* 43:231.

Gomez-Pompa, A. and C. Vazquez-Yanes, 1981, Successional studies of a rain forest in Mexico. In D. C. West, H. H. Shugart, and D. B. Botkin, eds., *Forest Succession: Concepts and Applications*. Springer-Verlag, New York, pp. 247–266.

Gorham, E., P. M Vitousek, and W. A. Reiners, 1979, The regulation of chemical budgets over the course of terrestrial ecosystem succession. *Annual Review of Ecology and Systematics* 10:53–84.

Green, D. G., 1982, Fire and stability in the post-glacial forests of southwest Nova-Scotia. *Journal of Biogeography* 9(1):29–40.

Green, D. G., 1989, Simulated effects of fire, dispersal and spatial pattern on competition within forest mosaics. *Vegetatio* 82(2):139–153.

Greenhill, A. G., 1881, Determination of the greatest height consistent with stability that a vertical pole or mast can be made, and of the greatest height to which a tree of given proportions can grow. *Proceedings of the Cambridge Philosophical Society IV*, Part II:65–73.

Hall, F. G., D. B. Botkin, D. E. Strebel, K. D. Woods, S. J. Goetz, 1991, Large-scale patterns in forest succession as determined by remote sensing. *Ecology*, 72:628–640.

Halpern, C. B., 1988, Early succession pathways and the resistance and resilience of forest communities, *Ecology* 69:1703–1715.

Hardin, G., 1960, The competitive exclusion principle, *Science* 131:1292–1297.

Hansen, J., I. Fung, A. Lacis, D. Rind, S. Lebedeff, R. Ruedy, and G. Russell, 1988, Global climate changes as forecast by Goddard Institute for Space Studies three-dimensional model. *Journal of Geophysical Research* 93:9341–9364.

Hansen, J., G. Russell, D. Rind, P. Stone, S. Lebedeff, R. Ruedy, and L. Travis, 1983, Efficient three-dimensional global models for climate studies. Models I and II. *Monthly Weather Review* 111:609–662.

Hanson, J. S., G. P. Malanson, and M. P. Armstrong, 1990, Landscape fragmentation and dispersal in a model of riparian forest dynamics. *Ecological Modelling* 49(3–4):277–296.

Harcombe, P. A. and P. L. Marks, 1983, Five years of tree death in a *Fagus-Magnolia* forest, Southeast Texas. *Oecologia* 57(1–2):49–54.

Hardin, G., 1960, The competitive exclusion principle. *Science* 131:1292–1297.

Harper, J. L., 1977, *Population Biology of Plants.* Academic Press, New York, 892 pp.

Harries, H., 1966, *Soils and Vegetation in the Alpine and Subalpine Belt of the Presidential Range.* Ph.D. Thesis, Rutgers University, New Brunswick, New Jersey, 542 pp.

Harrison, E. A. and H. H. Shugart, 1990, Evaluating performance of an Appalachian oak forest dynamics model. *Vegetation* 86(1):1–13.

Hatton, T. J., 1989, Spatial-analysis of a subalpine heath woodland. *Australian Journal of Ecology* 14(1):65–75.

Heinselman, M. L., 1970, Landscape evolution, peatland types, and the environment in the Lake Agassiz peatlands natural area, Minnesota. *Ecological Monographs* 40:235–261.

Heinselman, M. L., 1973, Fire in the virgin forests of the Boundary Waters Canoe Area, Minnesota. *Journal of Quaternary Research* 3:329–382.

Heinselman, M. L., 1973, Fire and succession in the conifer forests of Northern North America. In D. C. West, H. H. Shugart, and D. B. Botkin, eds., *Forest Succession, Concepts and Applications*, Springer-Verlag, New York, pp. 374–405.

Hemstrom, M. and V. Adams, 1982, Modeling long-term forest succession in the Pacific Northwest, pp. 14–23. In J. E. Means, ed., *Forest Succession and Stand Development in the Northwest.* Oregon State University, Corvallis, pp. 14–23.

Henderson, L. J., 1913, *The Fitness of the Environment.* Macmillan, Boston.

Hendrickson, O. Q., 1990, Asymbiotic nitrogen fixation and soil metabolism in three Ontario forests. *Soil Biology and Biochemistry* 22:967.

Hibbs, D. E., 1983, Forty years of forest succession in Central New England. *Ecology* 64(6):1394–1401.

Holdridge, L. R., 1947, Determination of world plant formations from simple climate data. *Science* 105:367–368.

Holling, C. S., 1965, The functional response of predators to prey density and its role in mimicry and population regulation. *Memoirs of the Entomological Society of Canada* 45:1–85.

Holling, C. S., 1974, Resilience and stability of ecological systems. *Annual Review of Ecology and Systematics* 4:1–24.

Hollinger, D. Y., 1987, Gas exchange and dry matter allocation responses to elevation of atmospheric CO_2 concentration in seedings of three tree species. *Tree Physiology.* 3:193–202.

Horn, H. S., 1971, *The Adaptive Geometry of Trees.* Princeton University Press, Princeton, New York.

Horn, H. S., 1974, The ecology of secondary succession. *Annual Reviews of Ecology and Systematics* 5:25–37.

Horn, H. S. 1981, Some causes of variety in patterns of secondary succession. In D. C. West, H. H. Shugart, and D. B. Botkin, eds., *Forest Succession: Concepts and Applications.* Springer-Verlag, New York, pp. 24–35.

Huntington, T. G., D. F. Ryan, and S. P. Hamburg, 1988, Estimating soil nitrogen and carbon pools in a northern hardwood forest ecosystem. *Soil Science Society of America Journal* 52:1162–1167.

Huston, M., D. DeAngelis, and W. Post, 1988, New computer-models unify ecological theory. Computer simulations show that many ecological patterns can be explained by interactions among individual organisms. *Bioscience* 38 (10):682–691.

Hutchinson, G. E., 1944, Limnological studies in Connecticut. VII. A critical examination of the supposed relationship between phytoplankton periodicity and chemical changes in lake water. *Ecology* 25:3–26.

Hutchinson, G. E., 1950, The Biogeochemistry of vertebrate excretion. *Bulletin of the American Museum of Natural History* 96:1–554.

Hutchinson, G. E., 1954, The biochemistry of the terrestrial atmosphere. In G. P. Kuiper, ed., *The Earth as a Planet*, University of Chicago Press, Chicago, pp. 371–433.

Hutchinson, G. E., 1958, Concluding remarks. *Cold Spring Harbor Symposium on Quantitative Biology* 22:415–427.

Hutchinson, G. E., 1978, *An Introduction to Population Ecology*, Yale University Press, New Haven, Conn.

Jacoby, G. C. and J. Hornbeck, eds., 1987, *Proceedings of the International Sysposium of Ecological Aspects of Tree-Ring Analysis.* CONF-8608144. National Technical Information Service, U.S. Department of Commerce, Springfield, Va.

Jarvis, P. G., 1981, Plant water relations in models of tree growth. *Studies of Forest Succession* 160:51–60.

Jurik, T. W., J. A. Weber, and D. M. Gates, 1984, Short-term effects of CO_2 on gas exchange of leaves of bigtooth aspen (*Populus grandidentata*) in the field. *Plant Physiology* 75:1022.

Kalm, P., 1770, *Travels in America: The America of 1750.* A. B. Benson, ed., English version 1770 (Dover, New York, 1966), pp. 300, 309.

Kasanaga, H. and M. Monsi, 1954, On the light-transmission of leaves, and its meaning for the production of matter in plant communities. *Japanese Journal of Botany* 14:304–324.

Keane, R. E., S. F. Arno, and J. K. Brown, 1990a, *FIRESUM–An Ecological Process Model for Fires Succession in Western Conifer Forests.* U.S. Department of Agriculture Forest Service Intermountain Research Station General Technical Report INT-266.

Keane, R. E., S. F. Arno, and J. K. Brown, 1990b, Simulating cumulative fire effects in Ponderosa pine Douglas-fir forests. *Ecology* 71(1):189–203.

Keeling, D. D., D. J. Moss, and T. P. Whorf, 1989, *Final Report.* Carbon Dioxide Information and Analysis Center, Martin-Marietta Energy Systems, Oak Ridge, Tenn.

Ker, J. W. and J. H. G. Smith, 1955, Advantages of the parabolic expression of height–diameter relationships. *Forest Chronicle* 31:235–246.

Kercher, J. R. and M. C. Axelrod, 1981, SILVA: A model for forecasting the effects of sulfur dioxide pollution on growth and succession in a western coniferous forest. Lawrence Livermore Report UCRL-53109, 80 pp.

Kercher, J. R. and M. C. Axelrod, 1984, Analysis of a model for forecasting the effects of sulfer dioxide pollution on growth and succession on western coniferous forests. *Ecological Modelling* 23:165–184.

Kershaw, A. P., 1976, A late Pleistocene and Holocene pollen diagram from Lynch's Crater, Northeastern Queensland, Australia. *New Phytologist* 77:469–498.

Kessell, S. R., 1981, The challenge of modeling post-disturbance plant succession. *Environmental Management* 5(1):5–13.

Kienast, F. and N. Kuhn, 1989, Simulating forest succession along ecological gradients in southern Central Europe. *Vegetatio* 79(1–2):7–20.

Kimball, B. A, 1983, Carbon dioxide and agricultural yield: An assemblage and analysis of 430 prior observations. *Agronomy Journal* 75:779–788.

Kittedge, J., 1948, *Forest Influences*. McGraw-Hill, New York.

Koch, K. E., L. H. Allen Jr., P. Jones, and W. T. Avigne, 1987, Growth of citrus rootstock (Carizzo orange) seedlings during and after long-term CO_2 enrichment. *Journal of the American Society of Horticultural Science* 112:77–82.

Kozlowski, T. T., 1949, Light and water in relation to growth and competition of Piedmont forest tree species. *Ecological Monographs* 19:207–231.

Kozlowski, T. T., ed., 1968, *Water Deficits and Plant Growth*, Vol. 1. Academic Press, New York.

Kozlowski, T. T., ed., 1970, *Water Deficits and Plant Growth*, Vol. 2. Academic Press, New York.

Kozlowski, T. T., 1971, *Growth and Development of Trees*. 2 vols. Academic Press, New York.

Kozlowski, T. T., ed., 1972, *Water Deficits and Plant Growth*, Vol. 3, Academic Press, New York.

Kozlowski, 1982a, Water supply and tree growth, Part I, Water deficits. *Forestry Abstracts*, Commonwealth Forestry Bureau 43:57–95.

Kozlowski, 1982b, Water supply and tree growth, Part II, Flooding. *Forestry Abstracts*, Commonwealth Forestry Bureau 43:145–162.

Kozlowski, T. T., P. J. Kramer, and S. G. Pallardy, 1991, *The Physiological Ecology of Woody Plants*. Academic Press, San Diego.

Kramer, P. J. and J. P. Decker, 1944, Relation between light intensity and rate of photosynthesis of loblolly pine and certain hardwoods. *Plant Physiology* 19:350–358.

Kramer, P. J. and T. T. Kozlowski, 1960, *Physiology of Trees*. McGraw-Hill, New York.

Kramer, P. J. and T. T. Kozlowski, 1979, *Physiology of Woody Plants*. Academic Press, New York, 811 pp.

Kramer, P. J. and N. Sionit, 1987, Effects of increasing carbon dioxide concentration on the physiology and growth of forest trees. In W. H. Shands and J. S. Hoffman, eds., *The Greenhouse Effect, Climate Change, and U.S. Forests*. The Conservation Foundation, Washington, D.C., pp. 219–246.

Krupa, S. and R. N. Kickert, 1987, An analysis of numerical models of air pollutant exposure and vegetation response. *Environmental Pollution* 44(2):127–158.

Lack, D., 1967, *The Natural Regulation of Animal Numbers*. Clarendon Press, Oxford, 343 pp.

Larcher, W., 1969, The effect of environmental and physiological variables on the carbon dioxide gas exchange of trees. *Photosynthetica* 3:167–198.

Ledig, F. T. and D. R. Korbobo, 1983, Adaptation of sugar maple populations along altitudinal gradients: Photosynthesis, respiration, and specific leaf weight. *American Journal of Botany* 70:256–265.

Leemans, R., 1989, *Description and Simulation of Stand Structure and Dynamics in Some Swedish Forests*. Ph.D. Thesis, Uppsala University, Uppsala, Sweden.

Leemans, R. and I. C. Prentice, 1985, Fiby Urskog: Description and simulation of natural forest structure. In R. Leemans, I. C. Prentice, and E. Van der Maarel, eds., *Theory and Models in Vegetation Science: Abstracts. Studies in Plant Ecology* 16:61–62.

Leemans, R. and I. C. Prentice, 1987, Description and simulation of tree-layer composition and size distributions in a primaeval *Picea-Pinus* forest. *Vegetatio* 69:147–156.

Leemans, R. and I. C. Prentice, 1989, FORSKA, a general forest succession model. *Meddelanden fran Vaxtbiologiska institutionen*, Uppsala, Sweden.

Lehman, J. T., D. B. Botkin, and G. E. Likens, 1975a, The assumptions and rationale of a computer model of phytoplankton population dynamics. *Limnology and Oceanography* 20:343–364.

Lehman, J. T., D. B. Botkin, and G. E. Likens, 1975b, Lake eutrophication and the limiting CO_2 concept: A simulation study. V. Sladecer, ed., In *Proceedings of the International Congress of Theoretical and Applied Limnology*, Obermiller, Stuttgart, pp. 300–307.

Leslie, P. H., 1945, The use of matrices in certain population mathematics. *Biometrika* 35:213–245.

Leslie, P. H., 1958, A stochastic model for studying the properties of certain biological systems by numerical methods. *Biometrika* 45:16–31.

Leverenz, J. W. and D. J. Lev, 1987, Effects of carbon dioxide induced climate changes on the natural ranges of six major commercial tree species in the western United States. pp. 123–156, In W. H. Shands and J. S. Hoffman, eds., *The Greenhouse Effect, Climate Change, and U.S. Forests.* Conservation Foundation, Washington, D.C., 1987.

Levins, R., 1966, The strategy of model building in population biology. *American Scientist* 54:421–431.

Liebig, J., 1840, *Chemistry in Its Application to Agriculture and Physiology.* Taylor and Walton, London.

Lieffers, V. J. and J. S. Campbell, 1984, Biomass and net annual primary production regressions for *Populus tremuloides* in northeastern New Brunswick. *Canadian Journal of Forest Research* 6:441–447.

Lieth, H. and R. H. Whittaker, eds., 1975, *Primary Productivity of the Biosphere.* Springer-Verlag, New York.

Likens, G. E., F. H. Bormann, R. S. Pierce, J. S. Eaton, and N. M. Johnson, 1977, *Biogeochemistry of a Forested Ecosystem*, Springer-Verlag, New York.

Lippe, E., J. T. de Smidt, and D. C. Glenn-Lewin, 1985, Markov models and succession—a test from a heathland in the Netherlands. *Journal of Ecology* 73(3):775–791.

Livingstone, D.A., 1975, Late quarternary climatic change in Africa. *Annual Review of Ecology and Systematics* 6:249–280.

Loomis, R. S., W. A. Williams, and W. G. Duncan, 1967, Community architecture and the productivity of terrestrial plant communities. In A. San Pietro, F. A. Greer, and T. J. Army, eds., *Harvesting the Sun.* Academic Press, New York, pp. 291–308.

Lorimer, C.G., 1983, A test of the accuracy of shade-tolerance classifications based on physiognomic and reproductive traits. *Canadian Journal of Botany* 61 (6):1595–1598.

Lotka, A., 1925, *Elements of Physical Biology.* Williams and Wilkins, Baltimore, Md.

Loucks, O. L., 1970, Evolution of diversity, efficiency, and community stability. *American Zoologist* 10:17–25.

Lovelock, J. E., 1979, *Gaia, a New Look at Life on Earth*. Oxford University Press, New York.

Lovelock, J. E. 1988, *The Ages of Gaia: A Biography of Our Living Earth*. W. W. Norton, New York.

Lovelock, J. E. and L. Margulis, 1974, Atmospheric homeostasis by and for the biosphere: The gaia hypothesis. *Tellus* XXVI:2–10.

Maguire, B. Jr., L. B. Slobodkin, H. J. Morowitz, B. Moore III, and D. B. Botkin, 1980, A new paradigm for the examination of closed ecosystems. In J. P. Giesy Jr., ed., *Microcosms in Ecological Research*, Savannah River Ecology Lab, Augusta, Ga. U.S. Technical Information Center, U.S. Department of Energy, Symposium Series 52 (CONF-781101).

Maguire, D. A. and D. W. Hann, 1990, Constructing models for direct prediction of 5-year crown recession in southwestern Oregon Douglas-fir. *Canadian Journal of Forest Research* 20(7):1044–1052.

Malanson, G. P., 1984, Linked Leslie-matrices for the simulation of succession. *Ecological Modelling* 21(1–2):13–20.

Manabe, S. and R. S. Stouffer, 1980, Sensitivity of a global climate model to an increase of CO_2 concentration in the atmosphere. *Journal of Geophysical Research* 85:5529–5554.

Manabe, S. and R. T. Wetherall, 1987, Large-scale changes in soil wetness induced by an increase in carbon dioxide. *Journal of Atmospheric Science* 44:1211–1235.

Mankin, J. B., R. V. O'Neil, H. H. Shugart, and B. W. Rust, 1975, The importance of validation in ecosystem analysis. In G. S. Innis, ed., *New Directions in the Analysis of Ecological Systems*, pp. 63–72. vol. 5, The Society for Computer Simulation (Simulation Councils, Inc.), La Jolla, Calif.

Marks, P. L., 1974, The role of pin cherry (*Prunus pennsylvanica*) in the maintenance of stability in northern hardwood ecosystems. *Ecological Monographe* 44:73–88.

Marquis, D. A., 1969, Thinning in young northern hardwoods: 5-year results. *U.S. Department of Agriculture Forest Service Research Paper NE-139*, Upper Darby, Pa., 22 pp.

Marsh, G. P., 1864, *Man and Nature*. (Originally published in 1864, reprinted and edited by D. Lowenthal, 1967.) Belknap Press, Cambridge, Mass., 472 pp.

May, R. M., 1973a, *Stability and Complexity in Model Ecosystems*. Princeton University Press, Princeton, N.J., 235 pp.

May, R. M., 1973b, Time-delay versus stability in population models with two and three trophic levels. *Ecology* 54:315–325.

May, R. M., ed., 1976, *Theoretical Ecology*. Blackwell Scientific Pub., Oxford, 317 pp.

May, R. M., 1981, *Theoretical Ecology, Principles and Applications*. Sinauer Assoc., Sunderland, Ma.

Mayfield, H., 1960, *The Kirtland's Warbler*. Cranbrook Inst. Science, Bloomfield Hills, Michi.

McAuliffe, J. R., 1988, Markovian dynamics of simple and complex desert plant communities. *American Naturalist* 131(4):459–490.

McFee, W. W. and E. L. Stone, 1966, The persistence of decaying wood in the humus layers of northern forests. *Soil Science Society of America Proceedings* 30:513–516.

McIntosh, R. P., 1974, Plant ecology 1947–1972. *Annuals of the Missouri Botanic Garden* 61:132–165.

McIntosh, R. P., 1985, *The Background of Ecology: Concept and Theory*. Cambridge University Press, New York.

Mielke, D. L., H. H. Shugart, and D. C. West, 1978, A stand model for uplands forests of southern Arkansas. Report ORNL/TM-6225, Environmental Sciences Div. Pub. No. 1134, Oak Ridge National Laboratory, Oak Ridge, Tenn.

Miller, P. C., 1969, Solar radiation profiles in openings in canopies of aspen and oak. *Science* 164:308–309.

Miller, P. R., 1980, *Effects of Air Pollutants on Mediterranean and Temperate Forest Ecosystems*. USDA Forest Service General Technical Report PSW-43, US Forest Service, Berkeley, Calif.

Miller, R. S. and D. B. Botkin, 1974, Mortality rates and survival of birds. *American Naturalist* 108:181–192.

Milliken, Governor W. G., 1980, Proceedings of the Governor William G. Milliken's Forestry Conference, Michigan Technological University, Houghton, Mich., 90 pp.

Mitchell, H. L. and R. F. Chandler, 1939, The nitrogen nutrition and growth of certain deciduous trees of Northeastern United States. *Black Rock Forest Bulletin* 11.

Mitchell, J. F. B. *Quarterly Journal of the Meteorological Society* 109:113, 1983.

Monteith, J. L., 1965, Evaporation and environment. *Proceedings of the Symposium of the Society on Experimental Biology* 19:205–236.

Moore, A.W., 1989, On the maximum growth equation used in forest gap simulation models. *Ecological Modelling* 45:63–67.

Moorhead, D. C., 1985, Development and application of an upland boreal forest succession model. Ph.D. thesis, University of Tennesse, Knoxville.

Morris, J. T., 1982, A model of growth responses by *Spartina alterniflora* to nitrogen limitation. *Journal of Ecology*, 70:25–42.

Morris, J. T., R. A. Houghton, and D. B. Botkin, 1984, Theoretical limits of below-ground production of *Spartina alterniflora*. *Journal of Ecological Modeling* 26:155–175.

Nohrstedt, H.-O. 1989, Changes in carbon content, respiration rate, ATP content, and microbial biomass in nitrogen-fertilized pine forest soils in Sweden. *Canadian Journal of Forest Research* 19:323.

Norton, S. and H. E. Young, 1976, Forest biomass utilization and nutrient budgets. H. E. Young, ed., IUFRO Biomass Studies S4.01, Oslo, Norway, pp. 55–73.

Norton, S. A., D. W. Hanson, and R. J. Campana, 1980, The impact of acidic precipitation and heavy metals on soils in relation to forest ecosystems. In P. R. Miller, ed., *Proceedings of The International Symposium on Effects of Air Pollution on Mediterranean and Temperate Forest Ecosystems*. U.S. Forest Service, Berkeley, Calif., General Technical Report PS10-43 Government Document A 13.88:PSW-43, pp. 152–157.

Oberbauer, S. F., N. Sionit, S. J. Hastings, and W. C. Oechel, 1986, Effects of CO_2 enrichment and nutrition on growth, photosynthesis, and nutrient concentrations of Alaskan tundra plant species. *Canadian Journal of Botany* 64:2993–2998.

Oberbauer, S. F., B. R. Strain, and N. Fetcher, 1985, Effect of CO_2 enrichment on seedling physiology and growth of two tropical tree species, *Physiol. Plant* 65:352–356.

Odum, E. P., 1969, The strategy of ecosystem development. *Science* 164:262–270.

Odum, H. T., 1957, Tropic structure and productivity of Silver Springs, Florida. *Ecological Monographs* 27:55–112.

Oechel, W. C. and G. H. Riechers, 1986, Response of vegetation to carbon dioxide. Progress Rept. 37, *Response of a Tundra Ecosystem to Elevated Atmospheric Carbon Dioxide*, USDOE Office of Basic Energy Sciences, Carbon Dioxide Research Division.

Oja, T. (1983). Metsa suktsessiooni ja tasandilise struktuuri imiteerimisest (On the stimulation of succession and plain structure of forest). *Year-Book of the Estonian Naturalists' Society*. 69:110–117.

Olson, G. W., 1976, *Criteria for Making and Interpreting a Soil Profile Description*. State Geological Survey of Kansas, University of Kansas, Lawrence, Kansas, 47 pp.

O'Neill, R. V., D. L. DeAngelis, J. B. Waide, and T. F. H. Allen, 1986, *A Hierarchical Concept of Ecosystems*. Monographs in Population Biology, Princeton University Press, Princeton, N.J., 253 pp.

Orloci, L. and M. Orloci, 1988, On recovery, Markov chains, and canonical analysis. *Ecology* 69(4):1260–1265.

Parrish, J. A. D. and F. A. Bazzaz, 1982, Competitive interactions in plant communities of different successional ages. *Ecology* 63:314–320.

Pastor, J., J. D. Aber, C. A. McClaugherty, and J. M. Melillo, 1984, Above ground production and N and P cycling along a nitrogen mineralization gradient on Blackhawk Island, Wisconsin. *Ecology* 65:256–268.

Pastor, J. and W. M. Post, 1985a, Development of a linked forest productivity–soil process model. Oak Ridge National Laboratory TN-9519, 162 pp.

Pastor, J. and W. M. Post, 1985b, Influence of climate, soil moisture, and succession on forest carbon and nitrogen cycles. *Biogeochemistry* 2:3–27.

Pastor, J., R. H. Gardner, V. H. Dale, and W. M. Post, 1987, Successional changes in nitrogen availability as a potential factor contributing to spruce declines in boreal North America. *Canadian Journal of Forest Research* 17:1394–1400.

Pastor, J. and W. M. Post, 1988, Response of northern forests to CO_2-induced climate change. *Nature* 334:55–58.

Patten, B. C., 1971, *Systems Analysis and Simulation in Ecology*. Academic Press, New York, Vol. I, 607 pp.

Pearlstine, L., W. Kitchens, and M. McKellar, 1985, Modelling the impacts of a river diversion on bottomland forest communities in the Santee River floodplain, South Carolina, *Ecological Modelling* 29:283–302.

Perlin, J., 1989, *A Forest Journey: The Role of Wood in the Development of Civilization*. W. W. Norton, New York, 445 pp.

Perry, T. O., H. E. Sellers, and C. O. Blanchard, 1969, Estimation of photosynthetically active radiation under a forest canopy with chlorophyll extracts and from basal area measurements. *Ecology* 50:39–44.

Peters, R. L. and J. D. S. Darling, 1985, The greenhouse effect and nature reserves. *Bioscience* 35:707–717.

Pianka, E. R., 1988, *Evolutionary Ecology*. Harper and Row, New York, p. 338.

Pielou, E. C., 1969, *An Introduction to Mathematical Ecology*. Wiley-Interscience, New York, 286 pp.

Pielou, E. C., 1975, *Ecological Diversity*. Wiley-Interscience, New York, 165 pp.

Pliny, *Natural History*, Book 31, Chapter 30, VIII, 4171. Quoted by G.P. Marsh in *Man and Nature* (*op. cit.*) p. 188.

Pownall, T., 1784, *A Topographical Description of the Dominion of the United States*. Reprinted by Lois Mulkean, Pittsburgh, 1949.

Prentice, I. C., 1986, The design of a forest succession model. In J. Fanta, ed., *Forest Dynamics Research in Western and Central Europe*. Purdue, Wageningen, Germany, pp. 253–256.

Prentice, I. C., J. T. de Smidt, and O. van Tongeren, 1987, Simulation of heathland vegetation dynamics. *Journal of Ecology* 75(1):203–219.

Purohit, A. N. and E. B. Tregunna, 1976, Effects of carbon dioxide on growth of Douglas fir seedlings. *Indian Journal of Plant Physiology* 19:164–170.

Rambler, M. B., L. Margulis, and R. Fester, eds., 1989, *Global Ecology: Towards a Science of the Biosphere*, Academic Press Pub., Boston.

Rashevsky, 1938, *Mathematical Biophysics*. University of Chicago Press, Chicago.

Reed, K. L. and S. G. Clark, 1979, Succession simulator: A coniferous forest simulator, model documentation. *Coniferous Biome Ecosystem Analysis Bulletin Number 11*, University of Washington, Seattle.

Reid, W. V. and K. R. Miller, 1989, *Keeping Options Alive: The Scientific Basis for Conserving Biodiversity*. World Resources Institute, Washington, D.C., p. 4.

Reiners, W. A., D. Y. Hollinger, and G. E. Lang, 1984, Temperature and evapotranspiration gradients of the White Mountains, New Hampshire, USA. *Arctic and Alpine Research* 16(1):31–36.

Riechert, S. E. and C. R. Tracy, 1975, Thermal balance and prey availability: Bases for a model relating web-site characteristics to spider reproductive success. *Ecology* 56:265–284.

Ritchie, J. C. and G. A. Yarrantow, 1978, The late quaternary history of the boreal forest of central Canada, based on standard pollen stratigraphy and principal components analysis. *Journal of Ecology* (66):199–212.

Roberts, D. W., 1989, Analysis of forest succession with fuzzy graph theory. *Ecological Modelling* 45(4):261–274.

Rogers, H. H., J. F. Thomas, and G. E. Bingham, 1983, Response of agronomic and forest species to elevated atmospheric carbon dioxide. *Science* 220:428–429.

Roise, J. P., D. R. Betters, and B. M. Kent, 1981, An approach to functionalizing key environmental factors—forage production in Rocky-Mountain aspen stands. *Ecological Modelling* 14(1–2):133–146.

Root, R. B., 1967, The niche exploitation pattern of the blue-gray gnatcatcher. *Ecological Monographs* 37:317–319.

Rosenfeld, A. H. and D. B. Botkin, 1990, Trees can sequester carbon, or die, decay, and amplify global warming: Possible positive feedback between rising temperature, stressed forests, and CO_2. *Physics and Society* 19:4 pp.

Roskoski, J. P., 1980, Nitrogen fixation in hardwood forests of the northeastern United States. *Plant and Soil* 54:33–44.

Runkle, J. R., 1981, Gap regeneration in some old-growth forests of the Eastern United States. *Ecology* 62(4):1041–1051.

Running, S. W., 1984, Documentation and preliminary validation of $H_2OTRANS$ and DAYTRANS, two models for predicting transpiration and water stress in western coniferous forests. *U.S. Forest Service Rocky Mountain Forest Range Experiment Station Research Report* RM-252:1–45.

Russell, E. W. B., 1983, Indian-set fires in northeastern USA. *Ecology* 64:78–88.

Schlesinger, M. E. and J. F. B. Mitchell, 1987, Model projections of the equilibrium climate response to increased carbon dioxide. *Review of Geophysics* 25:760–798.

Schlesinger, M. E. and Z.-C. Zhao, 1988, *Seasonal Climate Changes Induced by Doubled CO_2 as Simulated by the OSU Atmospheric GCM/Mixed-Layer Ocean Model*. Climate Research Institute Report, Oregon State University, Corvallis, 84 pp.

Schlesinger, W. H., 1991, *Biogeochemistry: An Analysis of Global Change*. Academic Press, San Diego.

Schneider, S. H., 1989a, The greenhouse effect: Science and policy. *Science* 243:771–781.

Schneider, S. H., 1989b, *Global Warming: Are We Entering the Greenhouse Century?* Sierra Club Books, San Francisco, 317 pp.

Scholander P. F., H. T. Hammal, E. A. Hemmingsen, and E. D. Bradstreet, 1964, Hydrostatic pressure and osmotic potential in leaves of mangroves and some other plants. *National Academy of Sciences* 52:119–125.

Schroedinger, E. 1952, *Science and Humanism*. Cambridge University Press, London, 68 pp.

Sellers, W. D., 1965, *Physical Climatology*. University of Chicago Press, Chicago.

Sharpe, P. J. H., H. I. Wu, J. Walker, and L. K. Penridge, 1985, A physiologically based continuous-time Markov approach to plant-growth modeling in semi-arid woodlands. *Ecological Modelling* 29(1–4):189—213.

Shugart, H. H., 1984, *A Theory of Forest Dynamics*. Springer-Verlag, New York.

Shugart, H. H., 1987, Dynamic ecosystem consequences of tree birth and death patterns. *Bioscience* 37:596–602.

Shugart, H. H. and W. R. Emanuel, 1985, Carbon dioxide increase: The implications at the ecosystem level. *Plant, Cell and Environment* 8:381–386.

Shugart, H. H., W. R. Emanual, and A. M., Solomon, 1984, Modeling long-term changes in forested landscapes and their relation to the earth's energy balance. In B. Moore and M. N. Dastoor, eds., *The Interactions of Global Biogeochemical Cycles*. Pub. No. 84-21, Jet Propulsion Lab., NASA, Pasadena, Calif., pp. 229–252.

Shugart, H. H., M. S. Hopkins, I. P. Burgess, and A. T. Mortlock, 1980, The development of a succession model for subtropical rain forest and its application for assess the effects of timber harvest at Wiangaree State Forest, New South Wales. *Journal of Environmental Management* 11:234–265.

Shugart, H. H. and I. R. Noble, 1981, A computer model of succession and fire response of the high-altitude *Eucalyptus* forest of the Brindabella Range, Australian Capital Territory. *Australian Journal of Ecology* 6:149–164.

Shugart, H. H. and S. W. Seagle, 1985, Modeling forest landscapes and the role of disturbance in ecosystems and communities. In S. T. A. Pickett and P. S. White, eds., *The Ecology of Natural Disturbances and Patch Dynamics*. Academic Press, Orlando, Fla., pp. 353–368.

Shugart, H. H. and D. C. West, 1977, Development of an Appalachian deciduous forest succession model and its application to assessment of the impact of the chestnut blight. *Journal of Environmental Management* 5:161–179.

Shugart, H. H. and D. C. West, 1979, Size and pattern of simulated forest stands. *Forest Science* 25:120–122.

Shugart, H. H. and D. C. West, 1980, Forest succession models. *BioScience* 30:308–313.

Shugart, H. H. and D. C. West, 1981, Long-term dynamics of forest ecosystems. *American Scientist* 69(6):647–652.

Siccama, T. G., 1968, Altitudinal distribution of forest vegetation in relation to soil and climate on the slopes of the Green Mountains. Ph.D. thesis, University of Vermont, Burlington.

Siccama, T. G., 1974, Vegetation, soil, and climate on the Green Mountains of Vermont. *Ecological Monographs* 44:325–349.

Siccama, T., 1988, Forest plot data sets and associated kinds of ecological data. In *Ecological Data Exchange*, Yale School of Forestry and Environmental Studies, New Haven, Conn., 17 pp.

Siccama, T. and H. W. Vogelmann, 1987, Forest plot data—1964: Green Mountain series. In *Ecological Data Exchange*, Yale School of Forestry and Environmental Studies, New Haven, Conn., 19 pp.

Sionit, N., B. R. Strain, and H. H. Helmers, 1985, Long-term atmospheric enrichment affects growth and development of *Liquidambar styraciflua* and *Pinus taeda* seedlings. *Canadian Journal of Forestry Research* 15:468–471.

Slobodkin, L. B., D. B. Botkin, B. Maguire Jr., B. Moore III, and H. J. Morowitz, 1980, On the epistemology of ecosystem analysis. In V. S. Kennedy, ed., *Estuarine Perspectives*, Academic Press, New York, pp. 497–507.

Smith, D. M., 1986, *The Practice of Silviculture*. Wiley, New York.

Smith, J. B. and D. A. Tirpak, eds., 1989, *The Potential Effects of Global Climate Change on the United States*. U.S. Environmental Protection Agency, Washington, D.C., EPA-203-05-89-0.

Smith, T. and M. Huston, 1989, A theory of the spatial and temporal dynamics of plant communities. *Vegetatio* 83(1–2):49–69.

Smith, T. M. and D. L. Urban, 1988, Scale and resolution of forest structural pattern. *Vegetatio* 74:143–150.

Solomon, A. M., 1986, Transient response of forests to CO_2-induced climate change: Simulation modeling experiments in eastern North America. *Oecologia* 68:567–579.

Smith, W. B. and G. J. Brand, 1983, Allometric biomass equations for 98 species of herbs, shrubs, and small trees. North Central Forest Experiment Station, St. Paul, Minn. Research Note NC-229.

Sokal, R. R. and F. J. Rohlf, 1969, *Biometry: The Principles and Practices of Statistics in Biological Research* W. H. Freeman, San Francisco.

Solomon, A. M., H. R. Delcourt, D. C. West and T. J. Blasing, 1980, Testing a simulation model for reconstruction of prehistoric forest-stand dynamics. *Quarternary Research* 14:275–293.

Solomon, A. M. and H. H. Shugart, 1984, Integrating forest-stand simulations with paleoecological records to examine long-term forest dynamics. In G. I. Agren, ed., *State and Change in Forest Ecosystems—Indicators in Current Research*. Swedish University Agricultural Sciences, Uppsala, Sweden, pp. 333–336.

Solomon, A. M., M. L. Tharp, West, G. E. Taylor, J. W. Webb, and J. L. Trimble, 1984, Response of unmanaged forests to CO_2-induced climate change: Available information, initial tests, and data requirements. U.S. Department of Energy, Technical Report. 009, 93 pp.

Solomon, A. M. and T. Webb III, 1985, Computer-aided reconstruction of late-quarternary landscape dynamics. *Annual Review of Ecology and Systematics* 16:63–84.

Solomon, A. M. and D. C. West, 1987, Simulating forest ecosystem responses to expected climate change in Eastern North America: Applications to decision making in the forest industry. In W. E. Shands and J. S. Hoffman, eds., *The Greenhouse Effect, Climate Change, and U.S. Forests*. The Conservation Foundation, Washington, D.C., pp. 189–217.

Sprugel, D. G., 1985a, Changes in biomass components through stand development in wave-regenerated balsam fir forests. *Canadian Journal of Forest Research* 15:269–278.

Sprugel, D. G., 1985b, Ecology of natural disturbance and patch dynamics. In S. T. A. Pickett and P. S. White, eds., *Ecology of Natural Disturbance and Patch Dynamics*, Academic Press, Orlando, Fla.

Star, J. L. and J. E. Estes, 1990, *Geographic Information Systems: An Introduction*. Prentice Hall, Englewood Cliffs, N.J., 304 pp.

Stephenson, N. L., 1990, Climatic control of vegetation distribution: The role of the water balance, *American Naturalist* 135:649–670.

Stevens, P. A., J. K. Adamson, B. Reynolds, and M. Hornung, 1990, Dissolved inorganic nitrogen concentrations and fluxes in three British Sitka spruce plantations. *Plant and Soil* 128:103.

Stone, C. D., 1974, *Should Trees Have Standing? Towards Legal Rights for Natural Objects*, William Kaufmann, Pub., Los Atlos, Calif.

Swain, A. M., 1978, Environmental changes during the past 2000 years in

North-Central Wisconsin: Analysis of pollen, charcoal, and seeds from varied lake sediments, *Quarternary Research* 10:55–68.

Talbot, L. M., 1990, *A Proposal for the World Bank's Policy and Strategy for Tropical Moist Forests in Africa.* Report to the World Bank, 10 pp.

Tamm, C. O. and E. B. Cowling, 1977, Acidic precipitation and forest vegetation. *Water Air Soil Pollution* 7:503–511.

Teskey, R. O. and R. B. Shrestha, 1985, A relationship between carbon dioxide, photosynthetic efficiency and shade tolerance. *Physiol. Plant* 63:126–132.

Thomas, W. L. ed., 1956, *Man's Role in Changing the Face of the Earth.* University of Chicago Press, Chicago.

Thoreau, H. D., 1860, The succession of forest trees. An address read to the Middlesex Agricultural Society in Concord, Sept., 1860, extracted from the 8th Annual Report of the Massachusetts Board of Agriculture (1860).

Thornthwaite, C. W., 1948, An approach toward a rational classification of climate. *Geographical Review* 38:55–94.

Thornthwaite, C. W. and J. R. Mather, 1957, Instructions and tables for computer potential evapotranspiration and the water balance. *Publications In Climatology* 10(3), Drexel Institute of Technology, Centerton.

Tilman, D., 1985, The resource-ratio hypothesis of plant succession. *American Naturalist* 125(6):827–852.

Tilman, D., 1990a, Mechanisms of plant competition for nutrients: the elements of a predictive theory of competition. In J. B. Grace and D. Tilman, eds., *Perspectives on Plant Competition.* Academic Press, San Diego, pp. 117–142.

Tilman, D., 1990b, Constraints and trade-offs toward a predictive theory of competition and succession. *Oikos* 58:2–15.

Tolley, L. C. and B. R. Strain, 1985, Effects of CO_2 enrichment and water stress on gas exchange of seedlings grown under different irradiance levels. *Oecologia* 65:166–172.

Tracy, C. R., B. J. Tracy, and D. S. Dobkin, 1975, The role of posturing in behavioral thermoregulation by black dragons (*Hagenius brevistylus* Selys: Odonata). *Physiological Zoology* 52:565–571.

Turner, M. G., 1989, Landscape ecology: The effect of pattern on process. *Annual Review of Ecology and Systematics* 20:171–197.

Urban, D. L. and H. H. Shugart Jr., 1986, Avian demography in mosaic landscapes: Modeling paradigm and preliminary results. In J. Verner, M. L. Morrison, and C. J. Ralph eds., *Modeling Habitat Relationships of Terrestrial Vertebrates.* University of Wisconsin Press, Madison, pp. 273–279.

Urban, D. L. and T. M. Smith, 1989, Microhabitat pattern and the structure of forest bird communities. *American Naturalist* 133:881–829.

Usher, M. B., 1981, Modeling ecological succession, with particular reference to Markovian models. *Vegetatio* 46(7):11–18.

Vanclay, J. K., 1989, A growth model for North Queensland rainforests. *Forest Ecology and Management* 27(3–4):245–271.

Van Miegroet, H., D. W. Johnson, and D. W. Cole, 1990, Soil nitrification as affected by N fertility and changes in forest floor C/N ratio in four forest soils. *Canadian Journal of Forest Research*, 20:1012.

Van Voris, P., 1976, Ecological stability: An ecosystem perspective. Oak Ridge National Laboratory Publ. # 900 ORNM/TM-5517, 34 pp.

Van Voris, P., D. Botkin, D. Woodby, J. Bergengren, R. Nisbet, T. Demarais, D. Millard, John Thomas, and James Thomas, 1990, *TERRA-Vision, Terrestrial Environmental*

Resource Risk Assessment—Computer Visioning System. Unpublished report, Pacific Northwest Laboratory, Richland, Wash., 99352, 21 pp.

Verhulst, P. F., 1838, Notice sur la loi que la population suit dans son accroissement. *Correspondence Mathematique et Physique* 10:113–121.

Vernadsky, V. J., 1926, *The Biosphere.* Leningrad.

Vitousek, P. M. and W. A. Reiners, 1975, Ecosystem succession and nutrient retention: A hypothesis. *BioScience* 25:376–381.

Vitousek, P. M. and P. S. White, 1981, Process studies in succession. In West, D. C., H. H. Shugart, and D. B. Botkin, eds., *Forest Succession: Concepts and Applications.* Springer-Verlag, New York, pp. 267–276.

Volterra, V., 1926, Variazioni e fluttuazioni del numero d'individui in specie animali conviventi. *Memòria Accadèmia Nazionale dei Lincei* Series VI, vol. 2.

Vowinckel, T., W. C. Oechel, and W. C. Boll, 1975, The effect of climate on the photosynthesis of *picea marina* at the subarctic treeline. I. Field measurements. *Canadian Journal of Botany* 53:604–620.

Waggoner, P. E. and W. E. Reifsnyder, 1981, Differences between net radiation and water use caused by radiation from the soil surface. *Soil Science* 91:246–250.

Walker, J., C. H. Thompson, I. F. Fergus, and B. R. Tunstall, 1981, Plant succession and soil development in coastal sand dunes of subtropical Eastern Australia. In West, D. C., H. H. Shugart, and D. B. Botkin, eds., *Forest Succession: Concepts and Applications.* Springer-Verlag, New York, pp. 107–131.

Waring, R. H. and J. F. Franklin, 1979, Evergreen coniferous forests of the Pacific Northwest. *Science* 204(29):1380–1386.

Waring, R. H. and W. H. Schlesinger, 1985, *Forest Ecosystems: Concepts and Management.* Academic Press, New York, 340 pp.

Waring, R. H., P. E. Schroeder, and R. Ore, 1982, Application of the pipe model theory to predict canopy leaf area. *Canadian Journal of Forest Research* 12:556–560.

Watt, K. E. F., 1962, Use of mathematics in population ecology. *Annual Review of Entomology* 7:243–252.

Weinstein, D. A. and H. H. Shugart, 1983, Ecological modeling in landscape dynamics. In H. Mooney and M. Godron, eds., *Disturbance and Ecosystems.* Springer-Verlag, Berlin and New York, pp. 29–45.

West, D. C., S. B. McLaughlin, and H. H. Shugart, 1980, Simulated forest response to chronic air pollution stress. *Journal of Environmental Quality* 1:43–49.

West, D. C., H. H. Shugart, and D. B. Botkin, eds., 1981, *Forest Succession: Concepts and Applications.* Springer-Verlag, New York, 517 pp.

White, E. H., 1974, Whole-tree harvesting depletes soil nutrients. *Canadian Journal of Forest Research* 4:530–535.

White, J., 1981, The allometric interpretation of the self-thinning rule. *Journal of Theoretical Biology* 89:475–500.

Whittaker, R. H., 1966, Forest dimensions and production in the Great Smoky Mountains. *Ecology* 47:103–121.

Whittaker, R. H., 1970, *Communities and Ecosystems.* Macmillan Co., London & Toronto, p. 73.

Whittaker, R. H. and P. L. Marks, 1975, Methods of assessing terrestrial productivity. In H. Lieth and R. H. Whittaker, eds., *Primary Productivity of the Biosphere.* Springer-Verlag, New York.

Whittaker, R. H. and G. M. Woodwell, 1968, Dimension and production relations of trees and shrubs in the Brookhaven Forest, New York. *Journal of Ecology* 56:1–25.

Whittaker, R. H. and G. M. Woodwell, 1969, Structure, production and diversity of the oak-pine forest at Brookhaven, New York. *Journal of Ecology* 57:157–174.

Wigley, T. M. L., P. D. Jones, and P. M. Kelly 1980, Scenario for a warm, high CO_2 world. *Nature* 283:17–21.

Wilson, E. O., 1988, The current state of biodiversity. In E. O. Wilson, ed., *Biodiversity*. National Academy Press, Washington, D.C., pp. 3–18.

Wilson, M. V. and D. B. Botkin, 1990, Models of simple microcosms: Emergent properties and the effect of complexity on stability. *American Naturalist* 135:414–434.

Woodby, D. A. 1991, *An Ecosystem Model of Forest Tree Persistence*. Ph.D. thesis, University of California, Santa Barbara.

Woods, K. D., A. Feiveson, and D. B. Botkin, 1991, Statistical error analysis for biomass density and leaf area index estimation. *Canadian Journal of Botany* 21:974–989.

Woods, T. and F. H. Bormann, 1977, Short-term effects of a simulated acid rain upon the growth and nutrient relations of *Pinus strobus*, *Water Air Soil Pollution* 7:479–488.

Woodwell, G. M. and D. B. Botkin, 1970, Metabolism of terrestrial ecosystem by gas exchange techniques: The Brookhaven approach. In D. E. Reichle, ed., *Analysis of Temperate Forest Ecosystems*. Springer-Verlag, New York, pp. 73–85.

Woodwell, G. M. and E. V. Pecan, eds., 1973, *Carbon and the Biosphere*, Brookhaven National Laboratory Symposium No. 24. Oak Ridge, Tenn., Technical Information Service.

Woodwell, G. M., R. H. Whittaker, W. A. Reiners, G. E. Likens, C. A. S. Hall, C. C. Delwiche, and D. B. Botkin, 1977, The biota and the world carbon budget. *Science* 199:141–146.

Wright, H. E. Jr., 1976, The dynamic nature of Holocene vegetation. *Quaternary Research* 6:581–596.

Wu, L. S. and D. B. Botkin, 1980, Of elephants and men: A discrete, stochastic model for long-lived species with complex life histories. *American Naturalist* 116:831–849.

Wu, H., E. J. Rykiel, and P. J. H. Sharpe, 1987, Age size population-dynamics— derivation of a general matrix methodology. *Computers & Mathematics with Applications* 13:759–766.

Yeatman, C. S., 1970, CO_2 enriched air increased growth of conifer seedlings. *Forest Chronicles*. 46:229–230.

Zabinski, C. and M. B. Davis, 1989, Hard times ahead for Great Lakes forests: A climate threshold model predicts responses to CO_2-induced climate change. Chapter 5 in J. B. Smith and D. A. Tirpak, eds., *The Potential Effects of Global Climate Change on the United States: Appendix D-Forests*. Office of Policy, Planning and Evaluation, U.S. Environmental Protection Agency, Washington, D. C., pp. 5-1–5-19.

Author Index

Subject Index